ENGINEERING GEOLOGY AND SOIL MECHANICS

工程地质与土力学

主编
殷杰 陈亮

副主编
苗永红 耿维娟 周恩全 沈圆顺 钱海

主审
陆建飞

镇江

图书在版编目(CIP)数据

工程地质与土力学 = Engineering Geology and Soil Mechanics：英文 / 殷杰，陈亮主编. —镇江：江苏大学出版社，2020.12
ISBN 978-7-5684-1288-9

Ⅰ.①工… Ⅱ.①殷… ②陈… Ⅲ.①工程地质－高等学校－教材－英文②土力学－高等学校－教材－英文 Ⅳ.①P642②TU43

中国版本图书馆 CIP 数据核字(2020)第 003159 号

工程地质与土力学
Engineering Geology and Soil Mechanics

| 主　　编/殷　杰　陈　亮 |
| 责任编辑/郑晨晖 |
| 出版发行/江苏大学出版社 |
| 地　　址/江苏省镇江市梦溪园巷 30 号(邮编：212003) |
| 电　　话/0511-84446464(传真) |
| 网　　址/http：press.ujs.edu.cn |
| 排　　版/镇江文苑制版印刷有限责任公司 |
| 印　　刷/镇江文苑制版印刷有限责任公司 |
| 开　　本/787 mm×1 092 mm　1/16 |
| 印　　张/20.25 |
| 字　　数/674 千字 |
| 版　　次/2020 年 12 月第 1 版 |
| 印　　次/2020 年 12 月第 1 次印刷 |
| 书　　号/ISBN 978-7-5684-1288-9 |
| 定　　价/68.00 元 |

如有印装质量问题请与本社营销部联系(电话：0511-84440882)

Preface

Engineering Geology and Soil Mechanics is a textbook edited for undergraduate students majoring in Civil Engineering and other related majors. The aim of this book is to provide students with a basic knowledge of engineering geology, the essential concepts of the physical properties of soils as a civil engineering material and the fundamental principles of soil mechanics.

The book is divided into 10 chapters. Chapters 1~3 describe the basic knowledge and theory of *Engineering Geology*, mainly including geological processes, rock cycle, geologic time and structures, and geologic hazards; Chapters 4-10 describe the basic knowledge and principle of *Soil Mechanics*, mainly including the engineering properties of soil, permeability and seepage of soil, stress in soil, compressibility of soil, shear strength of soil, earth pressure and slope stability. In order to cultivate the students' ability in solving practical engineering problems, each chapter of the textbook begins with an introductory case to arouse the enthusiasm and interests of the students. At the end of each chapter, there are knowledge expansions, and some questions. These are convenient for students to review after class and use network expanding their knowledge by self-study. In the appendix, we offer the list of symbols, terminologies translated in Chinese and references for students to check.

The following editorial board members (unspecified participants are all teachers in Jiangsu University) who participated in the compiling of the textbook: Chapter 1 was edited by Yin Jie and Geng weijuan; Chapter 2 was edited by Yin Jie and Professor Wu Xiaoliang (Zhenjiang Institute of Survey and Investigation); Chapter 3 was edited by Yin Jie and Zhao Jinshuai; Chapter 4 was edited by Chen Liang, Miao Yonghong and Geng weijuan; Chapter 5 was edited by Chen Liang, Shen Yuanshun and Yin Jie; Chapter 6 was edited by Chen Liang, Miao Yonghong and Yin Jie; Chapter 7 was edited by Yin Jie and Shen Yuanshun; Chapter 8 was edited by Geng weijuan, Yin Jie and Qian Hai; Chapter 9 was edited by Zhou Enquan and Yin Jie; Chapter 10 was edited by Zhou Enquan, Wang Fei and Yin Jie. Editor-in-chief Yin Jie (Associate Professor at Jiangsu University) and Chen Liang (Professor at Hohai University) made the outline of the book and compiled the integrated manuscript. Professor Lu Jianfei from Jiangsu University reviewed the textbook chiefly.

We are indebted to all who have contributed to the publication of this textbook.

Thanks for the graduate students of the authors who made a great contribution to the book. Thanks to Zheng Chenhui and her colleagues from Jiangsu University Press for their tireless efforts for the publication of this textbook. Thanks to the overseas education college (OEC) at Jiangsu University who rated the course of Engineering Geology and Soil Mechanics as an excellent course in Jiangsu University and gave financial support in publication of this textbook for this course. Besides, some of the images or photographs were collected from the websites on the internet, which are highly appreciated.

Given volume limitation, this textbook only covers the most important contents of relevant knowledge about Engineering Geology and Soil Mechanics. Criticism and suggestions from the readers will be highly welcomed and appreciated. Due to the limited knowledge of the authors, defects in the textbook are unavoidable. We deeply appreciate your criticism and correction.

Yin Jie & Chen Liang

June, 2020

CONTENTS

Chapter 1　The Earth and Rock Structures ··· 1
　1.1　Introductory case ·· 1
　1.2　Structure of the Earth ··· 2
　　1.2.1　Shape of the Earth ·· 2
　　1.2.2　Inner structure of the Earth ··· 2
　1.3　Geologic processes ··· 4
　　1.3.1　Internal geologic processes ·· 4
　　1.3.2　External geologic processes ··· 5
　1.4　Rock types ·· 8
　　1.4.1　Igneous rocks ·· 9
　　1.4.2　Sedimentary rocks ··· 16
　　1.4.3　Metamorphic rocks ·· 21
　1.5　Rock cycle ··· 25

Chapter 2　Geologic Time and Structures ··· 29
　2.1　Introductory case ·· 29
　2.2　Geologic time ··· 29
　　2.2.1　Relative geologic time ··· 30
　　2.2.2　Absolute geologic time ·· 32
　　2.2.3　Geologic time scale ·· 33
　2.3　Geologic structures ··· 35
　　2.3.1　Attitude of stratum ··· 35
　　2.3.2　Determination of stratum attitude ·· 36
　　2.3.3　Representation of stratum attitude ··· 36
　2.4　Contact of the stratum ··· 37
　　2.4.1　Nonconformity ··· 38
　　2.4.2　Disconformity ··· 38
　　2.4.3　Angular unconformity ··· 38
　2.5　Fold structure ·· 38
　　2.5.1　Definition of fold structure ·· 38
　　2.5.2　Types of fold structures ··· 39

2.6 Fracture structure ·· 41
  2.6.1 Joints ··· 41
  2.6.2 Faults ··· 43
2.7 Geologic map ··· 48
  2.7.1 Contents of a geologic map ·· 49
  2.7.2 Method of reading a geologic map ·································· 50

Chapter 3 Geologic Hazards ··· 52

3.1 Introductory case ·· 52
3.2 Land subsidence ··· 53
  3.2.1 Definition of subsidence ··· 53
  3.2.2 Classification of subsidence ·· 53
  3.2.3 Subsidence due to groundwater extraction ······················· 54
  3.2.4 Groundwater control ··· 58
3.3 Landslide ·· 62
  3.3.1 Definition of landslide ·· 62
  3.3.2 Classification of landslide ··· 64
  3.3.3 Development of landslide ··· 65
  3.3.4 Causes of landslide ··· 67
  3.3.5 Landslide prevention and mitigation ································· 68
3.4 Debris flow ·· 69
  3.4.1 Definition of debris flow ··· 69
  3.4.2 Formation conditions of debris flow ································· 69
  3.4.3 Development characteristics of debris flow ······················· 71
  3.4.4 Classification of debris flow ·· 71
  3.4.5 Debris flow prevention and mitigation ······························ 73
3.5 Karst ·· 74
  3.5.1 Definition of karst ·· 74
  3.5.2 Morphological characteristics of karst ······························ 75
  3.5.3 Formation conditions of karst ··· 77
  3.5.4 Engineering problems in karst area ································· 79
3.6 Earthquake ··· 81
  3.6.1 Basic concepts of earthquake ·· 81
  3.6.2 Types of earthquake ··· 85
  3.6.3 Methods of reducing earthquake hazards ························ 86

Chapter 4 Engineering Properties of Soil ·· 92

4.1 Introductory case ·· 92
4.2 Soil formation and structure ··· 92
  4.2.1 Soil formation ··· 93

  4.2.2 Soil structure ··· 99
 4.3 Physical and index properties ··· 103
  4.3.1 Direct measured physical properties ·· 104
  4.3.2 Other index properties ·· 105
  4.3.3 Basic relations among physical and index properties ··································· 106
 4.4 Consistency of cohesive soil ··· 109
  4.4.1 Soil cohesion ·· 109
  4.4.2 Plasticity ··· 110
  4.4.3 Atterberg limits ··· 110
 4.5 Relative density of non-cohesive soil ··· 114
 4.6 Compaction characteristics of soil ··· 116
  4.6.1 Soil compaction tests ··· 116
  4.6.2 Factors affecting soil compaction ·· 117
 4.7 Classification of soil ··· 119
  4.7.1 Soil classification for engineering purpose ·· 119
  4.7.2 Soil classification standards in China ·· 122

Chapter 5 Permeability and Seepage ··· 132
 5.1 Introductory case ··· 132
 5.2 Permeability of soil and Darcy's Law ··· 134
  5.2.1 Permeability of soil ··· 134
  5.2.2 Darcy's Law ·· 134
  5.2.3 Influencing factors of soil permeability ·· 135
  5.2.4 Permeability of layered soil ·· 136
  5.2.5 Method for determination of soil permeability coefficient ··· 138
 5.3 Basic equations of seepage and net characteristics ··································· 142
  5.3.1 Continuity equation of flow ·· 143
  5.3.2 Basic equation of seepage field ·· 144
  5.3.3 Description of seepage field by flow net ·· 144
 5.4 Seepage force and seepage deformation ·· 146
  5.4.1 Seepage force ··· 146
  5.4.2 Basic form of seepage deformation ·· 148
  5.4.3 Determination of flowing soil and piping ·· 148
  5.4.4 Effective stress principle ·· 152

Chapter 6 Stresses in Soil ··· 159
 6.1 Introductory case ··· 159
 6.2 Self-weight stress in the ground ··· 161
  6.2.1 Natural state of ground ·· 161
  6.2.2 Calculation of self-weight stress ·· 162

6.2.3 Discussion on self-weight stress ······ 163
6.3 Contact pressure under foundations ······ 165
  6.3.1 Distribution of contact pressure ······ 166
  6.3.2 Contact pressure distribution under rigid foundation ······ 167
6.4 Induced stress calculation for a space problem ······ 171
  6.4.1 Induced stress due to a point load ······ 171
  6.4.2 Induced stress below a rectangular base under uniform pressure ······ 173
  6.4.3 Induced stress below a rectangular base under triangular pressure ······ 177
  6.4.4 Induced stress below a rectangular base under horizontal pressure ······ 181
  6.4.5 Induced stress below a circular base under uniform pressure ······ 182
6.5 Induced stress calculation for a plane problem ······ 183
  6.5.1 Induced stress due to an uniformly distributed line load ······ 183
  6.5.2 Induced stress due to uniformly distributed pressure on a strip foundation ······ 183
  6.5.3 Induced stress due to a vertical triangular distributed pressure ······ 186
  6.5.4 Induced stress due to a horizontally uniform pressure ······ 187

Chapter 7 Soil Consolidation and Settlement ······ 192

7.1 Introductory case ······ 192
7.2 Components of total settlement ······ 194
  7.2.1 Immediate (elastic) settlement ······ 195
  7.2.2 Consolidation settlement ······ 198
  7.2.3 Secondary compression settlement ······ 200
7.3 Compressibility characteristics ······ 201
  7.3.1 Consolidation test ······ 201
  7.3.2 One-dimensional compression curve ······ 202
7.4 Analysis of consolidation ······ 207
  7.4.1 Piston-spring analogy ······ 208
  7.4.2 Terzaghi's consolidation theory ······ 208
  7.4.3 Solution of Terzaghi's theory ······ 210
  7.4.4 Settlement and time ······ 213

Chapter 8 Shear Strength of Soil ······ 219

8.1 Introductory case ······ 219
8.2 Shear strength and Mohr-Coulomb failure criterion ······ 220
  8.2.1 Coulomb's law of shear strength ······ 220
  8.2.2 Mohr's circle ······ 221

8.2.3	Mohr-Coulomb failure criterion	222
8.3	Laboratory tests for soil shear strength	226
8.3.1	Type of shear tests and drainage conditions	226
8.3.2	Direct shear test	228
8.3.3	Triaxial compression test	229
8.3.4	Pore pressure coefficients in triaxial compression test	232
8.3.5	Unconfined compression test	235
8.4	Shear behavior of sand	237
8.4.1	Internal friction angle of sand	237
8.4.2	Stress strain characteristics of sand	238
8.4.3	Sand liquefaction	239
8.5	Shear behavior of clay	239
8.5.1	Shear strength of normally consolidated clay	240
8.5.2	Shear strength of overconsolidated clay	242
8.5.3	Stress strain behavior of normally and overconsolidated clay	243
8.5.4	Discussion on shear strength problems	244
8.5.5	Creep of clay during shear	246

Chapter 9   Earth Pressure on Retaining Structure 251

9.1	Introductory case	251
9.2	Calculation of earth pressure at rest	253
9.3	Rankine's earth pressure theory	256
9.3.1	Basic assumptions and principles	256
9.3.2	Calculation of Rankine's active earth pressure	257
9.3.3	Calculation of Rankine's passive earth pressure	264
9.4	Coulomb's earth pressure theory	267
9.4.1	Basic assumptions and principles	267
9.4.2	Calculation of Coulomb's active earth pressure	267
9.4.3	Calculation of Coulomb's passive earth pressure	276
9.5	Discussion on earth pressure	277
9.5.1	Influence of the wall movement	277
9.5.2	Differences between Rankine's and Coulomb's theory	278

Chapter 10   Slope Stability 284

10.1	Introductory case	284
10.2	Stability analysis of non-cohesive soil slope	287
10.2.1	For non-cohesive soil slope without seepage flow	287
10.2.2	For non-cohesive soil slope with seepage flow	288
10.3	Stability analysis of cohesive soil slope	289
10.3.1	Friction circle method	289

　　　　10.3.2　Slices method ······ 292
　　　　10.3.3　Assumptions of commonly used slice methods ······ 294
　　10.4　Swedish circle method ······ 294
　　　　10.4.1　Basic assumptions and equations ······ 294
　　　　10.4.2　Safety factor for layered soil slope with surcharge load ······ 295
　　　　10.4.3　Safety factor with seepage by unit weight substitution method ······ 296
　　10.5　Bishop method of slices ······ 300
　　　　10.5.1　Effective stress analysis ······ 300
　　　　10.5.2　Total stress analysis ······ 301
　　10.6　Slope stability analysis of non-circular failure surface ······ 304
　　　　10.6.1　Janbu method ······ 304
　　　　10.6.2　Imbalance thrust force method ······ 306

References ······ 311

Chapter 1　The Earth and Rock Structures

1.1　Introductory case

　　The demand for energy resources is increasing with the vast consumption as a result of human activities and engineering constructions. Fossil energy sources, including oil, coal and natural gas, are non-renewable resources that formed when prehistoric plants and animals died and were gradually buried by layers of rock. Oil and gas can move through the porous rocks (rocks with gaps between the grains). The oil and gas move upwards from the source rock where they were formed. When they met a layer of cap rock (a rock with no spaces between the grains) the oil and gas are trapped. For example, shale gas is a natural gas that is trapped within shale formations, as shown in Figure 1-1. Shales are fine-grained sedimentary rocks that can be rich sources of petroleum and natural gas. The shales were deposited as fine silt and clay particles at the bottom of relatively enclosed bodies of water. Hydraulic fracturing is usually the most efficient and economical technology to extract the shale gas. Basically, it is natural gas-primarily methane-found in shale formations, some of which were formed 300 million to 400 million years ago during the Devonian period of Earth's history. Besides fossil energy, a renewable resource that can be harvested for human use is geothermal energy. Geothermal energy is the heat produced deep in the Earth's core. It is a clean, renewable resource that can be harnessed for use as heat and electricity.

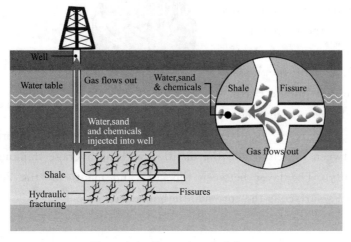

Figure 1-1　Extraction of shale gas

Engineering Geology and Soil Mechanics

1.2 Structure of the Earth

1.2.1 Shape of the Earth

The Earth is one of the nine planets revolving around the sun, and it's a rotating ellipsoid. Based on the satellite observations and changes of satellite orbit, the equatorial radius of the Earth is 6378.137km, the polar radius is 6356.752km, the mass is 5.9724×10^{24} kg, the surface area is 5.1007×10^{8} km^2 and the volume is 1.0832×10^{12} km^3. As shown in Figure 1-2, it's like a "pear-shaped" body.

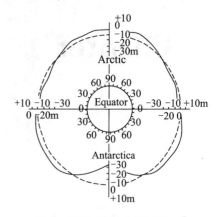

Figure 1-2 Shape of the Earth

1.2.2 Inner structure of the Earth

The structure of the Earth can be described by two ways: including mechanical properties and chemical properties. Mechanically, it is composed of lithosphere, asthenosphere, mantle, outer core, and inner core. Chemically, the earth is composed of crust, upper mantle, lower mantle, outer core, and inner core (Figure 1-3). The thickness of core is 3473km, which accounts for 16.2% of the Earth's total volume and 31.4% of the total mass. The core of the Earth mainly composed of iron and nickel with an average density of 10.5g/cm^3. The mantle is the transition layer between the crust and the Earth's core with a thickness of about 2900km. The volume of the mantle is about 82.5% of the total volume of the earth and the mass of mantle accounts for 67.8% of the mass, which is the main part of the Earth. On the surface of crust, human does

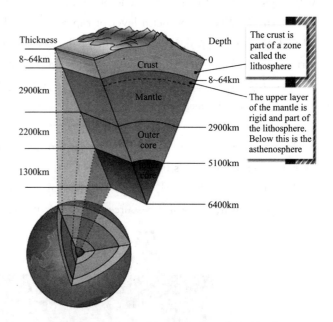

Figure 1-3 Inner layer structure of the Earth

activities, where is the research subject of engineering geology. In engineering geology, we call the layer between the crust and the mantle, with an average thinkness of 33km, as Moho discontinuity. The layer between the core and mantle at the depth of 2885km beneath the Earth's surface is the core-mantle boundary.

1.2.2.1 Earth's crust

The crust is located outside the Moho and is a hard shell as the Earth's surface, consisting of solid rocks. The crust varies in its thickness, with an average thickness of 17km, which is 1/400 of the Earth's radius. The volume of crust accounts for 1.5% of the total volume of the Earth, and the mass of crust accounts for 0.8% of the total mass of the earth. There are obvious differences in structure between continental crust and oceanic crust. The thinkness of continental crust is longer than that of oceanic, with the thickest of 70km, and the average is about 35km; the oceanic crust is thinner, with the thinnest of less than 5km, and the average is 6km.

The rocks that comprise the crust are mostly igneous rocks except the surface sedimentary rocks on the surface of Earth's crust. According to the genesis or origin of the rock, the crust can be divided into two layers: the upper silicon aluminum layer (granite layer) and the lower silicon and magnesium layer (basaltic rock). The oceanic crust directly exposes to ocean basin.

There are more than one hundred chemical elements in the crust, but the content of element are extremely uneven. There are nine elements including oxygen, silicon, aluminum, iron, calcium, sodium, potassium, magnesium and hydrogen, which account for more than 98% of the total mass of the crust.

1.2.2.2 Earth's mantle

Earth's mantle is a layer of silicate rock between the crust and the outercore, with a thickness of about 2900km. According to the seismic data, the mantle can be categorized by upper mantle and lower mantle.

I. Upper mantle

The rocks with higher Fe and Mg contents and less Si content than in crust are called mantle rocks.

The low velocity zone (asthenosphere) refers to the zone at the depth of 60~250km in the upper mantle. The seismic wave is strongly attenuated with the P-wave velocity at 7.7~8.1km/s and the S-wave velocity at 4.0~4.2km/s. The rocks that comprise the low velocity zone are with large plasticity, which are called "asthenosphere" or "soft flow layer". It is generally believed that the asthenosphere is the origin of magmatism. The material in asthenosphere can flow slowly, as they are molten. The upper mantle and the "soft flow layer" in crust composed of lithosphere. The lithosphere is a solid thin shell consisted of rocks, which is the surface of the Earth.

II. Lower mantle

It is generally believed that the chemical composition of layer mantle is similar to that of the upper mantle, but with more Fe components, which corresponds to stony iron meteorites. In addition, the lower mantle is a high-density crystalline structure.

1.2.2.3 Earth's core

The Earth's core is the very hot, very dense center of our planet. The ball-shaped core lies beneath the cool, brittle crust and the mostly-solid mantle. According to the seismic research, it can be divided into three layers: the outer core, the transition layer and the inner core.

- Outer core: make up of liquid material (shear wave can't pass through outer core).
- Transition layer: its material in the transition state is from liquid state to solid state.
- Inner core: the inner core material is characterized by a solid state consisting mainly of iron and nickel elements.

1.3 Geologic processes

The Earth has not ever stopped moving since the formation. The internal structure, material composition and surface morphology of the Earth's crust are changing constantly with the evolution of the Earth. It is called geologic process that caused by the dynamic change, with the development of the Earth, crustal composition, internal structure and crustal formation. According to the dynamic sources, geologic process can be categorised by two types: internal geologic process and external geologic process.

1.3.1 Internal geologic processes

It is caused by the internal energy of the Earth, such as the rotation energy, the gravitational energy and the heat that is produced by the decay of radioactive elements. This process includes crustal movement, magmatism, metamorphism, and earthquake action.

1.3.1.1 Crustal movement

Crustal movement is also called tectonic movement. Crustal movement is the physical movement of the Earth's crust caused by the internal motion of the earth. It can push the evolution of the lithosphere, the formation and end of the landforms and ocean, and the formation of trenches and mountains. It can also lead to the earthquake and volcano eruption. According to the direction of movement, it can be categorised by horizontal and vertical motion.

I. Horizontal motion

Horizontal motion, that is, the crustal movement is roughly along the tangent of the earth's surface. Horizontal motion is characterized by horizontal compression or stretching of the lithosphere, forming huge fold peaks and graben, rift and so on.

II. Vertical motion

Vertical motion, that is, the crustal movement perpendicular to the Earth's crust along the radius of the Earth. The vertical motion is characterized by a large area of ascending and descending motion, forming large uplifts and depressions, resulting in regression and transgression.

1.3.1.2 Magmatism

Magma is a mixture of molten and semi-molten rock, volatiles and solids that is found below the surface of the Earth. Due to the large pressure, the process that active magma intrudes through the weak crust or splits out of the surface, cooling, and forming into rock is called magmatism. The rock formed by magmatism is called igneous rock.

The compositions of igneous rock are different from each other due to different tectonic activities. The main minerals of crystal precipitation are as following: Mg, Fe silicate olivine, pyroxene, hornblende, biotite, K, Na, portland-plagioclase, microcline and orthoclase, quartz, etc.

1.3.1.3 Metamorphism

The geologic process involves composition, structure and tectonic activities for rocks that formed the Earth's crust in high temperature and pressure, which is called metamorphism, where the rocks formed are called metamorphic rocks. The change occurs primarily due to heat, pressure, and the chemically activated fluids.

Metamorphism includes contact metamorphism, dynamic metamorphism, gas-liquid metamorphism, combustion metamorphism, high-temperature metamorphism, thermal metamorphism, burial metamorphism, ocean floor metamorphism, etc.

1.3.1.4 Earthquake action

Earthquakes are the phenomena of rapid crust movement. Generally, it is caused by tectonic movement. In addition, volcanic eruption, underground excavation and collapse of underground mine can also cause earthquakes.

1.3.2 External geologic processes

The energy of external geological process comes from solar radiation, solar and lunar gravity, and human activities, etc.

1.3.2.1 Weathering

Weathering is the decomposition of rocks, soil and minerals as well as artificial materials affected by the Earth's atmosphere, water and biological organisms. Weathering occurs with little or no movement, thus should not be confused with erosion, which involves the movement of rocks and minerals by agents such as water, ice, snow, wind, waves, gravity and being transported and deposited in other locations. The important classifications of weathering processes are physical weathering, chemical weathering and biological weathering.

Ⅰ. **Physical weathering (Figure 1-4)**

It's also called mechanical weathering, refers to the rock in weathering process that mechanically weakened, where the chemical decomposition does not occur. The main factors causing physical weathering include the actions on the rocks by abrasion, frost chattering, temperature fluctuations and salt crystal growth. For example, in cold and high

Figure 1-4 Physical weathering

mountain areas, the water frozen at temperatures below 0℃ seeping into the rock void, and the ice volume is about 110% larger than that of water, and the expansion pressure is up to 96MPa. As the temperature changes, the water freezes and melts over time, causing the rock fissures to deepen and expand, and finally crack.

II. Chemical weathering (Figure 1-5)

Chemical weathering refers to the chemical reaction of rock under the influence of water, oxygen, carbon dioxide and acids, resulting in the decomposition of rocks and the formation of new minerals. Water is the most important agent in chemical weathering, since it is a natural solvent. The main ways of chemical weathering are dissolution, hydrolysis, hydration, oxidation, and carbonization.

Figure 1-5   Chemical weathering

III. Biological weathering (Figure 1-6)

It refers to the physical or chemical weathering of rocks caused by organisms-animals, plants, fungi and microorganisms such as bacteria. For example, plant roots in the rock fissures that lead rocks to disintegrate, and plant roots continue to grow, squeeze in the rock voids, and secrete acids to react with minerals and absorb nutrients.

Figure 1-6   Biological weathering

1.3.2.2   River (Figure 1-7)

River or stream refers to the water flowing along the land surface, mainly results from atmospheric precipitation, following recharge sources of melting snow and groundwater. Rivers are powerful and dynamic geological agents.

The geologic processes of rivers include three aspects: erosion, transportation and deposition.

Figure 1-7   Geologic process of river

I. Erosion

River erosion is the effect of wear and tear on the land surface of river water. It's the particular way in which water modifiers the landscape. There are vertical erosion and lateral erosion according to the direction of river erosion. Vertical erosion refers to water flow to the bottom of the river, deepening the river bed. The erosion depends on the energy and the geological conditions at the river bed. Generally, it begins with the river of high flow velocity, large flow rate and less erosion resistance of river bed, and gradually expands upwards, which is called the erosion source of river bed. The lateral erosion of rivers indicates that rivers flush both sides of the river and widen the effect of the river bed, which mainly occurs in the middle and lower reaches of the river. Due to the lateral erosion forms the gully shape, which lead to the formation of the oxbow lake.

II. Transportation

The process of carrying the materials to the downstream of a river is called the transportation. The water transport capacity is related to the flow rate, so the mountain rivers can carry huge rock blocks during the flood (dozens of tons). There are different sizes of debris or sediment in the riverbed, with soil particles gradually finer from upstream to downstream, which is called sorting. Particles collide and rub with each other during their movement, which makes the stones become rounded. The particle size gradually reduces, and the particles are finally grinded into spherical- and elliptical-shape. The process is called grinding effect.

III. Deposition

In the process of mass movement, energy of the river losses continuously. When the sediment, gravel and other substances along the river exceed the transport capacity of the flow, the transported materials gradually deposit under the gravity, forming river alluvium. Since the dissolved solvent in the river is far from saturated, the chemical deposition does not occur. The alluvium of the river is characterized by well separation, roundness grinding and clear-bedding.

1.3.2.3 Groundwater

Groundwater is the water in the pores of rocks and soils below the Earth's surface. The geologic processes of groundwater include erosion, transportation and deposition.

I. Erosion of groundwater

The velocity of groundwater flow is slow, thus the effect of groundwater on rock is mainly chemical erosion: limestone is a soft rock that dissolves in water, which forms the karst caves under the long-term erosion of groundwater. In a large karst cave, the rate of groundwater flow is relatively fast, which can cause erosion to the rock, resulting in the collapse of ground and safety problems of the ground structures.

II. Transportation of groundwater

The transportation of groundwater is mainly chemical transportation. The main chemical substances transported is bicarbonate. When the flow rate of groundwater is large, it can carry gravels and sands along the stream of underground river.

III. Deposition of groundwater

Ground water dissolves minerals and rocks into ions. Groundwater deposits those ions into different types of structures. Chemical deposition will occur when the pressure reduces or temperature changes and release of CO_2, forming exotic stalactites, stalagmite, stone pillars and other landforms.

1.3.2.4 Lake

I. Mechanical deposition

In the rainfall season, the surface water carrys large amounts of sediment collected along the lake. When the precipitation falls into the lake, the flow rate of lake drops abruptly, sediment deposits gradually, where coarse particles are deposited near the lake shore, fine particles are deposited in the middle of lake for days and months.

II. Chemical deposition

In the arid area, when the infiltration of the lake is greater than the precipitation, the salt content of the lake increases. When it is saturated, the dissolved salts crystallize. The salt crystallizes in the following order: calcite, dolomite, gypsum, mirabilite and halite due to the different solubility of salts.

III. Biological deposition

Lakes and swamps in humid areas contain a lot of organic matters, which are decomposed to form peat, coal, fertilizer and other materials in anoxic condition.

1.3.2.5 Sea

Coastal features are the results of deposition of sediments caused by wave action. The wave can erode the shoreline and affect the rock of shoreline to form different marine topography. Storms and tidal waves can also cause the erosion of shoreline rocks. Small particles and dissolved material are continuously deposited in shallow water under the wave action. The sedimentary particles become smaller as closer to water front, depositions of ocean floor sediments are biological and/or chemical substances.

1.3.2.6 Wind

Wind is a geologic agent that shapes landforms by erosion, transportation, and deposition of fine-grained materials. Wind can carry soil particles in suspension, or roll and bounce them along the ground surface. Carrying capacity of wind depends on wind strength and the size of soil particles. The sediments get dropped and deposited forming two types of aeolian deposits, sand dunes and loess.

1.3.2.7 Glacier

Glaciers can be categorized by continental glaciers and mountain glaciers. The ice can erodes bed rock, resulting in a variety of glacial types. Moraine is the landform that is deposited directly by glaciers. As the glacier recede, the original moraines will deposit and form glacial soil, which is characterized by boulders, peddles, gravels, sands and clay mixed composition, that is poor sorting and no layer structured.

1.3.2.8 Mass movement

Mass movement, also called mass wasting, refers to bulk movements of soil and rock debris down slopes in response to the pull of gravity, or gradual sinking of the Earth's surface.

1.4 Rock types

Rock are classified by their mineral and chemical composition, by the texture of the constituent particles and by the processes that formed them. These indicators separated rocks into igneous rocks, sedimentary rocks and metamorphic rocks.

The earth's surface is mainly composed of sedimentary rocks, accounting for about 75% of the continental area and the majority of the ocean floor. The crust is dominated by igneous rocks and metamorphic rocks, accounting for about 95% of the total volume.

1.4.1 Igneous rocks

1.4.1.1 Some terms

I. Magma

It is generally believed that the magma is formed in the upper mantle at the depths of the earth's crust with silicate as the main element. It is of high temperature, and is rich in volatile components as a molten body. Magma is a melting body formed in the deep or flow of the lithosphere. The main components are silicates and volatile substances. The magma is mainly flow through the earth's crust, intrude into the crust, and cooling down to condensation. The activity of magma can be classified by two kinds: intrusion and extrusion. If the magma intrudes from the weak zone to the cooled surface, it is an intrusion activity; whereas if magma flows through the volcanic channel, the activity on the Earth's surface is an extrusion.

According to the geological survey and experimental study, the basic characteristics of magma are as follows:

1) Composition of the magma. The magma is mainly composed of oxygen, silicon, aluminum, iron, calcium, sodium, potassium, magnesium, titanium, phosphorus, manganese, etc.

2) Temperature of the magma. The temperature of magma crystallization refers to the temperature where the magma is transformed to a solid state from a molten state, which is the temperature of the final crystallization of the magma. In addition to the temperature of the magma, the crystallization temperature is related to the temperature interval between the crystallization and the cooling.

3) Viscosity of the magma. The viscosity of magma is mainly related to the composition element of magma (mainly SiO_2), temperature, pressure and volatile contents.

The higher the content of SiO_2 in magma is, the greater the viscosity of magma will be. Viscosity of magma decreases as the temperature increases. When the temperature of magma drops rapidly, the viscosity of magma will increase dramatically. Generally, the greater the pressure is, the greater the viscosity of the magma will be. The viscosity at same temperature will drop differently according to different components of magma. The higher the volatile content of magma is, the smaller the viscosity will be.

II. Igneous rocks

Igneous rocks (derived from the Latin word for fire, ignis) can have very different mineral backgrounds, but they all share one thing in common: they formed by the cooling and crystallization of a melt. This material may have been lava erupted at the Earth's surface, or magma (unerupted lava) at depths of up to a few kilometers, known as magma in deeper bodies. Rock formed of lava is called extrusive, rock from shallow magma is called intrusive (<3km), and rock from deep magma is called plutonic (>3km). The deeper the magma, the slower it cools, and it forms larger mineral crystals.

1) Intrusive rocks or Plutonic rocks. When magma never reaches the surface and cools to form intrusions (dykes, sills, etc.) the resulting rocks are called plutonic. Depending on their silica content, they are called (in ascending order of silica content) gabbro, diorite, granite and pegmatite. By quantity, these are the by far most common rock types. Most magma actually never reach the surface of the earth.

2) Extrusive rocks or Volcanic rocks. When magma does reach the surface during a volcanic eruption, the rocks that form there are called lavas or volcanic rocks. The basic classification is the same as for plutonic rocks: with increasing silica content, they include: basalt, andesite, dacite, rhyolite, pumice and obsidian.

The most widely used and simplest classification of igneous rocks is according to the silica (SiO_2) content in the bulk rock composition. The common igneous rocks are shown in Table 1-1 and Figure 1-8.

Table 1-1 Main types of igneous rocks

Mass fraction of SiO_2 /%	Plutonic rock type	Volcanic rock equivalent
45~53	Gabbro	Basalt
53~63	Diorite	Andesite
63~68	Granodiorite	Dacite
68~75	Granite	Rhyolite

Figure 1-8 Common igneous rocks

III. Minerals

Minerals form in igneous rock as magma or lava cools. The minerals will form from the available chemicals in the magma as their crystallization temperature is reached. There are currently more than 4000 known minerals, according to the International Mineralogical Association, which is responsible for the approval of and naming of new mineral species found in nature. Of these, perhaps 150 can be called "common", 50 are "occasional", and the rest are "rare" to "extremely rare". Minerals may be classified according to chemical composition, for example-the silicate class, the carbonate class, element class, etc. A mineral can be identified by several physical properties such as crystal structure, hardness, color, luster, specific gravity, etc.

1) Color. Many minerals come in a wide variety of colors. Different minerals can be the same color. It is difficult to use just color to identify a mineral.

2) Streak. Streak is the color of a mineral in powdered form. You can see a mineral's streak by rubbing a sample across an unglazed ceramic plate and observing the powder left behind. Sometimes a mineral's streak is very different from the color of the sample.

3) Hardness. Hardness is the resistance of a mineral to scratching. Geologists use the Mohs hardness scale (Figure 1-9) to identify and describe mineral hardness. We can use the Mohs hardness scale by testing an unknown mineral against one of the standard minerals. Whichever one scratches the other is harder, and if both scratch each other they are the same hardness.

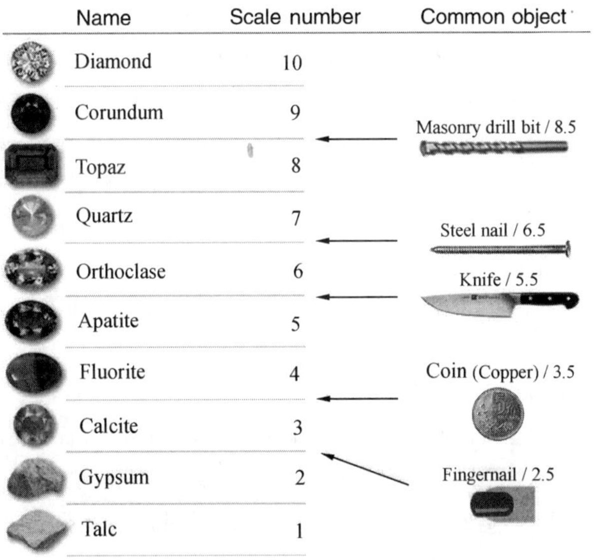

Name	Scale number	Common object
Diamond	10	
Corundum	9	Masonry drill bit / 8.5
Topaz	8	
Quartz	7	Steel nail / 6.5
Orthoclase	6	Knife / 5.5
Apatite	5	
Fluorite	4	Coin (Copper) / 3.5
Calcite	3	
Gypsum	2	Fingernail / 2.5
Talc	1	

Figure 1-9 Mohs hardness scale

4) Specific gravity. Specific gravity is the ratio of the weight of a given volume of a mineral to the weight of an equal volume of water. Higher specific gravity means the mineral is heavier.

5) Cleavage. Cleavage is how a mineral breaks. Some minerals break in smooth, flat surfaces at identifiable angles, such as calcite. Others fracture and produce no flat surfaces, such as quartz.

6) Fracture. Fracture is how a mineral breaks when no cleavage surfaces form. For example, quartz breaks in a pattern known as conchoidal fracture. Conchoidal fracture looks like smooth, curved surfaces.

7) Luster. Luster is how a mineral reflects light or how it shines. Some ways to describe luster include glassy or vitreous, metallic, dull, and pearly.

8) Crystal form. Crystal form describes the geometric shape of a crystal. There are seven

main groups of crystal shapes, including cubic, tetragonal, hexagonal, trigonal, orthorhombic, monotonic, and triclinic crystals.

9) Transparency. Transparency describes how a mineral transmits light. Some minerals are transparent (you can see through them); others are translucent (some light passes through a sample) or opaque (no light passes through a sample).

10) Magnetism. Magnetism is a special property of some minerals, especially magnetite. Samples are attracted by a magnet. Lodestone, a special form of magnetite, is a magnet itself.

11) Reaction to acid. Some minerals react to acid. Calcite especially will fizz and bubble when it comes in contact with an acid such as hydrochloric acid at room temperature.

1.4.1.2 Composition of the igneous rocks

The composition of igneous rocks includes chemical composition and mineral composition.

I. Chemical composition

The geochemical study shows that most of the elements found in the igneous rocks are found in the earth's crust. The eight most abundant elements are: O, Si, Al, Fe, Ca, Na, K, Mg. Those elements are account for 98% of the total amount of igneous rocks, with P, H, N, C, Mn, account for other 2%. The content of O is highest, accounting for more than 46% by weight of igneous rocks. Therefore, igneous rocks are often expressed as a percentage of oxide.

The chemical composition of igneous rocks is mainly composed of SiO_2, Al_2O_3, CaO, Na_2O, FeO, MgO, K_2O, Fe_2O_3 and other oxides, among which SiO_2 is the most important component. It reflects the nature of magma and affects the variation of the minerals in igneous rocks. It is of great significance to study the chemical composition of igneous rocks to accurately identify the igneous rocks.

II. Mineral composition

The most common minerals of igneous rocks are: feldspar, quartz, hornblende, pyroxene and olivine. For igneous rock, the composition is divided into four groups: felsic, intermediate, mafic, and ultramafic. These groups refer to differing amounts of silica, iron, and magnesium found in the minerals that make up the rocks. It is important to realize these groups do not have sharp boundaries in nature, but rather lie on a continuous spectrum with many transitional compositions and names that refer to specific quantities of minerals. As an example, granite is a commonly-used term but has a very specific definition which includes exact quantities of minerals like feldspar and quartz.

1) Felsic group. It refers to a predominance of the light-colored (felsic) minerals feldspar and silica in the form of quartz. These light-colored minerals have more silica as a proportion of their overall chemical formula. Minor amounts of dark-colored (mafic) minerals like amphibole and biotite mica may be present as well. Felsic igneous rocks are rich in silica (in the 65%~75% range, meaning the rock would be 65%~75% weight percent SiO_2) and poor in iron and magnesium.

2) Intermediate group. It is a composition between felsic and mafic. It usually contains roughly-equal amounts of light and dark minerals, including light grains of plagioclase feldspar

and dark minerals like amphibole. It is intermediate in silica in the 55%~60% range.

3) Mafic group. It refers to an abundance of ferromagnesian minerals (with magnesium and iron, chemical symbols Mg and Fe) plus plagioclase feldspar. It is mostly made of dark minerals like pyroxene and olivine, which are rich in iron and magnesium and relatively poor in silica. Mafic rocks are low in silica, in the 45%~50% range.

4) Ultramafic group. It refers to the extremely mafic rocks composed of mostly olivine and some pyroxene which have even more magnesium and iron and even less silica. These rocks are rare on the surface, but make up peridotite, the rock of the upper mantle. It is poor in silica, in the 40% or less range.

We can classify an igneous rock based on its mineral or chemical composition. For example, it is common to find a felsic rock composed almost entirely of the mineral plagioclase, but in chemical terms, such a rock is a subsilicic mafic rock. Another example is an igneous rock consisting solely of pyroxene. Mineralogically, it would be termed ultramafic, but chemically, it is a mafic igneous rock with a silica content of about 50%.

1.4.1.3 Texture and fabric of igneous rocks

The texture of an igneous rock normally is defined by the size and form of its constituent mineral grains and by the spatial relationships of individual grains with one another and with any glass that may be present. Texture can be described independently of the entire rock mass, and its geometric characteristics provide valuable insights into the conditions under which the rock was formed.

I. Texture of igneous rocks

Among the most fundamental properties of igneous rocks are crystallinity and granularity, two terms that closely reflect differences in magma composition and the differences between volcanic and various plutonic environments of formation.

1) Crystallinity. Generally it is described in terms of the four categories, entirely crystalline, crystalline material and subordinate glass, glass and subordinate crystalline material, and entirely glassy.

① Pegmatite texture. The rock is an intrusive rock. This rock is formed below the earth's surface, but close to the earth's surface under conditions of low temperature with large amount of water mixed with magma. The water assists the ions to move around to form large crystals. In this case, the rock formed consists of very large crystals without any matrix of smaller crystals around them.

② Porphyritic texture. A rock of this texture may be extrusive or intrusive. This rock is created by slow cooling followed by fast cooling of magma. A magma undergoes cooling slowly and due to certain environmental changes, it is pushed up out to the surface and hence subjected to fast cooling. Consequently the rock shows some large crystals mixed with crystals of small size which cooled fast. This texture showing large sized crystals within a matrix of small crystals is the porphyritic texture.

③ Glassy texture. This texture is created when an extrusive rock cools extremely fast from a lava flow. As the name implies this texture is that of glass and slag which has amorphous structure without definite crystals. This results when a magma is chilled so quickly that mineral crystals have no opportunity to form. This texture is most commonly seen in the solidification of lava having a high silica content.

2) Granularity. The general grain size ordinarily is taken as the average diameter of dominant grains in the rock. For the pegmatites, which are special rocks with extremely large crystals, it can refer to the maximum exposed dimensions of dominant grains (more than 10mm in diameter).

According to the size of average diameter, the grain size can be categorized by:
- Cryptocrystalline, particle diameter smaller than 0.02mm;
- Microcrystalline, particle diameter in the range of 0.02~0.2mm;
- Fine-grained, particle diameter in the range of 0.2~1mm;
- Medium-grained, particle diameter in the range of 1~5mm;
- Coarse-grained, particle diameter in the range of 5~20mm;
- Very coarse-grained, particle diameter greater than 20mm.

① Phaneritic texture. This is the texture of an intrusive rock whose crystals are large and can be seen with the naked eye. This is a coarse grained texture in which all the leading mineral constituents can be easily seen. This rock is formed at great depths where the magma cools very slowly.

② Aphanitic texture. This is the texture of an extrusive rock. This texture is created when the molten lava cools very fast. The mineral crystals do not have enough time to grow to large size. The individual grains are commonly less than 0.5mm in diameter and cannot be distinguished with the naked eye. The rock is crystalline, but so fine grained that it appears homogeneous. Felsite (composed of feldspar and quartz) generally has an aphanitic texture.

II. Fabric of igneous rocks

The fabric of igneous rocks refers to the mutual relationship between the grains in rocks. The most common structures of igneous rocks are:

1) Block structure. The minerals that are composed of rock are not arranged in one direction. They are evenly distributed in rocks. It is the common fabric of intrusive rocks.

2) Rhyotaxitic structure. That is, a texture where the materials are displayed in parallel arrangement of stripes with different colors. This texture is mostly found in acid lava, and the direction of the stripe is in accordance with the flow direction of molten magma.

3) Vesicular structure and amygdaloidal structure (Figure 1-10 and Figure 1-11). The two types are common textures in the extrusive rocks. When volatile components volatilize from the molten magma, a large number of bubbles can form. The bubbles in the rocks are retained to form an irregular void, due to quick condensation. The consequent rock has a vesicular texture; if the pores are filled with materials, the rock has an amygdaloid structure.

Figure 1-10 Vesicular structure   Figure 1-11 Amygdaloidal structure

1.4.1.4 Common igneous rocks

Igneous rocks are categorized according to the prevalence, texture, mineralogy, chemical composition, and the geometry of the igneous frame. The type of the various styles of distinct igneous rocks can provide us with essential records approximately the conditions underneath which they fashioned. Two critical variables used for the category of igneous rocks are particle size, which largely depends on the cooling records, and the mineral composition of the rock. Feldspars, quartz or feldspathoids, olivines, pyroxenes, amphiboles, and micas are all essential minerals inside the formation of virtually all igneous rocks, and they may be primary to the type of these rocks. Within the four groups of mineral composition, rocks are named depending on whether they display coarse-grained or fine-grained textures. The coarse-grained textures indicate intrusive rocks whereas the fine-grained textures generally indicate extrusive rocks. Some common igneous rocks are detailed below.

1) Gabbro. Crushed gabbro is commonly used as concrete aggregate, railroad ballast and road metal. This coarse-grained igneous rock is formed intrusively and composed of layers of minerals such as feldspar and augite. Occasionally it will contain olivine, a green crystalline mineral. It is generally dark in color. Gabbro can be cut and polished to form what is known as black granite.

2) Basalt. It is one of the most common types of igneous rocks in the world. The majority of the ocean floor is composed of basalt. This smooth, black igneous rock forms due to erupting magma beneath the ocean floor that cools as it comes in contact with the ocean water. Basalt is a finely grained rock that contains small crystals. It is composed of plagioclase and pyroxene and commonly used for aggregate.

3) Granite. It is a medium to coarse-grained igneous rock that is formed intrusively. Granite is commonly seen in ornamental stonework, monuments, architecture and construction. Light in color, granite is composed of quartz and feldspar minerals. It is high in silica, potassium and sodium, but low in iron, calcium and magnesium.

4) Rhyolite. It is a high-silica volcanic rock that is chemically the same as granite but is extrusive rather than plutonic. Rhyolite is often pink or gray and has a glassy groundmass. This type of igneous rock is typically found in continental settings where magmas have incorporated granitic rocks from the crust as they rise from the mantle. It tends to make lava domes when it erupts.

5) Diorite. It is a plutonic rock that is between granite and gabbro in composition. It consists mostly of white plagioclase feldspar and black hornblende. Unlike granite, diorite has no or very little quartz or alkali feldspar. Unlike gabbro, diorite contains sodic—not calcic—plagioclase. Typically, sodic plagioclase is the bright white variety albite, giving diorite a high-relief look.

6) Andesite. It is an extrusive igneous rock that is higher in silica than basalt and lower than rhyolite or felsite. In general, color is a good clue to the silica content of extrusive igneous rocks, with basalt being dark and felsite being light. Andesite is less fluid than basalt and erupts with more violence because its dissolved gases cannot escape as easily. Andesite is considered the extrusive equivalent of diorite.

7) Granodiorite. It is a plutonic rock that is among the most abundant intrusive igneous rocks. It composed of black biotite, dark-gray hornblende, off-white plagioclase, and translucent gray quartz. Granodiorite differs from diorite by the presence of quartz, and the predominance of plagioclase over alkali feldspar distinguishes it from granite.

8) Dacite. It is a fine-grained igneous rock that is normally light in color. It is often porphyritic. Dacite is found in lava flows, lava domes, dikes, sills, and pyroclastic debris. It is a rock type usually found on continental crust above subduction zones, where a relatively young oceanic plate has melted below.

9) Pumice. While most common types of igneous rocks are very hard, pumice is the exception to the rule, as it is very brittle. Pumice is an extrusive igneous rock that is formed due to cooling lava. Gas and air bubbles formed during the cooling process create tiny holes and crevasses within the rock. Pumice is a very common igneous rock and is used as an exfoliate. It is used to make emery boards and hand soap.

1.4.2 Sedimentary rocks

Sedimentary rocks are types of rock that are formed by the deposition of small particles and subsequent cementation of materials at the Earth's surface and within bodies of water. Sedimentation is the collective name for processes that cause minerals or organic matters (detritus) to settle in place. The particles that form a sedimentary rock by accumulating processes are called sediment. Before being deposited, the sediment was formed by weathering and erosion from the source area, and then transported to the place of deposition by water, wind, ice, mass movement or glaciers. Sedimentation may also occur as minerals precipitate from solutions or shells of aquatic creatures settling out of suspension. Sedimentary rocks are the most widely distributed rocks in the earth's crust, covering about 75% of the land surface area.

1.4.2.1 Composition of sedimentary rocks

The composition of sedimentary rocks mainly depends on the parent sediments lithology of sedimentary rocks and environment conditions at the site. It consists of two parts: sediment particles and cementing agents. Sediment particles can be grouped to the following categories according to their genetic types:

I. Rock debris

The rock fragment is mainly composed of rock detritus, such as anti-weathering quartz, feldspar, white mica, etc. In addition, It includes other forms of debris, such as volcanic ash.

II. Clay minerals

The minerals produced during the weathering process of clay minerals are mainly kaolinite, illite, montmorillonite, etc. The particles of this type of minerals are very fine (<0.005mm).

III. Chemical deposited minerals

Chemical deposited minerals are formed by sedimentation after chemical deposition and biochemical deposition, such as calcite, dolomite and gypsum.

IV. Organic and biological matters

Organic and biological matters mainly refer to the minerals formed though biological or organic processes, such as shells, peat, oil, etc.

1.4.2.2 Texture and fabric of sedimentary rocks

I. Texture of sedimentary rocks

The texture of sedimentary rock is determined by the formation, particle size and content. Sedimentary rocks can usually be divided into the following four types:

1) Clastic texture (Figure 1-12). Fragmented materials are cemented to form a structure. According to the particle size, it is divided into conglomeratic clastic texture (particle size $>$ 2mm), sandy texture ($0.05 \sim 2$mm), silty texture ($0.005 \sim 0.05$mm). It can also be divided into siliceous cementation, iron cementation, calcareous cementation, argillaceous cementation according to the composition of the cementing agents.

Figure 1-12　Clastic texture

2) Clay texture. It is composed of clay mineral particles with particle size less than 0.005mm, which is a unique texture of mudstone and shale.

3) Crystalline texture. The texture formed by chemical or biochemical precipitates or evaporites from a solution. It is the main texture of limestone, dolomite and siliceous rocks.

4) Bioclastic texture. This texture is formed by biological remains or fragments, such as shell, coral and bioclastic. Bioclastic texture is the unique texture of sedimentary rocks.

II. Fabric of sedimentary rocks

The fabric of sedimentary rocks refers to the spatial distribution of components and their arrangement. Bedding is one of the most important features of sedimentary rocks.

Bedding is the stratification of sedimentary rocks during the formation processes due to the depositional environment, which alters the grain size, composition, shape and color of sediments. Bedding is the main characteristic that distinguishes sedimentary rocks from igneous and metamorphic rocks. According to the different bedding structure, it can be classified by horizontal bedding, imbricate bedding and oblique bedding, as shown in Figure 1-13. The layered patterns indicate different depositional environments and transportation conditions.

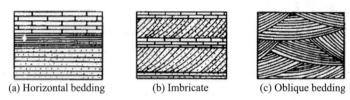

Figure 1-13　Common bedding types of sedimentary rocks

1.4.2.3　Common sedimentary rocks

According to the origin, mineral composition and texture, sedimentary rocks can be grouped to clastic sedimentary rock, clay sedimentary rock, chemical and biochemical sedimentary rock. Some common sedimentary rocks are shown in Figure 1-14.

Figure 1-14　Common sedimentary rocks

Ⅰ. Clastic rock

1) Pyroclastic rocks. The pyroclastic rocks are referred to the rocks that are formed during the cause of volcanic eruption. Due to nature of volcanic eruption and formation, the classification of pyroclastic rocks are as followed:

① Volcanic breccia. It accounts for more than 90% volcanic debris. The range of particle size is 2~100mm with mostly angular shaped. It is often with ash of dark grey, blue grey and brown gray in color.

② Tuff. It is formed by the cementation of volcanic ash and fine debris with particles size that are less than 2mm. The tuff has a rough surface with greyish color. The tuff is of large porosity and is easy to be weathered.

③ Volcanic agglomerate. It is mainly composed of coarse pyroclastic materials with particle size that are greater than 100mm. The cementation materials are mainly volcanic ash or lava, sometimes calcium carbonate, silica or argillaceous matter.

2) Clastic sedimentary rocks. The rocks are formed by the transport, deposition and

compression processes of debris from weathered rocks.

① Conglomerate and breccia. They are clastic rocks with more than 50% particle size greater than 2mm. If the gravels are well rounded, it is called conglomerate; whereas breccias have angular shaped gravels. The lithological composition of breccia is relatively complicated and is composed of various rock and mineral particles. Cementing agents can be calcium, iron, silicon, etc. The type of cementation material has a great influence on the physical and mechanical properties of the rock.

② Sandstone. They are clastic rocks with particle sizes ranging from 0.05 to 2mm, which contain more than 50% detritus particles. It can be subdivided by quartz, feldspar and lithic according to the mineral composition of the sand grains, and can be classified by coarse grained, medium grained, fine grained and mudrock according to the grain size; according to the cementation, it can be divided into siliceous sandstone, iron sandstone, calcareous sandstone and argillaceous sandstone. Sandstones have different weathering resistance and color, such as siliceous sandstone is white, with hard texture, while mudrock is often gray yellow and gray brown, which has water absorption, ability that makes it easy to soften, of low strength and poor stability.

③ Siltstone. They are clastic rocks with particle size range from 0.005 to 0.05mm, which contain more than 50% detritus particles. The mineral in clastic rock is mainly quartz. The grain size distribution is poor, with a powder texture, and most of them contain a clear horizontal bedding. Most of the cementing agents are calcium and iron. They often appear in brown or red brown color, with loose texture. The stability and intensity are low.

II. Clay rocks

Clay rocks, also known as argillaceous rocks, are the most common rock found in sedimentary rocks. They are a transitional type between clastic rock and chemical rock, consisting of cement minerals. Clay rock is sandwiched between hard rock layers, forming weak intercalated layer. It is easy to soften and slide upon wetting.

1) Claystone. It is formed by dehydration and solidification of clay. It is thickly stratified. It is easy to expand after absorbing water with black or brown color. Claystone is mainly composed of microcrystalline kaolinite with white, rose or light green color. The claystone, mainly composed of kaolinite, which is usually in grey or yellowish white color.

2) Shale. It is a rock formed by compression process. Shale has thin layers that can split into thin slices along the laminar surface. This feature is called foliation. Shale is roughly grouped to clayey shale, siliceous shale, calcareous shale, sandy shale and carbonaceous shale. Exception in the high strength of the siliceous shale, other types of shale are weak and fragile with low strength and easy to soften upon wetting and lose stability.

III. Chemical and biochemical rocks

The minerals from rock weathering are transported to low land in the form of ions in solutions or colloids, forming chemical rocks by evaporation or cementation, such as organic agents forming biochemical rocks.

1) Limestone. The mineral composition of limestone is dominated by calcite, with a small amount of dolomite and cementing agents. It is often gray and light gray, while pure limestone is white with dense block texture, where the dilute hydrochloric acid produces bubbles. Karst is easy to form in limestone areas, which is an unfavorable engineering material.

2) Dolomite. It is mainly composed of mineral dolomite with a small amount of calcite and clay minerals. The color is mostly gray or light gray with cryptocrystalline or fine grained texture. Dolomite and limestone are very similar in the surface features where dolomite can be dissolved by weak acidic solutions, but would not rapidly dissolve in dilute hydrochloric acid as limestone does. Since the appearance of dolomite is similar to that of limestone, thus it is difficult to distinguish them in the wild, but can be identified by hydrochloric acid.

1.4.2.4 Characteristics of sedimentary rocks

1) Sedimentary rocks are found over the largest surface area (95%) of the globe. However, they constitute only 5 percent of the composition of the crust whereas 95% of the crust is composed of igneous and metamorphic rocks.

2) Sedimentary rocks are typically deposited in layers or strata under cool surface conditions, but these are seldom crystalline rocks.

3) Sedimentary rocks are formed of sediments derived from the older rocks, plant and animal remains and thus these rocks contain fossils of plants and animals.

1.4.2.5 Sedimentary rocks identification

Sedimentary rocks such as limestone or shale are hardened sediment with sandy or clay-like layers. They are usually brown to gray in color and may have fossils and water or wind marks. When identifying a sedimentary rock, we need to check its grain size and hardness. Coarse grains are visible to the naked eye, and the minerals can usually be identified without using a magnifier. Fine grains are smaller and usually cannot be identified without using a magnifier. Hardness of the minerals contained within a rock can be measured with the Mohs scale (Figure 1-9). In simple terms, hard rock scratches glass and steel, usually signifying the minerals quartz or feldspar, which has a Mohs hardness of 6 or higher. Soft rock does not scratch steel but will scratch fingernails (Mohs scale of 3 to 5.5), while very soft rock won't even scratch fingernails (Mohs scale of 1 to 2). Clastic sedimentary rocks contain clasts. Chemical sedimentary rocks are identified by identifying the mineral from which they are composed. Biologic sedimentary rocks are which form as the result of the accumulation of organic material or biologic activity.

I. Clastic rock

We can observe the size of clastic particles first when identifying clastic rocks, then analyze the cements properties and main mineral components. It can be distinguished from conglomerate, sandstone or siltstone according to the size of clastic particles. Then, we can identify the type of the rocks according to the nature of the cementing agents and the main mineral components.

II. Clay rock

Common clay rocks include shale and claystone, both of which have the similar characteristics. However, the stratification of the shale is clear, and the layers can be separated to thin slices, with fragmented weathering, while claystone is not clear in layers and does not have fragmented weathering.

III. Chemical and biochemical rock

Common chemical rocks include limestone, dolomite and mudstone, etc. They have similar appearances, but differ in the content of calcite, dolomite and clay minerals. When identifying the chemical rocks, attention needs to be paid on the reactions. Limestone reacts with small drop of hydrochloric acid. However, the color is cloudy due to the high content of clay minerals in limestone. There is often a muddy spot after drying. The dolomite would not react with weak hydrochloric acid, but it is marked by a significant bubbling phenomenon.

1.4.3 Metamorphic rocks

Metamorphic rocks form when rocks are subjected to high heat, high pressure, and hot mineral-rich fluids or, more commonly, some combination of these factors. Conditions like these are found deep within the earth or where tectonic plates meet.

1.4.3.1 Metamorphism and related factors

Metamorphism is under the influence of extreme factors that making the rock to change its composition, structure and characteristics. The main factors that cause metamorphism are high temperature, high pressure or hot mineral-rich fluids.

I. High temperature

Most of metamorphism processes are under high temperature condition. The increase of temperature can make rocks to recrystallize and to form a new crystalline structure, such as the change of grain size after the recrystallization of limestone becoming marble. Another function of high heat is to promote chemical reactions between minerals, to produce new metamorphic minerals.

There are three types of heat sources that cause metamorphism: the first is the heat generated by tectonic movement; the second is the high temperature from the earth's crust; and the third is the heat from the hot magma.

II. High pressure

There are two types of high pressure that cause the metamorphism of rocks: one is the lateral pressure caused by tectonic movement or igneous activity; the other is the static pressure caused by the weight of overlying strata.

III. Hot mineral-rich fluids

In the process of metamorphism, the new chemical composition mainly comes from hot liquid and volatile gas containing complicated chemical reactions from igneous activities.

Under the combined condition of high temperature and high pressure, these components are easy to react with surrounding rocks to produce new metamorphic minerals.

Usually, the metamorphism of rocks result from the combination of these factors. The metamorphic rocks are formed from the rocks with different properties due to different degrees of metamorphism.

1.4.3.2 Mineral composition of metamorphic rocks

The minerals that make up metamorphic rocks are basically the same igneous rocks or sedimentary rocks, such as quartz, feldspar, mica, amphibolite, pyroxene, calcite, etc. The different component are produced after the metamorphism of metamorphic minerals, such as kyanite, andalusite, skarn, wollastonite, corundum, chlorite, epidote, sericite, talc, garnet, graphite, etc. These minerals can indicate the chemical composition of the original rocks and the physical and chemical environment where the metamorphic minerals are formed.

1.4.3.3 Texture and fabric of metamorphic rocks

I. **Texture of metamorphic rocks**

The texture of metamorphic rocks refers to the size, shape and arrangement of mineral particles in metamorphic rocks. According to different metamorphism, the texture of metamorphic rocks can be grouped to three categories: residual texture, crystalline texture and crushing texture.

1) Residual texture. Because of recrystallization and metamorphic reaction did not complete, some of the original rock structure feature is part of the retention, this form of structure is called a residual texture.

2) Crystalline texture. It is formed by recrystallization or recombinant cooperation and the mineral is more oriented and arranged, which is the most important structure of metamorphic rocks.

3) Crushing texture. The original rock is subjected to strong stress (dynamic metamorphism), resulting in the fragmentation, dislocation and recrystallization of mineral particles, thus forming a crushing texture. There are common breccia structure, fracture structure, porphyry structure, mylonitic structure and so on.

II. **Fabric of metamorphic rocks**

The fabric of metamorphic rocks refers to the spatial distribution and arrangement of minerals in metamorphic rocks.

1) Schistose structure. The structure is formed by the tectonic movement, where the components of the rock are crystallized into the new crystal and aligned. Because of the dense and repetitive arrangement of flake minerals, it would form a strip or slice surface along the laminar zone.

2) Lamellar structure. Lamellar structure is formed by the orientation of crystals, schistose minerals and granular minerals, as shown in Figure 1-15.

Figure 1-15　Lamellar structure

3) Phyllitic structure (Figure 1-16). Phyllitic structure is formed by the metamorphism of soil or tuff. The components in the rock were recrystallized into microcrystalline and aligned. The fracture surface is well developed.

4) Tabular structure (Figure 1-17). The rock presents a foliated fracture surface, some of the microcrystalline schistose minerals such as sericite and chlorite can be seen on the surface.

Figure 1-16　Phyllitic structure　　　　Figure 1-17　Tabular structure

1.4.3.4　Common metamorphic rocks

Metamorphic rocks maybe classified on the basis of foliation into foliated and non-foliated rocks. Foliated rocks includes the rock that can split into thin sheets. Non-foliated rocks do not have lineations, foliations, or other alignments of mineral grains. Figure 1-18 shows some common metamorphic rocks.

I. Foliated rocks

1) Slate. The metamorphic degree of slate is very low, the recrystallization is not obvious. The hardness increases after dehydration. The surface of slate is dense and cryptocrystalline, where the mineral composition is difficult to identify. In some case, there is a small amount of mica, chlorite and other new minerals on the surface. Slate is generally made of mudstone, siltstone and acid tuff. A hard and elastic slate can be sliced and peeled along the surface, it is used as a building material such as tiles or floor panels.

2) Phyllite. The metamorphism grade is slightly higher than that of slate. The mineral composition of the original rock has been completely recrystallized, consisting mainly of sericite, chlorite, quartz, sodium feldspar. The nature of phyllite rocks is weak and fragile.

3) Schist. Schist is a medium-grade metamorphic rock with medium to large, flat, sheet like grains in a preferred orientation. It is mainly composed of schistose minerals (mica, chlorite, talcum, etc), lamellar minerals (actinolite, diorite, amphibolite, etc) and granular minerals (feldspar, quartz, etc).

4) Gneiss. The particle size of gneiss is generally greater than 1mm, and it is a coarse grained structure. Its mineral compositions are mainly feldspar, quartz, mica, amphibolite, pyrolite. In same cases, it include the characteristic of metamorphic minerals such as barite, cyanite, garnet, etc.

II. Non-foliated rocks

1) Marble. Marble is a metamorphic rock formed by process of contact thermal metamorphism or regional metamorphism of carbonate rocks (limestone and dolomite). Generally, it is of white color, with impurities that can present different colors and patterns. The mineral composition are mainly calcite and dolomite. Marble has the characteristic of metamorphic minerals such as serpentine, diatomite, and talc.

2) Quartzite. Quartzite is a metamorphic rock with granular crystal structure. Generally, it has a non-foliated structure and sometimes with an orientated structure. Quartzite is formed by the metamorphism of quartz sandstone or siliceous rocks by regional metamorphism or contact metamorphism. It can contain a small amount of feldspar, mica, chlorite, amphibole, pyrolite. Quartzite is widely distributed on the earth's surface, and it is a good building material.

III. Tectonic fractured rocks

1) Tectonic breccia. It is a rock formed by fracture in the fracture zone that causes the original rock to break into breccia (>2mm) and is cemented by fine debris, mud and partial exsolution. Tectonic breccia is an important symbol to identify the existence and properties of fracture zone.

2) Mylonite. It is a fined-grained, compacted metamorphic rock that formed in ductile fault zone. The fragments in the rock are spherical or lenticular and are aligned. Mylonite is present in the fault zone in the deeper depths of underground (>10km).

Figure 1-18 Common metamorphic rocks

1.4.3.5 Metamorphic rocks identification

The texture of the rock is first observed when identifying a metamorphic rock. According to the texture, the metamorphic rocks can be classified by two types: foliated and non-foliated. Then, determine the type of the rock according to the characteristics of the band and the main mineral components.

The metamorphic rocks are commonly found as marble and quartzite. Both are minerals with variable crystal structures, and the rocks are generally light in color. However, marble is mainly composed of calcite with low hardness, which can dissolve in hydrochloric acid. In contrast, quartzite is composed almost of quartz with high hardness, which cannot dissolve in hydrochloric acid.

1.5 Rock cycle

The rock cycle is a concept of geology that describes the transition of rocks between the three rock types: igneous, sedimentary, and metamorphic. It was first suggested by James Hutton, the 18th-century founder of modern geology. As shown in Figure 1-19, the rock cycle outlines how each rock type can be converted to another rock type through geologic processes. In essence, geologic processes can transform one type of rock into another. Igneous rocks can be transformed into either metamorphic rocks or sedimentary rocks, while sedimentary rocks can transform into both metamorphic and igneous rocks. Naturally, metamorphic rocks can transform into sedimentary or igneous rocks.

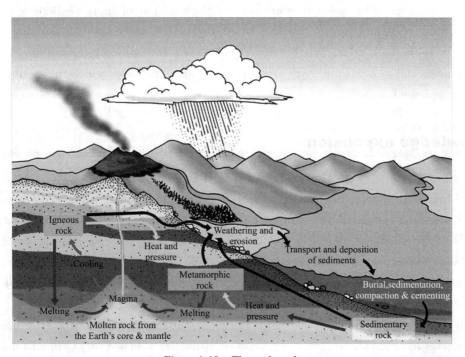

Figure 1-19 The rock cycle

The transition between rock types can happen due to a variety of processes. Metamorphic and sedimentary rocks can become igneous rocks by being turned into magma at high temperatures, which usually takes millions of years to happen. Metamorphic rocks are created through temperature and pressure, as long as the temperature isn't too high. Metamorphic rocks can be created through either contact metamorphism or regional

Engineering Geology and Soil Mechanics

metamorphism. Regional metamorphism happens as geologic regions smash together, like at mountain ranges, which creates foliated metamorphic rocks. Contact metamorphism happens when rocks come close to hot igneous rocks, which ends up altering the properties of the rock and creates crystals. Finally, surface processes like precipitation and weathering can transform metamorphic and igneous rocks into sedimentary rocks.

The primary forces behind the rock cycle are geological. These forces are subsidence and uplift. Uplift refers to the pushing of rock out of the Earth's surface, creating mountains. Subsidence is when one rock plate submerges beneath another plate and becomes magma deep underneath the Earth. Both subsidence and uplift are caused by plate tectonics. The continental plates which lie on the Earth's mantle are moving. While the plates usually only move at a few centimeters a year, over millions of years this leads to massive changes in the Earth's surface. As these plates collide and slip past one another, certain regions collide and are pushed up while other regions are pushed below.

The phenomenon of seafloor spreading also plays a role in the rock cycle. The tectonic plates which collide with each other are by necessity pulling away from each other in other areas. This means that magma is released from the Earth where the plates are dividing, deep in the ocean. This phenomenon leads to the creation of more seafloor along with metamorphic rocks and igneous rocks. Besides, surface processes like weathering also contribute to the rock cycle. Wind and rain can wear away and rocks over time, breaking apart igneous and metamorphic rock structures into small particles which then join together to create sedimentary rocks.

Knowledge expansion

Uses of Rocks in Civil Engineering

Most civil engineering projects involve some excavation of soils and rocks, or involve loading the Earth by building on it. In some cases, the excavated rocks may be used as constructional material, and in others, rocks may form a major part of the finished product, such as a highway cutting or the site for a reservoir. For example, blocks of stones are used in foundations, walls, bridge pier, abutments, lighthouses, aqueducts, and retaining walls. Rocks are used for masonry work, lintels, and vertical columns, covering floors of the building. Broken or crushed rocks are used as aggregates in concrete, in road constructions. Broken or crushed rocks are also used as railway ballast. Limestone is the basic material for the manufacture of lime concrete and cement. Rocks are used as blocks along the river and canal banks for preventing erosion called riprap. Base material for water and sewage filters, in case of waterworks and sewage treatment plants. In the construction of the masonry dam, stones of good quality and durability are of vital significance. Rocks are also used for ornamental works in and outside the buildings, like blocks, slabs, and chips.

Following are some rocks used in building construction in Civil Engineering:

1) Igneous rocks

① Granite is an igneous-plutonic rock, medium to coarse-grained that is high in silica, potassium, sodium and quartz but low in calcium, iron and magnesium. It is widely used for architectural construction, ornamental stone and monuments. ② Pumice is an igneous-volcanic rock, it is a porous, brittle variety of rhyolite and is light enough to float. It is formed when magma of granite composition erupts at the earth's surface or intrudes the crust at shallow depths. It is used as an abrasive material in hand soaps, emery boards, etc. ③ Gabbro is an igneous-plutonic rock, generally massive, but may exhibit a layered structure produced by successive layers of different mineral composition. It is widely used as crushed stone for concrete aggregate, road metal, railroad ballast, etc. Smaller quantities are cut and polished for dimension stone (called black granite). ④ Basalt is an igneous volcanic rock, dark gray to black, it is the volcanic equivalent of plutonic gabbro and is rich in ferromagnesian minerals. Basalt can be used in aggregate.

2) Sedimentary rocks

① Coal is a sedimentary rock, formed from decayed plants, is mainly used in power plants to make electricity. ② Limestone is a sedimentary rock, it is used mainly in the manufacture of Portland cement, the production of lime, manufacture of paper, petrochemicals, insecticides, linoleum, fiberglass, glass, carpet backing and as the coating on many types of chewing gum. ③ Shale is a sedimentary rock, well stratified in thin beds. It splits unevenly more or less parallel to bedding plane and may contain fossils. It can be a component of bricks and cement. ④ Conglomerate is a sedimentary rock with a variable hardness, consisted of rounded or angular rock or mineral fragments cemented by silica, lime, iron oxide, etc. Usually found in mostly thick, crudely stratified layers. Used in the construction industry. ⑤ Sandstone is a sedimentary rock more or less rounded. Generally thick-bedded, varicolored, rough feel due to uneven surface produced by breaking around the grains. Used principally for construction, it is easy to work.

3) Metamorphic rocks

① Schist is a metamorphic uneven-granular, medium to coarse grained, crystalline with prominent parallel mineral orientation. It goes from silvery white to all shades of gray with yellow to brown tones depending on the mineral concentration. Some schists have graphite and some are used as building stones. ② Gneiss is metamorphic uneven granular medium to coarse grained crystalline with more or less parallel mineral orientation. Colors are too variable to be of diagnostic value. Due to physical and chemical similarity between many gneisses and plutonic igneous rocks some are used as building stones and other structural purposes. ③ Quartzite is a metamorphic rock with crystalline texture, consists of rounded quartz grains cemented by crystalline quartz, generally white, light gray or yellow to brown. Quartzite has the same uses as sandstone. ④ Marble is a metamorphic even-granular grain to medium-grained and may be uneven granular and coarse-grained in calc-silicate rock. The normal color is white but accessory minerals act as coloring agents and may produce a variety of colors. Depending upon its purity, texture, color and marbled pattern it is quarried for use as dimension stone for statuary, architectural and ornamental purposes.

Exercises

[1-1] What is igneous rock? What is sedimentary rock? What is metamorphic rock?

[1-2] What is the metamorphism? What are the main minerals in metamorphic rocks?

[1-3] What's the difference between a rock and a mineral?

Chapter 2　Geologic Time and Structures

2.1　Introductory case

Geologic structure plays an important role in the stability of surrounding rocks. When the tunnel passes through the layered rock mass between soft and hard phase, it is easy to deform or slump at the contact surface. If the axis of the tunnel and the direction of the strata are close to normal, the length of the tunnel through the weak strata can be shortened. If the direction of the strata is close to parallel to the axis and cannot be completely constructed through the hard rock and the cross section passes through different strata, the cave should be properly adjusted, so that

Figure 2-1　Tunnel construction through a mountain

the surrounding rock would be more stable. The tunnel should be set in the hard rock, or in the layer as hard as the rock.

When the tunnel is near or in the fracture zone of a fault, if the fault zone width is larger, the smaller the angle of the strike to the tunnel axis, the longer it appears in the tunnel and the greater its impact on the stability of the surrounding rock. In conclusion, it is very important to understand the relevant knowledge of geological structure for tunnel construction (Figure 2-1).

2.2　Geologic time

The Earth is very old—4.6 billion years or more—according to the recent estimates. The vast span of time, called geologic time named by earth scientists, is difficult to understand in the traditional time units as months or years, or even centuries. How do scientists define geologic time, why do they believe the Earth is so old? A great part of the Earth's age is locked up in the rocks, and our centuries-old search for the beginning and nourished the growth of geologic science.

The crust is made up of rocks of different ages, and the rock is a product from different times and different geologic processes. In geological history, the earth's crust has experienced many times of intense tectonic movements, magmatic activities and the prosperity of sea and land changes, biological and extinction, the crust of weathering and erosion. To find out geological events or geological formation of the era and order is necessary. In addition, for engineering geology exploration of a region, we need to get all kinds of geologic structure information and stratigraphic contact relationship, reading and analysis of geological data, maps and other data. We must have the basic knowledge of geological time, thus to have a basic understanding of the geological time.

Two scales are used to define the episodes and to measure the age of the earth: a relative time scale, based on the sequence of layered rocks and the evolution of lives, the absolute or radiometric time scale, based on the natural radioactivity of chemical elements in the rocks.

2.2.1 Relative geologic time

Since the formation of the earth, geologic processes throughout geological history have been continuously carried out. In each of geological history stage, there are the formation and development of minerals, rocks, biological and geological structures, as well as the destruction and extinction of them. Combing various geological period of rock formation, and things buried in them like fossils, we can get the relationship of different ages according to the relative time.

The relative geologic time is mainly determined by the sedimentary sequence, biological evolution and geological structure of rock strata. The following items are usually used to determine the relative geologic time including stratigraphy, palaeontology, comparative lithology antitheses, stratigraphic contact relation and tectonics.

2.2.1.1 Stratigraphy

Sedimentary rocks are always formed by layers. After the formation of sedimentary rocks, the bottom strata are usually older than the upper layers, which means that the strata of the original output have the new regularity of the old. By connecting all the layers of rock in one area, the rock formations at different times can be divided into stratum, which is associated with time. The stratigraphic sequence method is the basic method to determine the relative geological time of the strata.

In 1669, Nicolaus Steno (1638—1686) proposed the principles of stratigraphy as follows:

1) Principle of superposition. In a sequence of strata, any stratum is younger than the sequence of strata on which it rests, and is older than the strata that rest upon it. At the time when any given stratum was formed, the matters resting upon were fluids, therefore, at the time when the bottom stratum was formed, there is no upper strata existed.

2) Principle of initial horizontality. Strata are deposited horizontally and then deformed to various attitudes. Strata that is either perpendicular to the horizon or inclined to the horizon was at one time parallel to the horizon direction.

3) Principle of strata continuity. Strata can be assumed to be continued laterally. Materials forming any stratum were continuous over the surface of the earth unless some other solid bodies stood in the way.

4) Principle of cross cutting relationships. The core of cross-cut layers are probably younger. If a body cuts across a stratum, it should have formed after that stratum.

2.2.1.2 Palaeontology

Creatures in geologic history are called paleontology. The remains of these ancient creatures can be found in sedimentary rocks, which are composed of calcium, silica, or metasomatism, fossil. In geological history, geological processes constantly change the surface of the earth's environment. In order to adapt to the change of the natural environment, constantly changing inside and outside the function of the organ itself, namely each geological time have to adapt to the natural environment characteristic biota. Generally, the evolutionary trend of biology is constantly evolving from simple to complex irreversibly. Therefore, the more primitive, simple and inferior organisms in the older strata of the stratum; Conversely, the more advanced, complex and advanced biology in the new stratum of the era. The more simple the structure of the biological fossils buried in the strata, the more complex the structure of the fossils is. During the same geological period, the formation of rock formations in the same geological environment often contained the same fossil or fossil assemblages. On the other hand, as long as the fossil species are the same, their geological time should be the same. To this end, according to the fossil species in the strata, the method of establishing the stratigraphic sequence and determining the geological time is called paleontological method.

It should be pointed out that fossils of decisive significance for studying geologic time should have the characteristics of rapid evolution, short duration, large quantity and wide distribution in geological history, such fossil known as index fossil. It is more accurate to use index fossil to determine stratigraphic record dating.

2.2.1.3 Comparative lithology antitheses

During the same period, the rocks formed under the same geological environment have the same lithology characteristics and sequence laws of color, composition, and structure. Therefore, we can determine the age of rock strata in a certain area according to the comparison of lithological characteristics.

2.2.1.4 Stratigraphic contact relation

In geological history, a region cannot be permanently deposited and is often affected by tectonic movements and magmatic activity. Weathering and denudation of rock formations, or tilting and folding of strata, result in discontinuities in sedimentation or uncoordinated formation of upper and lower strata. Therefore, any stratum sections presented in the region will be missing strata in some times, resulting in incomplete geological records, so that sedimentary rocks, igneous rocks and their different types of contact with each other, which can determine the relationship between the formation of rocks of different ages.

2.2.1.5 Tectonics

Method of structural geology tectonics and magmatic activities can make the different age of strata, break and thrust deep into the relation between rock mass. This relationship can be used to determine the order of these strata (or rock) and geological time, which is called tectonics.

2.2.2 Absolute geologic time

The relative time can only show the relative age relation of rocks and strata, rather than the exact date of the formation of rock or rock formation. To determine the exact age of rocks or minerals, other methods can be used to determine the age of their formation.

There are radioactive isotopes in nature, which are found in substances in nature. Geologists use the decay of the radioactive isotopes to determine the age of minerals or rocks, which are known as the isotopic age or absolute age. This method has been widely used in geological field.

There are many radioactive isotopes in nature, their radioactive decay are different, nor their half-life. Most radioactive isotopes decay rapidly with short half-life, and only a small amounts of radioactive elements decay slowly with half-life in billions of years, as shown in Table 2-1. Half-life is the time with unit a (year) required for one-half of parent istope in a sample to decay into a stable daughter isotope.

Table 2-1 Radioactive isotopes used to determine the geologic time

Parent isotope	Daughter isotope	Half life/a	Decay constant λ
U^{238}	Pb^{206}	4.47×10^9	1.55×10^{-10}
U^{235}	Pb^{207}	7.04×10^8	9.85×10^{-10}
Th^{232}	Pb^{208}	1.04×10^{10}	4.95×10^{-11}
Rb^{87}	Sr^{87}	4.89×10^{10}	1.42×10^{-11}
K^{40}	Ar^{40}	1.25×10^9	5.54×10^{-10}
Sm^{147}	Nd^{144}	1.06×10^{10}	6.54×10^{-12}
C^{14}	N^{14}	5.73×10^3	1.21×10^{-4}

The mathematical expression that relates radioactive to geologictime is:

$$t = \frac{1}{\lambda} \times \ln\left(1 + \frac{D}{P}\right) \tag{2-1}$$

where, t—the age of rock or mineral specimen;

D—the number of atoms of a daughter product today;

P—the number of atoms of the parent product today;

λ—the appropriate decay constant (Table 2-1).

2.2.3 Geologic time scale

The geologic time scale (GTS) is a system of chronological measurement that relates stratigraphy to time, and is used by geologists, paleontologists, and earth scientists to describe the timing and relationships between events that have occurred throughout earth's history.

The geologic time scale is a "calendar" for events in Earth history. It subdivides all time into named units of abstract time called—in descending order of duration—**eons**, **eras**, **periods**, **epochs**, and **ages**. The enumeration of those geologic time units is based on stratigraphy, which is the correlation and classification of rock strata. Table 2-2 shows the geologic time scale.

Table 2-2 Geologic time scale

Million years ago	Eon	Era		Eon	Era	Period		Epoch		Million years ago
0										
	Phanerozoic	Cenozoic								
		Mesozoic				Quaternary		Holocene		0.01
500		Paleozoic						Pleistocene	Upper	0.13
									Middle	0.8
									Lower	
		Neoproterozoic					Neogene	Pliocene	Late	2.59
1000					Cenozoic				Early	3.6
										5.3
	Proterozoic							Miocene	Late	11.2
1500		Mesoproterozoic							Middle	16.4
1600						Tertiary			Early	23.03
							Paleogene	Oligocene	Late	28.4
2000		Paleoproterozoic							Early	33.9
								Eocene	Late	41.3
2500									Middle	49
	Precambrian								Early	55.8
2800		Neoarchean						Paleocene	Late	61
				Phanerozoic					Early	65.5
3000		Mesoarchean			Mesozoic	Cretaceous		Late		99
3200								Early		145.5
						Jurassic		Late		159
3500	Archean	Paleoarchean						Middle		180
3600								Early		199.6
		Eoarchean				Triassic		Late		227
4000								Middle		242
								Early		252
						Permian		Late		256
								Early		299
						Carboniferous		Pennsylvanian		323
								Mississippian		359.2
					Paleozoic	Devonian		Early		370
								Middle		391
4500	Hadean							Late		416
About 4600						Silurian		Early		423
								Late		443.7
						Ordovician		Late		458
								Middle		470
								Early		488.3
						Cambrian				541

Eons, a long span of geologic time. In formal usage, eons are the longest portions of geologic time (eras are the second-longest). Three eons are recognized as following: the Phanerozoic Eon (dating from the present back to the beginning of the Cambrian Period), the Proterozoic Eon, and the Archean Eon. Less formally, eon often refers to a span of one billion years. Formal geologic time begins at the start of the Archean Eon (2.5 billion to 4.0 billion years ago) and continues to the present day. Modern geologic time scales additionally often include the Hadean Eon, which is an informal interval that extends from 4.0 billion years ago to about 4.6 billion years ago (corresponding to Earth's initial formation), as shown in Table 2-2.

Eras, a long span of geologic time; in formal usage, the second longest portions of geological time (eons are the longest). Ten eras are recognized by the International Union of Geological Sciences: the Eoarchean Era (3.6 billion to 4.0 billion years ago), the Paleoarchean Era (3.2 billion to 3.6 billion years ago), the Mesoarchean Era (2.8 billion to 3.2 billion years ago), the Neoarchean Era (2.5 billion to 2.8 billion years ago), the Paleoproterozoic Era (1.6 billion to 2.5 billion years ago), the Mesoproterozoic Era (1.0 billion to 1.6 billion years ago), the Neoproterozoic Era (541 million to 1.0 billion years ago), the Paleozoic Era (252 million to about 541 million years ago), the Mesozoic Era (roughly 65.5 million to 252 million years ago), and the Cenozoic Era (65.5 million years ago to the present). Because of the difficulties involved in establishing accurate chronologies, the eras contained within Precambrian time (approximately 541 million to 4.6 billion years ago) are classified independently. An era is composed of one or more geological periods.

Period, in geology, the basic unit of the geologic time scale; during these spans of time specific systems of rocks were formed. Originally, the sequential nature of defining periods was relative, originating from the superposition of corresponding stratigraphic sequences and the evidence derived from paleontological studies. With the advent of radiometric dating methods, absolute ages for various periods can be determined.

Epoch, unit of geological time during which a rock series is deposited. It is a subdivision of a geological period, and the term is capitalized when employed in a formal sense (e.g., Pleistocene Epoch). Additional distinctions can be made by appending relative time terms, such as early, middle, and late. The use of epoch is usually restricted to divisions of the Paleogene, Neogene and Quaternary periods.

Ages, a geologic age is a subdivision of geologic time scale divides an epoch into smaller parts. A succession of rock strata laid down in a single age on the geologic time scale is a stage.

Geological history in each geologic time has a corresponding stratigraphic formation, known as the chronostratigraphic units, including **eonothem**, **erathem**, **system**, **series**, **stage**. Table 2-3 shows the units in geochronology and stratigraphy.

Table 2-3 Units in geochronology and stratigraphy

Segments of rock (strata) in chronostratigraphy	Eonothem	Erathem	System	Series	Stage
Time spans in geochronology	Eon	Era	Period	Epoch	Age

In Table 2-3, **Eonothem** is the totality of rock strata laid down in the stratigraphic record deposited during a certain eon of the continuous geologic timescale. **Erathem** is the total stratigraphic unit deposited during a certain corresponding span of time during an era in the geologic time scale. **System** in stratigraphy is a unit of rock layers that were laid down together within the same geological period. **Series** are subdivisions of rock layers based on the epoch of the rock and formally defined by international conventions of the geologic time scale. A series is a sequence of strata defining a chronostratigraphic unit. **Stage** is a succession of rock strata laid down in a single age on the geologic time scale, which usually represents millions of years of rock deposition.

2.3 Geologic structures

Geologic structures are usually the result of the tectonic movements that occur on the earth. The movements fold and break rocks, forming deep faults and mountains. Geologic structures such as faults and folds are the architecture of the earth's crust. Geologic structures influence the shape of the landscape, determine the degree of landslide hazard, bring old rocks to the surface, bury young rocks, trap petroleum and natural gas, shift during earthquakes, etc.

2.3.1 Attitude of stratum

Geological structures formed by the movement of the earth's crust, which are always made up of fracturing surfaces in rocks with a certain amount and space. Therefore, a basic content to study of geological structure is to determine the spatial state of the strata and fracture surfaces and their characteristics on the ground.

Determining the output status of rock strata is the basis of studying geological structures. The concept of attitude of stratum is often used in geology.

The formation of strata in the earth's crust is the result of spatial orientation of the strata. Elements of attitude that include strike, dip and dip angle (Figure 2-2) are used to determine the occurrence of rock formations.

In Figure 2-2, strike refers to the line formed by the intersection of a horizontal plane and an inclined surface. This line is called a strike line, and the direction the line points in (either direction, as a line points in two opposite directions) is the strike direction. For a horizontal

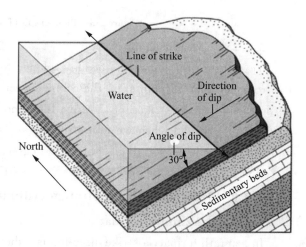

Figure 2-2 Element of attitude

plane, strike is undefined. The dip direction will be orthogonal to the strike direction. Water will flow down the plane in this direction. Angle of dip is the angle between that horizontal plane and the inclined surface measured perpendicular to the strike line down to the inclined surface. A horizontal plane has a dip angel of 0°, and a vertical plane a dip angel of 90°. The larger the dip angel, the steeper is the plane.

2.3.2 Determination of stratum attitude

The measurement of rock formations is an important work in geological survey, which is measured at the level of rock strata in filed.

2.3.2.1 Measuring strike

When measuring the strike of a bedding plane, a Brunton compass shown in Figure 2-3 is used. To measure the strike, place the side or edge of the compass against the plane of the outcrop. Sometimes it is easier to put your field book against the outcrop and then the compass against the book to get a smoother and/or a larger surface. Rotate the compass while keeping the lower side edge of the compass fixed, until the bulls-eye level bubble is centered (the round tube; not the long narrow one). When the bubble is centered, the compass is horizontal against the plane and parallel to the line of strike. Then, with the bulls-eye bubble centered, record the number that either end of the compass needle shown.

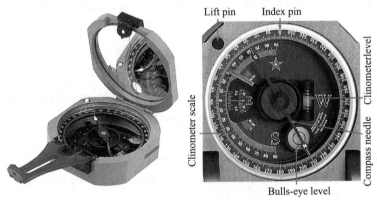

Figure 2-3 Photograph of a Brunton compass

2.3.2.2 Measuring dip

To measure the dip of the bedding plane, take your compass and put its side against the rock so that it points in the same direction as the line of dip (the dip line is perpendicular to the strike line). Move the clinometer until the clinometer level bubble is centered. As we did when we found the strike, record where the white tipped end of the clinometer needle is pointing. Note the degrees and the direction. Recall that the dip direction must always be perpendicular to the strike direction.

2.3.3 Representation of stratum attitude

2.3.3.1 Azimuth format

In azimuth format, a strike direction is indicated by an angle 0°~360° with North at 0° (or 360°), East at 90°, South at 180°, and West at 270°. Since dip direction is assumed to be clockwise from strike bearing (right hand rule), the strike of 0° means the bed is dipping east. Besides, a dip angle of 0° means its flat and 90° for a vertical bed. Therefore, the stratum attitude can be captured using azimuth format by direction and dip angle. For example, 90°/30° indicates the plane is inclined 30° and towards the east (90°).

2.3.3.2 Quadrant format

In quadrant format, the direction is determined by an angle ≤ 90° relative to a North or South compass direction (N40°E or S80°W). Note that N40°E quadrant is the same as 40° azimuth, and S80°W is equivalent to 260° azimuth.

2.4 Contact of the stratum

A unconformity is a buried erosional or non-depositional surface that separates two different-age rock masses or strata, indicating that the deposition of sediments was not continuous. The older layer was generally exposed to erosion for a period of time before the younger layer was deposited, but the term is used to describe any break in the sedimentary geological record.

Unconformities are gaps in the geologic record that may indicate episodes of crustal deformation, erosion, and sea level variations. They are characteristic of stratified rocks thus usually found in sediments (it can also be found in stratified volcanics). They are surfaces that form a substantial break (hiatus) in the geological record between rock bodies. Unconformities represent times when deposition stopped, some of the previously deposited rock was removed by erosion and resumed deposition. There are three types of unconformities: nonconformity, angular unconformity, and disconformity, as shown in Figure 2-4.

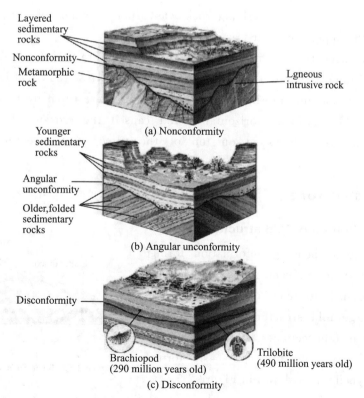

Figure 2-4 Three types of unconformities

2.4.1 Nonconformity

A nonconformity is the contact that separates a younger sedimentary rock unit from an igneous intrusive rock or metamorphic rock unit. A nonconformity suggests that a period of long-term uplift, weathering, and erosion occurred to expose the older, deeper rock at the surface before it was finally deposited below the younger rocks. A nonconformity is the aged erosional surface on the underlying rock.

2.4.2 Disconformity

Disconformities are usually erosional contacts that are parallel to the bedding planes of the upper and lower rock units. Since disconformities are hard to recognize in a layered sedimentary rock sequence, they are often discovered via the fossils. A gap in the fossil record indicates a gap in the depositional record, and the length of time the disconformity represents can be calculated. Disconformities are usually a result of erosion but can occasionally represent periods of non-deposition.

2.4.3 Angular unconformity

An angular unconformity is the contact that separates a younger, gently dipping rock unit from older underlying rocks that are tilted or deformed layered rock. The contact is more obvious than a disconformity because the rock units are not parallel and at first appear cross-cutting. Angular unconformities generally represent a longer time hiatus than disconformities because the underlying rock was usually metamorphosed, uplifted, and eroded before the upper rock unit was deposited.

The Phanerozoic strata in most of the Grand Canyon are horizontal. However, near the bottom horizontal strata overlies tilted strata. It is known as the Great Unconformity is an example of angular unconformity. The lower strata were tilted by tectonic processes that disturbed their original horizontality and caused the erosion of strata. Then, horizontal strata were deposited on top of the tilted strata creating the angular unconformity.

2.5 Fold structure

2.5.1 Definition of fold structure

The continuous bending deformation formed under the tectonic movement is called fold structure. The size difference of fold structure is very large, large fold structure extends tens of kilometers, small fold tectonics also can be seen in the specimen. The fold has two basic forms: anticline and syncline, as shown in Figure 2-5.

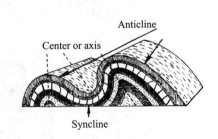

Figure 2-5 Basic forms of fold structure

1) Anticline. It is a series of up-arched strata. Sides (limbs) dip in opposite directions from central fold which is split by axial plane or fold axis. An eroded surface indicates a

pattern of progressively younger rocks away from the fold axis.

2) Syncline. It is a series of down-arched strata dipping towards the fold axis on both sides. Formations become progressively older from fold axis on an eroded surface.

Fold is one of the most common geologic structures found in rocks. When a set of horizontal layers are subjected to compressive forces, they bend either upward or downward. In a series of folds it is evident like waves. They consist of alternate crests and troughs. The crest of the fold is anticline while the trough is called syncline. An anticline and syncline constitute a fold (Figure 2-6). To describe the shape characteristics of the fold, some elements are given in Figure 2-6:

1) Core. It refers to the strata in the center of a fold.

2) Limb or flank. It is sloping side from the crest to the trough.

3) Axial plane. An imaginary plane bisects the vertical angle between equal slopes on either sides of the crest line.

4) Hinge line. It refers to the axis of the fold that divides the section of the fold.

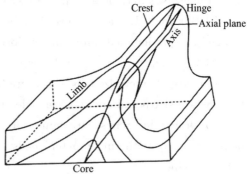

Figure 2-6 Elements of a fold

2.5.2 Types of fold structures

Folds show a great variety of forms; some may be quite simple, whereas others may be highly complex and complicated in their geometry and morphology. In fact, in most cases folds may be simple or complex modifications of two basic types of folds: anticline and syncline. Based on the geometrical appearance in cross-section, the types of folds are described below (with some types shown in Figure 2-7).

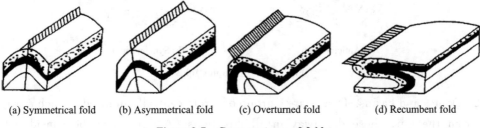

(a) Symmetrical fold   (b) Asymmetrical fold   (c) Overturned fold   (d) Recumbent fold

Figure 2-7 Common types of folds

1) Symmetrical folds. These are also called normal or upright folds. In such a fold, the axial plane is essentially vertical. The limbs are equal in length and dip equally in opposite directions.

2) Asymmetrical folds. All those folds, anticlines or synclines, in which the limbs are unequal in length and these dip unequally on either side from the hinge line are termed as asymmetrical folds. The axial plane in an asymmetrical fold is essentially inclined.

3) Overturned folds. These are folds with inclined axial planes in which both the limbs

are dipping essentially in the same general direction. The amount of dip of the two limbs may or may not be the same. Over folding indicates very severe degree of folding.

4) Recumbent folds. These may be described as extreme types of overturned folds in which the axial plane acquires an almost horizontal attitude.

5) Isoclinal folds. These are group of folds in which all the axial planes are essentially parallel, meaning that all the component limbs are dipping at equal amounts.

6) Conjugate folds. A pair of folds that are apparently related to each other may have mutually inclined axial planes. Such folds are described as conjugate folds. The individual folds themselves may be anticlinal or synclinal or their modifications.

7) Box fold. It may be described as a special type of fold with exceptionally flattened top and steeply inclined limbs almost forming three sides of a rectangle.

8) Chevron fold. Usually the crest and troughs of a fold are rounded, but sometimes the folds are characterized by sharp crests and troughs. Such folds where the crests and troughs are sharp and angular are called Chevron Folds.

9) Drag fold. These are minor or small folds formed when competent beds (weak beds) moves over the incompetent beds. The axial planes of these folds are inclined to the bedding planes.

10) Plunging fold. Any fold in which fold axis is not horizontal.

11) Non-plunging fold. Any fold in which the axis of fold is essentially horizontal.

Based on the mode of occurrence, we have the following types:

1) Anticlinorium and synclinorium (Figure 2-8). They are called respectively to exceptionally large sized fold. An Anticlinorium is a fan shaped structure and the trend if the fold is anticline character. Synclinorium is a regional structure of general synclinal form that includes a series of smaller folds.

(a) Anticlinorium (b) Synclinorium

Figure 2-8 Complex folds

2) Domes and basins. Domes are a group of strata centrally uplifted in such a way that seen from the top, these dip away in all directions. A dome may be considered as a compound anticline. Basins are the reverse of the domes and may be defined as a group of strata that are centrally depressed in such a way that the involved layers dip towards a common central point from all the sides. A basin may be called a compound syncline.

Based on the folding of rocks with depth, we have the following types:

1) Similar folding. In this folding the bedding planes are similar having the dame shape downwards or upwards so the beds near the crest are thicker and the beds at the limbs are thinner.

2) Parallel folding. In this type of folding the bedding plane remains parallel through

but because of this anticlines, which are sharp, becomes rounded and more broad. Similar Synclines, which are broad and rounded, becomes sharper with depth.

3) Diaper fold. This type of folding is common on sedimentary beds where more mobile beds are found at the center. There are anticlines where more mobile core has broken through the overlying brittle rocks.

4) Disharmonic folding. In this type, folding is not uniform throughout the stratigraphic column.

5) Suprataneous folding. When folding and sedimentation are contemporaneous, suprataneous folding is formed. The beds near the crest are thinner and near the troughs are thicker.

6) Decollement fold. In this type of folding a sheet of sedimentary bed breaks loose from the underlying fold independently without affecting the lower beds.

2.6 Fracture structure

The rock strata is subjected to tectonic movement. When the tectonic stress is more than the rock strength, the continuous integrity of the rock is destroyed resulting in fracture, which is called fracture structure. According to the fracture whether there was obvious relative displacement, two types of fracture structure can be obtained: joints and faults.

2.6.1 Joints

Joint is a fracture structure when the rock strata is disconnected and there was no relative displacement along the fracture surface. Joint fracture surface is called joint surface. Joint distribution is common, almost all of the rock has joint development. The extension of the joint varies greatly from a few centimeters to tens of meters. The state of the joints in the space is called the joint formation, and its definition and measurement methods are similar to those of the rock formation. Joints often divide the rock strata into different shapes and sizes. The strength of small pieces of rock with or without joints is significantly different. The instability of rock slopes and collapse of tunnel roof are often associated with joints.

2.6.1.1 Joint classification

The joints can be classified according to the causes, the mechanical properties, the relationship with the formation of rocks and the degree of opening.

I. Classification according to cause of formation

According to the cause of formation, joints can be divided into primary joints and secondary joints. Secondary joints are subdivided into structural joints and non-structural joints.

1) Primary joints (Figure 2-9). A joint formed during the formation of a rock.

2) Structural joints (Figure 2-10). The joints were formed by tectonic movement. Structural joints are often grouped together, where a set of parallel fractures in one direction are called a set of joints. In this process, the relationship between the two groups of tectonic stress is formed, and it is combined in a certain pattern.

Figure 2-9　Primary joints　　　　　　Figure 2-10　Structural joints

3) Non-structural joints (Figure 2-11). The joints that are formed by unloading, weathering and blasting are called unloading joints, weathered joints and blasting joints. These joints belong to secondary non-structural joint. The surface joints are generally distributed in the shallow layer of the ground.

II. Classification by mechanical properties (Figure 2-12)

1) Shear joint. It is usually composed of structural joints, formed by the shear forces

Figure 2-11　Non-structural joints

acting on the rocks. Generally, it intersects with the principal stress $(45°-\varphi/2)$ direction, where φ is the rock internal friction angle. Shear joints generally appear in pairs, cross each other as the X-shaped. Shear joints are mostly straight, often in a closed state or have a small opening.

2) Tension joint. It can be formed due to tensile forces acting on the rocks. The most common location of such joints in folded sequence is the outer margins of crests and troughs. The opening of tension joints is large, with good water permeability.

III. Classification according to the relationship between rock formations (Figure 2-13)

1) Strike joint. The joint strike is parallel to the strike of the strata.

2) Dip joint. The joint strike is vertical to the strike of the strata.

3) Oblique joint. The joint strike is oblique to the strike of the strata.

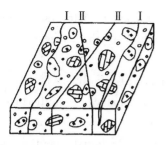

Ⅰ—Tension joint; Ⅱ—Shear joint

Figure 2-12 Tension and shear joints in the rock

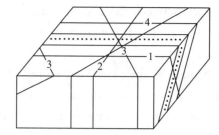

1—Strike joint; 2—Dip joint; 3—Oblique joint; 4—Rock strata direction

Figure 2-13 Joints formed during rock formation

Ⅳ. Classification by degree of opening

1) Wide open joint. The joint width is greater than 5mm.

2) Open joint. The joint width is 3~5mm.

3) Micro tension joint. The joint width is 1~3mm.

4) Closed joint. The joint width is less than 1mm.

2.6.1.2 Joint investigation

Joint is a kind of geological structure that is widely developed, the investigation of which should include the following contents:

① The cause of the joints.

② The number of joints. The density of joints are generally expressed as linear density or volume joint number. The volume joint number (J_v) is expressed as the joint number per unit volume.

③ The degree of opening, length and roughness of the joint.

④ Filling material and thickness, water content.

⑤ Joint development extent.

In addition, many rock formations in the field can be observed with dozens or hundreds of joints. Sedimentary rocks especially those of plastic nature and rich in moisture in the initial stages (clays, shales, limestones and dolomites etc.) undergo some contraction on drying up which might have resulted into irregular jointing. Igneous rocks which form by cooling and crystallisation from an originally hot and molten material (magma or lava) also necessarily undergo considerable contraction during the cooling process giving rise to tensile forces strong enough to break the congealing masses into jointed blocks. Such contraction or shrinkage is generally accepted to be the cause of the vertical type of joints in granites and the so well-known columnar joints of basalts and other effusive rocks.

2.6.2 Faults

In geology, a fault is a planar fracture or discontinuity in a volume of rock, where there was significant displacement as a result of rock-mass movement. Large faults within the Earth's crust attribute to the action of plate tectonic forces, with the largest forming the boundaries between the plates, such as subduction zones or transform faults. Energy release is associated with rapid movement on active faults, which is the cause of most earthquakes.

A fault plane is the plane that represents the fracture surface of a fault. A fault trace or fault line is a place where the fault can be seen or mapped on the surface. A fault trace is the line commonly plotted on geologic maps to represent a fault.

Since faults do not usually consist of single and clean fracture. Geologists use the term "fault zone" when referring to the zone of complex deformation associated with the fault plane.

2.6.2.1 Fault elements

To clarify the spatial distribution of faults and the movement characteristics of the strata on both sides of fault, the components of the fault is called the fault element, as shown in Figure 2-14.

I. Fault plane

It is the surface that the movement has taken place within the fault. The dip and strike can be measured on this surface. Faults may be vertical, horizontal, or inclined at any angle. Although the angle of inclination of a specific fault plane tends to

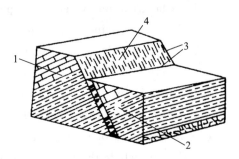

1—Footwall; 2—Hanging wall;
3—Fault line; 4—Fault plane

Figure 2-14 Fault elements

be relatively uniform, it may differ considerably along its length from place to place.

II. Hanging wall

Hanging wall, or head wall, is the upper or overlying block resting on the fault plane.

III. Footwall

It is the lower block or rock mass beneath the fault plane.

IV. Slip

The displacement that occurs during faulting is called the slip. It describes the movement parallel to the fault plane. Dip slip describes the up and down movement parallel to the dip direction of the fault. Strike slip describes the movement that is parallel to strike of the fault plane. Oblique slip is a combination of strike slip and dip slip.

2.6.2.2 Common classification of faults

I. According to the relative motion of two walls

1) Normal fault. It refers to the fault of the hanging wall droping down in relation to the footwall (Figure 2-15). Normal fault is generally affected by the Earth's crust tension or formed by gravity, with dip angle mostly more than 45°. The normal fault can be exposed separately or in multiple consecutive sections, such as graben and horst (Figure 2-16). It is called a graben when the middle section is a common descending wall. When the middle ground is a common ascending wall, it is called the horst. A normal fault on either side of the ground or the base can be a single output or a series of faults that are

similar to each other.

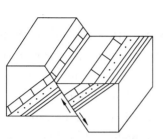

Figure 2-15 Normal fault

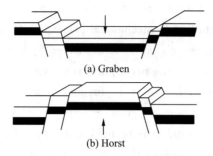

(a) Graben

(b) Horst

Figure 2-16 Graben and horst

2) Reverse fault. It refers to the fault of the hanging wall sliding up over the footwall, see Figure 2-17. The reverse fault is mainly formed by the compression of the Earth's crust.

Although reverse faults are also dip-slip faults, they behave the opposite way that a normal fault does. If the reverse faults that dip less than 45°, Thrust faults are relatively common in mountain belts that were created by continent-continent collisions. While reverse faults with a very low-angle fault plane and a very large total displacement are called overthrust fault.

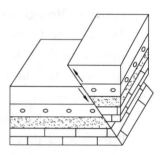

Figure 2-17 Reverse fault

3) Strike-slip fault. Faults which move along the direction of the dip plane are dip-slip faults and described as either normal or reverse, depending on their motion. In contrast, faults that move horizontally are known as strike-slip faults (Figure 2-18). Strike-slip faults include transform (which end at another plate boundary) and transcurrent (which end before reaching another plate boundary) fault lines. They are also classified as either right-lateral or left-lateral. A fault that moves to the left is a left-handed, left-lateral, or sinistral strike-slip fault, and a fault that moves to the right is a right-handed, right-lateral, or dextral strike-slip fault.

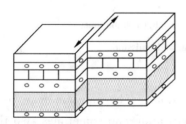

Figure 2-18 Strike-slip fault

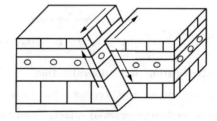

Figure 2-19 Oblique fault

4) Oblique fault. Faults which show both dip-slip and strike-slip motion are known as oblique-slip faults (Figure 2-19). Oblique-slip faults have simultaneous displacement up or down the dip and along the strike. The displacement of the blocks on the opposite sides of the fault plane usually is measured in relation to sedimentary strata or other stratigraphic

markers, such as veins and dikes. The movement along a fault may be rotational, with the offset blocks rotating relative to one another.

II. According to mechanical properties of fault

1) Compression fault. It is formed by compressive stress with direction perpendicular to the direction of the principal stress. Compression fault belongs to reverse fault.

2) Tension fault. It is formed under the action of tensile stress. The strike is perpendicular to the direction of tensile stress. Tension fault belongs to normal fault.

3) Shear fault. It is formed under the action of cutting with the direction of principal compressive stress less than 45°. Shear stress causes bodies of rock to slide sideways to form strike-slip fault.

2.6.2.3 Recognition of faults

To recognize faults in the field, a number of criteria are used. The faults may be directly seen in the field, particularly in artificial exposures such as river-cuttings, road cuttings, etc. But in majority of cases, faults are recognized by stratigraphic and physiographic evidences as below.

I. Discontinuity of structure

If there is an abrupt termina-tion of any geologic structure, a fault can be expected near about the point of termination. Sudden termination of dykes and veins, etc. also suggests the existence of a fault.

II. Repetition and omission strata

Sometimes a bed may suddenly terminate but occurs again somewhat off from the place where it terminated besides sometimes rock beds forming the country are found to be repeated and/or omitted (Figure 2-20) indicates the occurrence of a fault.

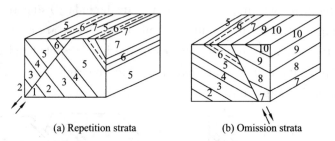

(a) Repetition strata   (b) Omission strata

Figure 2-20 Strata repetition and omission

III. Silicification and mineralization

Faults are often the avenues for moving solutions. The solution may replace the country rock with fine-grained quartz, causing silicification and sometimes also they form mineral deposits at that site. It points to the occur-rence of a fault in that area.

Ⅳ. Features characteristic of fault plane

These features are produced due to friction of the blocks on either side of the fault plane and include features like slickenside (Figure 2-21), grooves, drag (Figure 2-22), fluccan (i. e., pulverized clayey matter), breccia (commonly known as crush breccia), mylonites, horses and slices etc.

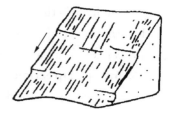

Figure 2-21 Slickenside feature

Figure 2-22 Drag features

Ⅴ. Difference in sedimentary facies

Different sedimentary facies of rocks of the same age may be brought into juxtaposition by large horizontal displacements, is suggestive of faulting.

Ⅵ. Physiographic evidences

The effects of faulting on outcrops constitute the physiographic evidences for faulting. They include features like: triangular facets (Figure 2-23), offset ridges, fault scrap, piedmont scrap, offset streams (Figure 2-24), fault springs (Figure 2-25) following a straight course (sometimes springs aligned along the fault planes also), lineament suggest the presence of faults in the field, alluvial fan, monocline, etc.

Figure 2-23 Triangular facets for fault cliff

Figure 2-24 Offset stream   Figure 2-25 Fault spring

To judge whether a fault exists, it is mainly based on two marks in repetition, deletion and discontinuity of strata. Other signs can only be used as a supplementary information, which cannot be used to determine the existence of fault.

2.6.2.4 Direction of fault movement

The property of the fault is determined by the shape of the fault surface, thus the hanging wall and footwall of the fault can be determined. The direction of the footwall movement is determined as nature of the fault. The direction of the footwall movement can be determined by the following conditions.

I. Stratigraphic age

On both sides of a fault, the rising wall is usually older, and the lower wall is relatively young. The stratum inversion is the opposite.

II. Stratigraphic boundary

When the cross section of the fault is transversal, the strata of the anticlinal ascending plate become wider and the strata of the core are narrowed.

III. Fault associated feature

The cut grooves are shallow, and the direction of the slickenside refers to the direction of the wall movement. The direction of the drag reflects the direction of the movement.

IV. Symbol recognition

On a geologic map, the faults are marked with bold lines, where the fault is represented by corresponding symbols shown in Figure 2-26. In normal fault and reverse fault, the arrow refers to the inclination of the fault plane, which is the angle of the fault plane. The direction of the short lines refers to direction of the upper wall movement. The long line is the line of strike. The direction of the arrows in the strike-slip fault is the direction of the movement.

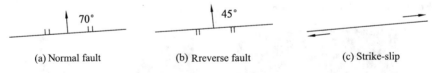

(a) Normal fault (b) Rreverse fault (c) Strike-slip

Figure 2-26 Fault symbol

2.7 Geologic map

A geologic map or geological map is a special-purpose map that aims to show geological features. Rock units or geologic strata are shown by color or symbols to indicate where they are exposed at the ground surface. Bedding planes and structural features such as faults, folds, foliations, and lineations are shown with strike, dip or trend and plunge symbols.

Stratigraphic contour lines may be used to illustrate the surface of a selected stratum illustrating the subsurface topographic trends of the strata. Isopach maps detail the variations in thickness of stratigraphic units. It is not always possible to properly show this when the strata are extremely fractured and mixed, or where they are disturbed. Rock units are typically represented by colors. Instead of (or in addition to) colors, certain symbols can be used. Different geologic mapping agencies and authorities use different standards for the colors and symbols for rocks with different types and ages.

2.7.1 Contents of a geologic map

The geologic map is a representation of comprehensive survey work on the topography. According to certain scale, with the symbol, color and line, landform, strata, geologic structure can be drawn on a map. Civil engineer can analyze the existing geologic maps and combine with the engineering conditions to conduct a preliminary work. Geological maps include geologic plans, geologic cross sections, bedrock map, and other specialty maps.

Geologic maps show various geologic data obtained in field, such as geomorphology, stratigraphy, geologic structure, mineral potential, hydrogeology, etc.

A geologic cross-section is a graphic representation of the intersection of the geologic bodies in the subsurface with a vertical plane of a certain orientation. It is a section of the terrain where the different types of rocks, their constitution and internal structure and the geometric relationship between them are represented. It is an approximate model of the real distribution of the rocks in depth, consistent with the information available on the surface and the subsurface. It can also represent the extension of the materials of the structures that have been eroded above the topographic surface.

The cross-sections are an indispensable complement of the geologic maps; maps and cross-sections are fruit of the interpretation of the arrangement of the rocks using diverse types of data, normally incomplete and with different degrees of uncertainty. Both are bi-dimensional representations of the geologic reality and jointly allow us to understand the tri-dimensional structure of the rocky volumes and, in consequence, the geologic history of a zone.

The geologic cross-sections have a very relevant economic and social importance. They are the basis for planning engineering works, fundamentally the lineal works that affect the surface and the subsurface (roads, tunnels, utilities) and for the exploration and production of geologic resources: water, stones, minerals and energy.

I. Strata

1) Horizontal strata. The strata line is parallel or overlapped with the contour line of the terrain, as shown in Figure 2-27.

2) Tilted strata. The dividing line is a "V" curve that intersects the terrain contour line.

3) Vertical strata. The boundary of rock strata is not affected by topographic contour lines. The boundary line is extended in a straight line.

II. Folds

It is common to identify the folds according to the symbol. If there is no symbol sign, it is necessary to determine the relationship between the young and old rocks distribution.

III. Fault

It is also common to identify faults according to symbols (Figure 2-26). If there is no symbol sign, it can be recognized by the phenomena such as repetition, loss, interruption, narrow change or dislocation.

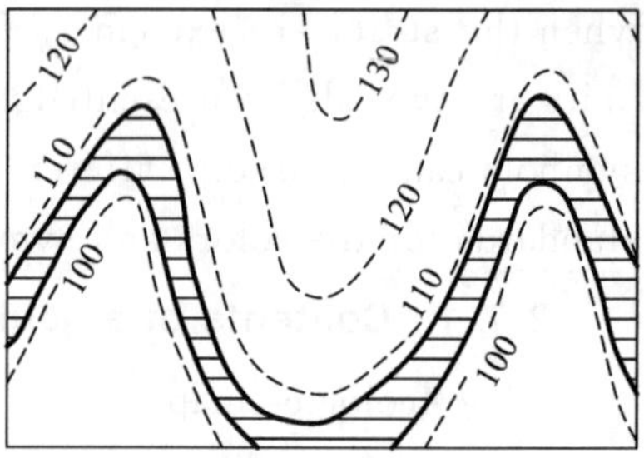

Figure 2-27 Horizontal strata in a geologic map

2.7.2 Method of reading a geologic map

I. Scale

Scale is the relationship between distance on the map and distance on the ground. For example, your map has a scale of 1 : 5000, which means that every 1cm on the map represents 5000cm (=50m) of measurement on the ground. The magnitude of the scale reflects the precision of the graph. The larger the scale, the higher the accuracy of the map, and the more detailed the geological conditions would be presented.

II. Legend

The legend to a geologic map is usually printed on the same page as the map and follows a customary format. The symbol for each rock or sediment unit is shown in a box next to its name and brief description. The map legend also contains a listing and explanation of the symbols shown on the map, such as the symbols for different types of faults and folds.

III. Geomorphology

It can be used to understand the topography, geomorphology, mountains and rivers, etc. It can be used to analyze the distribution of quaternary strata.

IV. Stratigraphic distribution and lithology

Stratigraphic age, lithology, occurrence, lithologic characteristics and the relationship with topography in the area should be addressed.

V. Geologic structure

Check the main geologic structure of the area and its relationship with the terrain. Such as the type, size, distribution and nature of faults and folds.

VI. Physical geologic phenomenon

Check the relationship between actual geological phenomena and topography, lithology, geologic structure and groundwater.

VII. Evaluation

According to the geologic conditions in the map, the preliminary evaluation of the construction site can be made. Suggestions for further investigation can be made in advance. Experienced civil engineers can predict future problems from reading and analyzing geologic map, and make reasonable site selection and design according to geologic conditions. However, the geologic survey and tests shall be performed in the design stages for the middle and large projects to provide the required information.

Knowledge expansion

Engineering Considerations of Folds

1) For a major project like a dam, tunnel, railway station, etc., a site which is highly folded should be avoided because the engineer may have to face many troubles sooner or later as folds are easily fractured even due to a slight disturbance.

2) If the project is of a scattered nature like electric or telephonic poles the work can be carried out without much of a risk.

3) Folds are also important to a water supply engineer especially when he has to select a suitable site for digging wells for water supply purpose. It has been observed that if the excavation of a well is done through impervious strata it will not yield any amount of water. If another well is excavated through previous strata it will yield abundance of water.

4) Synclinal folded rocks may yield hard and tough quality stones; whereas anticlinal folded rocks will yield weaker stones.

5) The anticlinal folds provide good prospects for stored petroleum, and hence in oil exploration, folds must not be overlooked.

Exercises

[2-1] What is a geologic time scale?

[2-2] What are geologic structures? How to characterize them?

[2-3] What is a geologic map? How to read it?

Chapter 3　Geologic Hazards

3.1 Introductory case

A geologic hazard is one of several types of adverse geologic conditions causing damages and loss of property and life. These hazards include sudden phenomena (e.g., earthquakes, landslides, rock falls, debris flows, etc) and long term phenomena (e.g., alluvial fans, ground subsidence, sinkholes, erosion, etc). Sometimes, the hazard is caused by humans through construction in which the conditions were not carefully taken into account. For example, naturally inactive landslides may be triggered by construction activities of roads or buildings that affect the stability of a hill slope. Other geologic hazards, such as earthquakes, rockfall, mudslides, and avalanches are naturally taking place, which can wreak havoc on buildings, roads, and other engineering structures. For example, the Wenchuan earthquake occurred in southwest China on 12 May, 2008, the earthquake's epicenter was located 80 kilometres northwest of Chengdu, with a focal depth of 19km.

Figure 3-1　A building after the Wenchuan earthquake

The magnitude of 2008 Wenchuan earthquake is 8.0. The earthquake lasted around 2 minutes, caused more than 80000 death and millions lost, almost 80% of buildings were destroyed, as shown in Figure 3-1. The earthquake could be felt in nearby counties and even in Beijing and Shanghai, where the office buildings swayed. Strong aftershocks, some of which exceeding 6 magnitude, continued to hit the area up to several months after the main quake, causing further casualties and damages. The earthquake made about 4.8 million people homeless, where the actual number could be as high as 11 million. Approximately there are 15 million people lived in the affected area. It was the deadliest earthquake that had ever hit in China since the 1976 Tangshan earthquake, which killed at least 240000 people. It was also the strongest in the country since the 1950 Chayu earthquake, which registered at 8.6 on the scale of richter magnitude. On November 6, 2008, the Chinese government announced that it would spend 1 trillion RMB (about US $146.5 billion) in three years to rebuild areas ravaged by the earthquake, as part of the Chinese economic stimulus program.

According to the China Earthquake Administration (CEA), the earthquake occurred along the Longmenshan Fault, a thrust structure along the border of the Indo-Australian plate and Eurasian plate. Seismic activities concentrated on its mid-fracture (known as Yingxiu-Beichuan fracture). The rupture lasted for 120 seconds, with the majority of energy released in the first 80 seconds. Starting from Wenchuan, the rupture propagated at an average speed of 3.1 kilometers per second toward 49° north east, rupturing a total of about 300km. Maximum displacement amounted to 9 meters. The focus was deeper than 10km.

3.2 Land subsidence

3.2.1 Definition of subsidence

Land subsidence is a widespread, well-recognized geologic hazard. Subsidence is the sudden sinking or gradual downward settling of the ground's surface with little or no horizontal motion. The definition of subsidence is not restricted by the rate, magnitude, or area involved in the downward movement.

Land subsidence causes many problems including:

① Changes in elevation and slope of streams, canals, and drains.

② Damage to bridges, roads, railroads, storm drains, sanitary sewers, canals, and levees.

③ Damage to private and public buildings.

④ Failure of well casings from forces generated by compaction of fine-grained materials in aquifer systems.

In some coastal areas, subsidence results in tides moving into low-lying areas, the levels of which were previously above high-tide levels.

3.2.2 Classification of subsidence

All types of subsidence can be divided into two major groups: endogenic subsidence and exogenic subsidence. Endogenic subsidence is mainly caused by processes originating within the planet, and exogenic subsidence is due to the forces originating on the Earth's surface and human activity.

Each of the major groups can be subdivided according to the nature of processes causing the subsidence. For example, endogenic subsidence can be subdivided into subsidence caused by volcanic activity, folding, faulting, and other endogenic processes. By definition, many forms of exogenic subsidence are directly caused by human activity. This subsidence is actually a surficial expression of compaction of material at depth due to removal of support (for example, oxidation of organic particles, thawing of permafrost, melting of ice, etc.), weakening of support (for example, hydrocompaction), or an increase in actual or effective loading. Examples of subsidence due to an increase in actual loading are collapses of karst cavities due to dewatering of unconfined aquifers. Typical examples of subsidence due to an increase in effective loading are subsidence related to piezometric decline of confined aquifer systems and subsidence due to withdrawal of crude oil and natural gas. Additionally, earth fissures are associated with land subsidence that accompanies extensive ground water pumping, with long, narrow cracks or openings in the earth.

3.2.3 Subsidence due to groundwater extraction

Compaction of soils in some aquifer systems can accompany excessive groundwater pumping and it is by far the single largest cause of subsidence. Excessive pumping of such aquifer systems has resulted in permanent subsidence and related ground failures, which has damaged buildings, buckled highways, and disrupted water supply.

3.2.3.1 Pore spaces in geomaterials

When large amounts of water are pumped, the subsoil compacts, thus reducing in size and number the open pore spaces in the rock or soil the previously held water. The pore spaces in geomaterials can be divided into three categories: void in soils, cracks or fractures in rocks, solutional openings formed by dissolution of soluble rocks. Figure 3-2 shows the typical occurrence of open pores in geomaterials including soils and rocks.

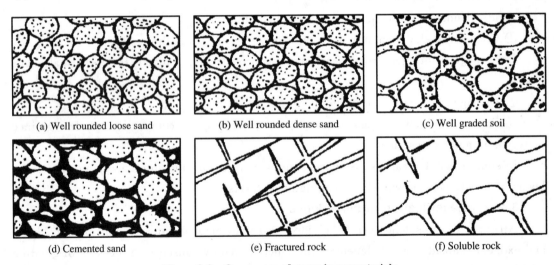

Figure 3-2 Occurrence of pores in geomaterials

Porosity is a parameter that describes the amount of open space in geomaterials. Porosity can be stated as a fractional value (e.g. 0.25) or percentage (25%) of open space (i.e. 25% of volume in the material is open space). Porosity values range from 0 to 50% typically. Open pores can be filled with water or air or a mixture of both.

Permeability describes how those pores are shaped and interconnected. This determines how easily water can flow from one pore to the next. Larger pores mean there is less friction between flowing water and the sides of the pores. Smaller pores mean more friction along pore walls, but also more twists and turns for the water to flow-through. A permeable material has a greater number of larger, well-connected pores spaces, whereas an impermeable material has fewer, smaller pores that are poorly connected. The characteristic of permeability of a geomaterial is quantified by permeability coefficient or hydraulic conductivity. The symbol used for permeability coefficient is k. Table 3-1 lists the variation of permeability coefficient for different types of geomaterials.

Table 3-1 Permeability of geomaterials

Permeability rating	Permeability coefficient $k/(m/d)$	Geomaterial type
Very good	>10	Gravel, coarse sand, karst rock, rock with big cracks
Good	10~1.0	Medium sand, fine sand, rock with cracks
Fair	1.0~0.01	Clayey silt, rock with fine cracks
Poor	0.01~0.001	Silty sand, silty clay, rock with microcracks
Impermeable	<0.001	Clay, shale, mudstone

Ⅰ. Voids in soils

Soil is a porous medium that has developed in the uppermost layer of Earth's crust. Soils are particles of weathered rock, which include tiny grains of clay and silt as well as larger sand particles and gravel. The porosity of soil is defined as the volume of voids in soil per the total volume of soil.

Ⅱ. Cracks in rocks

The cracks or fractures in rocks are affected by the movement of the Erath's crust and other internal and/or external geological processes. The porosity of rock is calculated as:

The porosity of rock fracture is calculated as:

$$K_r = \frac{V_r}{V} \times 100\% \tag{3-1}$$

where, K_r—porosity of rock;

V_r—volume of pore spaces in rock;

V—total volume.

Ⅲ. Openings in soluble rock

The openings in soluble rocks are formed by the long-term dissolution of rock in groundwater. in karst areas. Karst is associated with soluble rock types such as limestone, dolomite, and gypsum. The porosity of soluble rock can be obtained following Eq. (3-1).

3.2.3.2 Properties of groundwater

Ⅰ. Physical properties of groundwater

The physical properties of groundwater include temperature, color, transparency, smell, taste, conductivity and radioactivity. Pure groundwater should be colorless, tasteless, odorless and transparent. The physical properties will change when certain chemical compositions and/or suspensions are presented.

II. Chemical composition of groundwater

Groundwater can dissolve soluble substances in rocks and have complex chemical composition. The main gases in the groundwater are O_2, N_2, CO_2, H_2S, CH_4, etc. The cations mainly include: H^+, Na^+, K^+, NH_4^+, Ca^{2+}, Mg^{2+}, Fe^{3+}, Fe^{2+}, etc. The anions mainly include: OH^-, Cl^-, SO_4^{2-}, NO_2^-, NO_3^-, HCO_3^-, CO_3^{2-}, SiO_3^{2-} PO_4^{3-}, etc. Generally, the main chemical composition of groundwater are the following seven ions: Na^+, K^+, Ca^{2+}, Mg^{2+}, Cl^-, SO_4^{2-} and HCO_3^-. They are the main ions to evaluate the chemical composition of groundwater.

3.2.3.3 Types of groundwater

Groundwater is a part of the natural water cycle. Some part of the precipitation that lands on the ground surface infiltrates into the subsurface. Figure 3-3 shows the schematic of groundwater infiltration. It can be seen that a zone of saturation occurs where all the interstices are filled with water. There is also a zone of aeration where the interstices are occupied partially by water and partially by air, which is also known as unsaturated zone. Groundwater continues to descend until, at some depth, it merges into a zone of dense rock. Water is contained in the pores of such rocks, but the pores are not connected and water will not migrate. The process of precipitation replenishing the groundwater supply is known as recharge. In general, recharge occurs only during the rainy season in tropical climates or during winter in temperate climates. Typically, 10% to 20% of the precipitation that falls to the Earth enters the water-bearing strata, which are known as aquifers.

The upper boundary of the zone of saturation determines whether the groundwater is free or confined. In fact it is this upper boundary of free groundwater which is known as the water table. As shown in Figure 3-3, the water table is the contact plane between free groundwater and the capillary fringe. The water table definitely depends on the nature of the groundwater bearing material. In fine textured clay, there will be no clear cut "plane" surface dividing the zone of saturation from the lowest layer of the capillary fringe. In the case of granite, the water table may vary considerably in height over short distances and

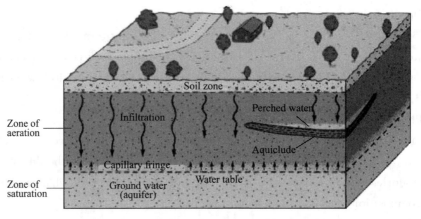

Figure 3-3 Schematic of groundwater infiltration

also may be inter-rupted in places. While in the case of an open textured rock such as well-jointed limestone, the groundwater will form a more horizontal surface.

There are three main classifications of groundwater (aquifer), defined by their geometry and relationship to topography and the subsurface geology. They are perched aquifer, unconfined aquifer and confined aquifer, as shown in Figure 3-4.

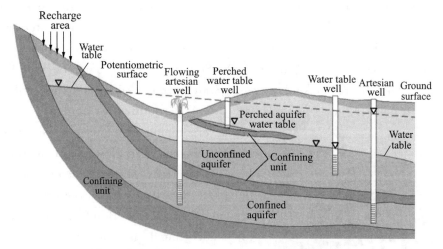

Figure 3-4　Common types of groundwater (aquifer)

I . Perched aquifer

Perched aquifers occur above discontinuous aquitards, which allow groundwater to "mound" above them. Thee aquifers are perched, in that they sit above the regional water table, and within the regional vadose zone (i. e. there is an unsaturated zone below the perched aquifer). The dimensions of perched aquifers are typically small (dictated by climate conditions and the size of aquitard layers), and the volume of water they contain is sensitive to climate conditions and therefore highly variable in time. Such aquifers are mainly found in clay parts of sedimentation deposits, which are flat in shape. From wells located in perched aquifers, water can be available in small quantity and for short periods.

II . Unconfined aquifer

An unconfined aquifer is defined as a body of water formed from groundwater, rain water runoff and streams with its water table, or the upper surface, open to the atmosphere. The water level in wells drilled into an unconfined aquifer will be at the same elevation as the water table. It is termed an unconfined aquifer because the aquifer formation extends essentially to the land surface. As a result, the aquifer is in pressure communication with the atmosphere. Unconfined aquifers are also known as water table aquifers because the water table marks the top of the groundwater system. Figure 3-4 shows that the unconfined aquifer is a layer of water that has a confining layer on the bottom and a layer of permeable soil above it.

III . Confined aquifer

A confined aquifer is basically a layer of water that is under pressure and is held between two confining layers (units) such as clay layers. It is isolated from pressure communication with overlying or underlying geologic formations and with the land surface and atmosphere. Confined aquifers differ from unconfined aquifers in two fundamental and important ways. First, confined aquifers are typically under considerable pressure, which may be derived from recharge at a higher elevation or from the weight of the overlying rock and soil (known as the overburden). In some cases, the pressure is high enough that wells drilled into the aquifer are free-flowing. This condition requires that the water pressure in the aquifer is sufficient to drive water up the wellbore and above the land surface, and such wells are called artesian wells (Figure 3-4). Second, confined aquifers typically remain saturated over their entire thickness, even as water is removed by pumping wells.

3.2.4 Groundwater control

Figure 3-5 Risks due to overpumping groundwater

Overpumping groundwater have irreversibly altered the aquifer and lowered of the water table. It will potentially cause the land subsidence and put great threats on infrastructures built on the ground (Figure 3-5). Besides, Excavation works for underground infrastructure such as deep basements or tunnels often encounter groundwater. Working below groundwater level can be difficult, as the excavation is at risk of instability and inundation.

Groundwater is an especial problem when excavating in water-bearing soil (such as sands and gravels) or fissured rock (such as chalk or sandstone). Without suitable control measures, inflows of groundwater can flood excavations or tunnels, and can also lead to instability when the soils or rock around the excavation weaken and collapse – either locally or on a large scale. How do civil engineers deal with groundwater is an important issue. The first thing that needs to be done to ensure that groundwater can be effectively dealt with is to gather information to understand the problem. This involves a site investigation, which might involve drilling and testing of boreholes, measurement of groundwater levels and tests to measure the permeability of the ground. High permeability soils and rocks tend to be water-bearing and are typical of the conditions where groundwater can cause problems for construction projects. There are two major methods for controlling groundwater including pumping technique and exclusion technique.

3.2.4.1 Pumping technique

Groundwater pumping, also known as dewatering, is an approach involves pumping groundwater from an array of wells or sumps around the excavation. The objective is to lower groundwater levels to below working level in the excavation. Examples of this group of techniques include sump pumping, wellpoints, deep wells and ejector wells.

Ⅰ. Dewatering by sump pumping

It is a reliable choice in a wide range of situations, and it's also referred to as the simplest, cheapest, and most effective dewatering method. A sump is a hole or an area in the ground (deeper than the basement floor) where water is collected and then pumped away for disposal (Figure 3-6). Drains and sumps are constructed at one or more sides or corners of the foundation pit. The drains collect the groundwater and convey it into the sump. From the sump, the water is continuously evacuated (either manually or mechanically). This method works well for most soil and rock conditions (can be applied in well-graded coarse soils or in hard fissured rock). When using this technique, there is a risk of collapse of the sides as "the groundwater flows towards the excavation with a high head or a steep slope". In fine-grained soils, such as silts and fine sands, there is also a risk of instability which may result in ground movements and settlement.

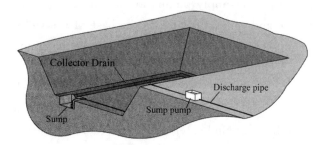

Figure 3-6　Schematic of sump pumping technique

Ⅱ. Dewatering by wellpoints

This method features easy installation, and it's relatively cheap and flexible, being practical and effective under most soil and hydrologic conditions. Wells are drilled around the construction area and pumps are placed into these wells. Wellpoint systems consist of a series of small-diameter wells, connected by a header pipe to a centrally located suction pump (Figure 3-7). Groundwater is abstracted via the wellpoints from a vacuum generated by the pump. The perforated pipe has a ball valve to regulate the flow of water (the ball

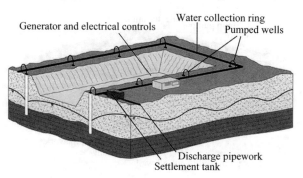

Figure 3-7　Schematic of wellpoints technique

valve also prevents the mud from entering into the pipe). Groundwater can be lowered about 6m by this method. This method is ideal for buildings with deep basements and is effective in sands and sandy gravels.

III. Dewatering by deep wells

When a deep excavation is needed and a large quantity of groundwater is required to be removed, dewatering may be done by constructing deep wells in soils or rocks where permeability is between moderate (e.g. sands) to high (e.g. gravels). Deep well dewatering system can drain out water up to 24 m depth. The capacity of the pumps as well as the number, depth, and spacing of deep wells may vary depending on the site conditions.

IV. Dewatering by ejector wells

The ejector system is a specialist technique used in low permeability soils such as very silty sands, silts, or clays. Ejectors are typically used to help stabilize the side slopes and soil in the excavation area. Unlike the wellpoint dewatering system, it uses high-pressure water in the riser units. Ejector supply pumps located at ground level feed high-pressure water to the ejector nozzle and venturi located at the base of the wells. The flow of water through the nozzle generates a vacuum in the well and draws in groundwater.

3.2.4.2 Exclusion technique

There are number of techniques by which ground water exclusion are obtained: ① forming impervious barriers by grouting with cement, clay suspension; ② chemical consolidation for controlling groundwater in excavation; ③ groundwater control by compressed air; and ④ freezing groundwater control.

I. Impervious barriers technique

In this technique, the permeability is reduced by creating an impervious barrier by using low permeability cut-off walls or injecting suspension material or fluids into the fissures of rocks or pore spaces. Figure 3-8 shows the schematic of cut-off walls. The cut-off walls are installed into the ground around the perimeter of the excavation. These walls act as barriers to groundwater flow, and effectively exclude groundwater from the excavation. The requirement to pump groundwater is limited to pumping out of the water trapped within the area enclosed by the cut-off walls. Examples of the techniques used to

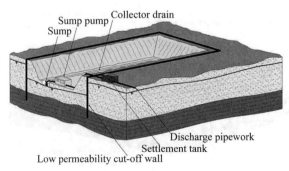

Figure 3-8　Schematic of low permeability cut-off walls

form cut-off walls include steel sheet-piling, concrete diaphragm walls, concrete bored piles and bentonite slurry walls.

Besides, grout barriers technique involves the injection into the ground of fluid grouts that set or solidify in the soil pores and rock fissures. The grout blocks the pathways for groundwater flow and can produce a continuous zone of treated soil or rock around the excavation that is of lower permeability than the native material. This reduces groundwater inflow in a similar way to cut-off walls. The most commonly used grouts are based on suspensions of cement in water.

II. Chemical consolidation technique

Chemical consolidation method is suitable for sandy gravels and fine grading sands. The most usual chemical material used for chemical consolidation is the sodium silicate. If the sodium silicate is mixed with other chemicals, moderately strong and insoluble silica gel can be produced. Two approaches have been practiced to conduct chemical consolidation, namely, two shot process and one-shot process. By and large, the latter process which is the most common one has replaced the former process. In two shot process, two pipes with spacing of 50cm are forced into the ground, then sodium silicate are driven to one pipe and calcium silicate injected into the other while they are pulled up gradually. As far as one-shot process is concerned, chemical grouts are usually created prior the injection process. So, the most important consideration in this technique is to postpone the formation of grout gel. This is because grout penetration would be easier and more efficient when its viscosity is low.

III. Compressed air technique

The compressed air method is used in pipe jacking to provide a positive pressure on the face of the tunnel. The pressure supplied by the compressed air prevents groundwater from entering the tunnel, thus building up in the pipe and damaging equipment. The compressed air method is one alternative that can be used to enable trenchless construction projects to be conducted safely and efficiently at depths below the groundwater table. In trenchless construction projects below the groundwater table, the water pressure from the surrounding ground can push large volumes of water into the tunnel. Compressed air supplied to the front face keeps the tunnel under a positive pressure and water out. When the tunnel is pressurized, workers can enter and work in the pipe jacked tunnels to remove large boulders that cannot be extracted by the screw conveyer. The compressed air method is an alternative to the pressure balance method where slurry is pumped to the face of the cutting head to provide a positive pressure on the tunnel face.

IV. Freezing technique

This is a technique that a very low temperature refrigerant (either calcium chloride brine or liquid nitrogen) is circulated through a series of closely-spaced boreholes drilled into the ground. The ground around the boreholes is chilled and ultimately frozen. Frozen soil or rock has a very low permeability, and will significantly reduce groundwater inflow into any excavation.

Controlling excavation ground water by freezing is not recommended to use unless all other methods fails to provide desired result or inappropriate to choose due to certain factors. This is because the cost of controlling ground water by freezing is significantly high due to large number of boreholes required to be drilled around the excavation area. However, there cases in which freezing is the only practical method to control ground water for example in extremely deep shaft excavation where the pressure of ground water is seriously high. To prevent the formation of unfrozen spaces in the frozen area, boreholes shall be exactly vertical and errors must be kept as minimum as possible in addition to provide small spacing between boreholes.

3.3 Landslide

3.3.1 Definition of landslide

A landslide is defined as the movement of a mass of soil and rock down a slope. Landslides are a type of "mass wasting", which denotes any down-slope movement of soil and rock under the direct influence of gravity. Landslides occur when gravitational and other types of shear stresses within a slope exceed the shear strength (resistance to shearing) of the materials that form the slope. Generally, the term "landslide" includes five modes of slope movement: slides, falls, topples, spreads, and flows. Specifically, the sliding can be rotational along a concave-upward set of shear surfaces (a rotational slide), as shown in Figure 3-9, or it can extend downward and outward along a broadly planar surface (a translational slide), as shown in Figure 3-10.

In a rotational slide (Figure 3-9), the axis of rotation is roughly parallel to the contours of the slope. The movement near the head of the slide is largely downward, exposing a steep head scarp, and movement within the displaced mass takes place along internal slip planes, each tending to tilt backward. Over time, upslope ponding of water by such back-tilted blocks can enlarge the area of instability, so that a stable condition is reached only when the slope is reduced to a very low gradient.

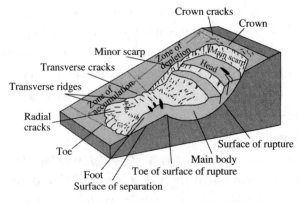

Figure 3-9 Schematic of a rotational landslide

A translational slide (Figure 3-10) typically takes place along structural features, such as a bedding plane or the interface between resistant bedrock and weaker overlying material. If the overlying material moves as a single, little-deformed mass, it is called a block slide. A translational slide is sometimes called a mud slide when it occurs along gently sloping, discrete shear planes in fine-grained rocks (such as fissured clays) and the displaced mass is fluidized by an increase in pore water pressure.

In detail, some components that form a landslide are marked on Figure 3-9 and Figure 3-10.

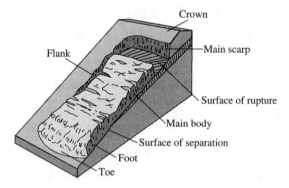

Figure 3-10   Schematic of a translational landslide

I. Crown

It is practically the undisplaced material still in place and adjacent to the highest parts of a landslide, sometimes with cracks.

II. Main Scarp

It is the steep slope at the upper edge of the landslide (at the head), caused by the movement of displaced material away from the undisturbed ground. It is the visible part of the slide surface.

III. Flank

It is the undisplaced material adjacent to the sides of the landslide. Flank usually describes the left and right lateral extents of the mass-wasting material.

IV. Surface of rupture

It is the lower boundary of the movement below the original ground surface. This is the surface along which material slides. The geomaterials below the slide surface do not move. It is marked on the sides by the flanks and at the end by the toe of the landslide.

V. Main body

It is the part of the landslide that overlies the surface of rupture, often with transverse and radial cracks at surface.

VI. Surface of separation

It is the part of the original ground surface that is now covered by the foot of the landslide.

Ⅶ. Foot

It is part of the landslide that overlies the original ground surface (i. e. right below the separation surface).

Ⅷ. Toe

It is the downhill end of the slide. The toe of the landslide marks the runout, or maximum distance traveled of the moving material. In rotational landslides, the toe is often a large, disturbed mound of geomaterial.

3.3.2 Classification of landslide

The various types of landslides can be differentiated by the kinds of material involved, the mode of movement, the landslide thickness, the landslide volume and the occurred time, etc. Table 3-2 lists the main types and features of landslide.

Table 3-2 Types and features of landslide

Divided basis	Category	Landslide features
Material composition	Soil landslide	Soil itself as main body, with rupture surface being soil or along the bedrock contact surface
	Loess landslide	Landside developed in the loess layers
	Debris landslide	Various types of deposits together as main body or slide along the bedrock
	Weathered rock landslide	Weathered rock sliding as main body, commonly occurred in igneous rocks (especially granite)
Relations between main body and strata	Homogeneous landslide	Occurred in the same geomaterial, with the smooth slip surface
	Translational landslide	Landslide with translational planes in hard rock and weak rock layers
	Rotational landslide	Sliding along slope bedding or bedrock interface or unconformity between bedrock or cutting beddings
Landslide thickness	Shallow landslide	Landslide body thickness within 10m
	Middle landslide	10~25m
	Deep landside	25~50m
	Super-deep landslide	Over 50m
Landslide volume	Small landslide	Landslide volume within $10 \times 10^3 \text{m}^3$
	Medium landslide	$100 \times 10^3 \sim 1000 \times 10^3 \text{m}^3$
	Large landslide	$1000 \times 10^3 \sim 10000 \times 10^3 \text{m}^3$
	Giant landslide	Over $10000 \times 10^3 \text{m}^3$
Landslide occurred time	New landslide	Landslide occurred within 5000 years
	Old landslide	5000~150000 years
	Ancient landslide	Over 150000 years

1) According to the compositions of geomaterial, landslide can be divided into soil landslide, loess landslide, debris landslide, weathered rock landslide and bedrock landslide.

2) According to the relationship between landslide main body and underlying geologic structure, landslide can be divided into homogenous soil landslide (Figure 3-11a), translational landslide along the inclined strata (Figure 3-11b), and rotational landslide along the slope of bedrock (Figure 3-11c) or cutting layers (Figure 3-11d).

3) According to the thickness of main body, landslide can be divided into shallow landslide, middle landslide, deep landslide, and super-deep landslide.

4) According to the volume of main body, landslide can be divided into small landslide, medium landslide, large landslide, and giant landslide.

5) According to the occurred time, landslide can be divided into new landslide, old landslide, and ancient landslide.

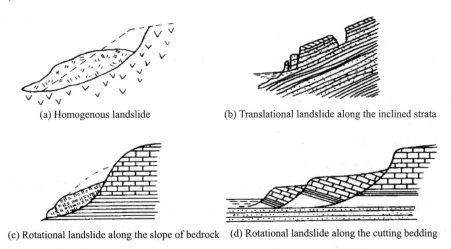

(a) Homogenous landslide   (b) Translational landslide along the inclined strata

(c) Rotational landslide along the slope of bedrock   (d) Rotational landslide along the cutting bedding

Figure 3-11　Schematics of different types of landslide

3.3.3　Development of landslide

Generally the occurrence of landslides is a long-term process. The development process of landslides can be divided into three stages: creep deformation stage, sliding failure stage and gradual stabilization stage. The study of landslide development is of great significance to recognize the landslide and choose available preventive treatments in engineering application.

Ⅰ. Creep deformation stage

The slope is usually stable before sliding. Sometimes, under the action of natural conditions and human involved factors, material strength on the slope can be reduced gradually (or the shear force increased continuously), causing the instability of slope. In a certain part of the slope, increased shear stress will cause a imperceptibly slow downward movement of material called creep. With further development of the creep deformation, the tension cracks will appear in the slope. Water infiltration make the creep deformation further developed and the crack widened. The shear cracks on both sides appear consecutively. The material near the foot of the slope is accumulated. The sliding surface has been formed almost. Slope deformation continues to develop with shear until the occurrence of landslide tensile cracks being continuously widened. From the slow deformation of the slope, cracks appear in the sliding surface to the complete formation of the sliding surface of the slope, this process is called creep deformation stage. The duration of this stage ranges from a few days to a few years.

II. Sliding failure stage

In this stage, the main body begins to move downward along the sliding surface. Main scrap appears with the rapid movement of sliding body. The main body can be split into small blocks and form step terrain on the slope surface. The trees on the main body were groggily tilted, as shown in Figure 3-12. The rapid movement of foot and toe of landslide will damage the buildings, roadways, and bridges on its way. The time of this stage depends on the sliding speed of landslide, and the speed is closely related to the reduction rate of the shear strength of the geomaterial. If the shear strength of the material decreases slowly, the main body will not slide sharply, and the sliding time will be longer. On the contrary, if the shear strength of the rock and soil decreases rapidly, the main body will slide at a speed of several meters or even tens of meters per second. Besides, the sliding is often accompanied by a loud noise and a huge air wave, which is more harmful.

III. Gradual stable stage

The process that a landslide stops moving and reaches a stable state again is called the gradual stable stage. This is mainly caused by friction of sliding surface. It should be noted that under the action of gravity, the loose rock and soil mass in the main body of landslide is gradually compacted, the cracks on the surface are filled, and the strength of soil and/or rock near the sliding surface is further enhanced due to compaction and consolidation. When the slope of the landslide becomes gentle, there would be no obvious crack on the surface. Besides, the trees on the main body surface become tilted due to the sliding. While after the sliding stops, the upper part of the trees turn to the upright after a long time, finally forming the bent trees (Figure 3-13).

Figure 3-12 Tilted trees

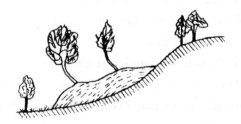

Figure 3-13 Bent trees

If the landslide stops, the landslide will no longer slip and be stable. If the main factor that causes the landslide is not completely eliminated, the stable landslide will slide again when the creep deformation accumulates to a certain extent.

3.3.4 Causes of landslide

The factors that cause a landslide is multiple. These factors can cause the change of slope appearance, the deterioration of the geomaterial property and the addition of load. The main causes of landslide are shown below.

I. Climate

Long-term climatic changes can significantly impact soil stability. A general reduction in precipitation leads to lowering of water table and reduction in overall weight of soil mass, reduced solution of materials and less powerful freeze-thaw activity. Water infiltration reduces the friction between the bedrock and the overlying sediment, and gravity sends the debris sliding downhill.

II. Earthquakes

Seismic activities have, for a long time, contributed to landslides across the globe. Any moment tectonic plates move, the soil covering them also moves along. When earthquakes strike areas with steep slopes, on numerous occasion, the soil slips leading to landslides.

III. Weathering

Weathering is the natural procedure of rock deterioration that leads to weak, landslide-susceptive materials. Weathering is brought about by the chemical action of water, air, plants and bacteria. When the rocks are weak enough, they slip away causing landslides.

IV. Erosion

Erosion caused by sporadic running water such as streams, rivers, wind, currents, ice and waves wipes out latent and lateral slope support enabling landslides to occur easily.

V. Volcanoes

Volcanic eruptions can trigger landslides. If an eruption occurs in a wet condition, the soil will start to move downhill instigating a landslide.

VI. Forest fires

Forest fires instigate soil erosion and bring about floods, which might lead to landslides. Plants help keep the soil stable by holding it together like glue with their roots. When this glue is removed, the soil loosens, and gravity acts upon it much more easily.

VII. Gravity

Landslides occur when gravity overcomes the force of friction. Steeper slopes coupled with gravitational force can trigger a massive landslide.

VIII. Human-related activities

Destruction of vegetation by droughts, fires, and logging has been associated with an increased risk for landslides. Mining activities that utilize blasting techniques contribute mightily to landslides. Vibrations emanating from the blasts can weaken soils in other areas susceptible to landslides. An excavation of slope or its toe, loading of slope or its crest will increase the risk for landslides. A drawdown of reservoirs will likely cause a landslide.

3.3.5 Landslide prevention and mitigation

The prevention of landslides is important before selection of site for engineering construction. When assessing and evaluating the site during the inspection one should watch out for the type of material on the slope to decide the appropriate mitigation technique. Generally, the contents of the slope are in the form of rocks or debris (soil, sand, etc), the slide of which is hampered by changing its trajectory or reducing its velocity. Some of the mitigation methods are summarized as below.

I. Facilitate drains

Groundwater is one of the major contributors in triggering a landslide. Therefore, to hamper the initiation of a landslide proper drainage system is required to prevent erosion and potential slumping of the surface. There are various drainage techniques like surface and subsurface drains that help in the slope stabilization process. For example, a circular intercepting ditch can be set at the stable section 5m away from the possible development boundary of the landslide to intercept and divert the surface water and groundwater outside the scope of the landslide, as shown in Figure 3-14. Besides, an effective form of subsurface drain includes a lateral trench which is constructed for an unstable slope.

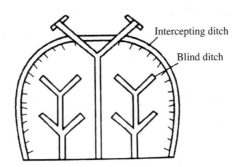

Figure 3-14 Schematic of the ditch arrangement

II. Plant vegetation

Another simple way to prevent landslides is to plant trees and small shrubs on the slope. As these trees and shrubs grow, their roots hold soil together, and help in reducing erosion of soil which is likely to make the slope unstable in course of time. In fact, there exist quite a few species of plants with shallow roots which are specifically used to protect the top layer of the soil in mountainous regions. This landslide prevention method works best on slopes that are not too steep or if the movement has not already begun.

III. Alter the slope gradient

It is possible to remove the material from upper part of the slope and put it near the base and reduce its gradient to for landslide mitigation. For example, rocks or debris from the slope can be removed by digging a sizable hole on the surface to improve the stability of the slope. It is mostly suitable for cuts into deep soil where rotational landslides may occur. The method falls short for translational failures on long, uniform, or planar slopes. Besides, reducing the height of the slope can minimize the driving force on the failure plane to increase the slope stability.

Ⅳ. Construct retaining walls

Retaining walls are structures designed to restrain the geomaterial. They are normally used in areas with steep slopes or where the landscape needs to be shaped severely for construction or engineering projects. There are various ways of constructing a retaining wall, such as gravity walls, piling walls, cantilever walls and anchored walls. For example, gravity walls manage to resist pressure from behind due to their own mass, as shown in Figure 3-15. Piling walls are usually used in tight spaces with soft soil having 2/3 of the wall beneath the ground. Cantilever walls have a large structural footing and convert horizontal pressure from behind the wall into vertical pressure on the ground below. Anchored walls use bolts or other stays anchored in the rock or soil behind to increase resistance, as shown in Figure 3-16.

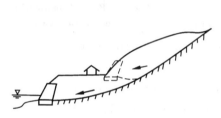
Figure 3-15   Gravity wall

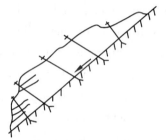
Figure 3-16   Anchor bolts

Ⅴ. Strengthen the slope

This method can be done by using soil reinforcement materials (cement, lime, geopolymers, etc.) that are available to increase the shear strength of the soil. Besides, lightweight plastic polymers with high-tensile strength can reinforce the soil by acting as a mesh or a grid which is stretched all across the slope.

3.4　Debris flow

3.4.1　Definition of debris flow

A debris flow is a moving mass of loose mud, sand, soil, rock, water and air that travels downwards a slope under the influence of gravity. In the debris flow development area, frequent debris flow destroyed the roads, railways, bridges and other traffic facilities. Large debris flow can have damage to factories, towns and farmland water conservancy projects, caused great losses to the people's lives and property and bringing all transportation to a halt.

3.4.2　Formation conditions of debris flow

The formation of debris flow is closely related to the geologic conditions of the region and human activities.

3.4.2.1 Geologic conditions

The debris flow development are with complex geologic processes, weak lithology, strong weathering, folds, fracture development, strong tectonic movement and frequent earthquakes. For these reasons, all kinds of bad geological phenomena such as rock fracture, collapse and landslide are generally developed, which provide abundant loose debris for the formation of debris flow.

3.4.2.2 Terrain conditions

Debris flows may originate when poorly sorted rock and soil debris became mobilized from slopes and channels by the addition of moisture. The essential conditions for debris flow are: abundant source of coarse-grained or fine-grained sediments, steep slopes, and plentiful supply of moisture along with space vegetation. The moisture is supplied by rainfall, snowmelt, and rarely by snow and ice during volcanic eruptions. These conditions can be found in mountainous areas in arid, semiarid, arctic

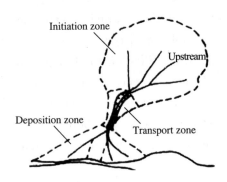

Figure 3-17 Schematic of typical debris flow basin

and humid areas. Small and steep drainage basins, where runoff can be concentrated and sediment source may be high, have the potential to transport large amounts of eroded material by debris flows. Typical debris flow can be categorized into three zones: initiation zone, transport zone and deposition zone, as shown in Figure 3-17.

I. Initiation zone

The initiation zone, also called the source area, is a slope failure in the headwall or side slope of a stream channel. It often has very steep slope (30°~60°), abundant supply of loose debris, a source of water and sparse vegetation. Such topographical conditions are conducive to the collection of water and solid materials on the surrounding uphill.

II. Transport zone

This is the middle part of a debris flow. Most of them are narrow and deep valleys or gullies with steep slope and large longitudinal gradient. The slope and valley deposits frequently supply a significant proportion of the volume of debris flows.

III. Deposition zone

Deposits of debris flows establish a debris fan or cone that can be referred as depositional zone. The deposition will start when the slope angle decreases. Debris flows can move for many kilometers from the source area, emerge from a channel and spread across an alluvial fan.

3.4.2.3 Hydrological and meteorological conditions

Water is not only the component of debris flow, but also the moving medium of debris flow. It will cause the reduction of friction resistance and the increase of sliding stress, thus resulting in flow. The formation of debris flow is closely related to the sudden flow of water in a short period of time. The sudden flow of water can form through: ① heavy rainfall; ② short-term intense melting of glaciers and snow; ③ sudden outburst of glacial lake, reservoir, etc. The sudden flow of water can be channeled over a steep valley filled with debris that is loose enough to be mobilized. The water soaks down into the debris, lubricates the material, adds weight, and triggers a flow.

3.4.2.4 Human factors

Improper engineering activities by human beings can promote the occurrence, development, resurrection or increase the damage of debris flow. These factors include unreasonable excavation of railway, highway and other engineering structures, abandonment of soil and slag for quarrying, increased soil erosion after deforestation, etc.

To sum up, there are three basic conditions in the formation of debris flow: ① a steep slope with loose unconsolidated deposits; ② abundant loose solid materials with plentiful supply of moisture accumulated in upstream; ③ the source of a sudden burst of water in a short term.

3.4.3 Development characteristics of debris flow

I. Regional features

Because of the regional regularity on the distribution of hydrological meteorology, topography and geologic conditions, the distribution of debris flow has regional characteristics.

1) In complex geologic structure, strong neotectonic movement, frequent seismic activity, rock crushing, sparse vegetation mountainous area will provide abundant loose debris for the formation and development of debris flow.

2) It is mainly distributed in temperate zone, semi-arid mountains, and glacier areas. In the dry seasons, the physical weathering of rock is strong, thus a lot of weathered deposits accumulated. It can easily trigger debris flow thereafter in the rainy season.

II. Time features

Since the characteristics of debris flow is periodic and varied in hydrometeorology, the accumulation of abundant loose solid materials would not complete in a short time period. Therefore, it takes time for the development of a debris flow. For example, the debris flows generally occur in China in some mountainous areas between June and September because of intense rainfall.

3.4.4 Classification of debris flow

The term debris flow is used for mass wasting in which the movement is occurring throughout the flow. Based on the formation environment, composition and fluid properties, debris flow can be classified according to the following criteria:

I. Formation environment

1) General debris flow. This is a typical debris flow with loose mud, sand, soil, rock, water and air. It has a fan-shaped drainage basin and an area of more than ten to tens of km^2. We can clearly distinguish the initiation zone, transport zone and deposition zone of debris flow.

2) Valley debris flow. The drainage basin is long and narrow with a sufficient water supply in upstream of the basin. The solid material mainly comes from the debris deposited in the middle stream. Along the river valley, there are both accumulation and erosion.

3) Hillside debris flow. The drainage basin is small, generally less than 1 km2. The basin is hopper-like shaped, with no obvious transport zone, and the initiation zone is directly connected with the deposition zone.

4) Debris avalanche. It is a large, very rapid to extremely rapid, flow of partially or fully saturated debris on a steep slope. It begins as a swallow surface slide when an unstable slope collapses and the fragmented debris continues to develop into a rapidly moving flow, but it does not move into a channel.

5) Volcanic mudflow. It is also called lahar, which originates on the slopes of the volcano. The debris of a lahar can be either hot or cold. They can originate from rainfall, crater lakes, condensation of erupted steam on volcano particles, or the melting of snow and ice at the top of high volcanoes.

II. Composition of debris

1) Mudflow. It is referred for fine-grained debris flows. Mudflow contains mainly clay and silt (accounting for $80\% \sim 90\%$), and only a small amount of rock debris with high viscosity. In general, the favorable conditions for a mudflow mainly include the availability of debris susceptible to saturation, discontinuous intense rainfalls as in the case of storms or sudden snow melting, lack of vegetation cover and steep slopes. These conditions are often found in mountainous regions, volcanic slopes with ash and dust deposits and in arid denuded areas. Moreover, mudflows can travel for long distances even over gently sloping terrain.

2) Normal debris flow. It combines loose soil, rock and sometimes organic matter (variety of grain sizes—from clay to boulders) and variable amounts of water to form a slurry that flows downslope. To be considered a debris flow, the moving material must be loose and capable of "flow," and at least 50% of the material must be sand-size particles or larger.

3) Water-rock flow. The solid material mainly contains some hard rocks, boulders, gravels and sands, but the silt and clay content is very small, generally less than 10%. This type of debris flow is mainly distributed in limestone, quartzite, dolomite, basalt and sandstone distribution area.

III. Fluid properties

1) Viscous debris flow. It contains a large number of fine materials (clay and silt). The solid content is usually in the range of $40\% \sim 60\%$, sometimes up to 80%. The debris can bond together and move as a whole at the same speed. The movement characteristics of this kind of debris flow are mainly viscous. When the viscous debris flow moves on the open accumulation fan, there is no diversion. After stopping deposition, it still maintains the structure of the movement, and the accumulation body is mostly long tongue shape.

2) Dilute debris flows. It is a mixture of water and solid matter, in which water is the main component. The solid content is usually in the range of 10%~40%. Since the clay and silt are less in solid matter, it cannot form a viscous flow. In the process of movement, the velocity of water and sand is much faster than that of rock. There is a significant difference between the velocities of solid and liquid, in which the rocks travel in the form of rolling or jumping, while the water is turbulent flow. In the accumulation fan area, the diluent debris flow is fan-shaped, with crisscross branches and frequent diversions. The accumulation fan is cut into deep gullies. The flow process of this kind of debris flow is smooth, and it is not easy to cause blockage. After stopping deposition, the main materials left are coarse particles and forming a fan-shaped and relatively flat surface.

3.4.5 Debris flow prevention and mitigation

The prevention and mitigation of debris flow includes a combination of land use planning and technical measures to reduce the loss. Debris flows are very difficult to stop once they have started because of their speed and intensity. However, methods such as modifying slopes and particularly preventing them from being vulnerable to debris-flow initiation through the use of erosion control are very useful in mitigation of debris flow hazard. Beside, afforestation and the prevention of wildfires can also help in the mitigation of debris flows.

In fact, construction activities make a slope prone to debris flow in several ways: ① undercutting the base of the slope along with removing the physical support of the upper part of the slope; ② removing the vegetation cover; ③ buildings on the upper part of slope add weight to the potential slide; ④ extra water may infiltrate to debris. In this case, preventive measures should be taken during constructions.

In general, the mitigation measures for debris flow hazards can be distinguished into two types: active and passive measures. Active measures aim at the hazard while passive measures aim at potential damages.

I. Active mitigation measures

Active debris flow mitigation measures influence the initiation, transport or deposition of debris flows. Some examples of constructions for mitigation debris flow are shown below.

1) Debris flow barriers. They are usually built at the base of slopes where debris flows are frequent. They are used especially in areas that are vulnerable to debris-flow damage and soil and debris are stopped from flowing into structures at the base of the slope. They should be planned to be able to hold the maximum flow volumes of an area and to prevent overtopping during a flow event.

2) Check dams. They are small, sediment-storage dams. They are typically built in the channels of steep slopes prone to erosion or gullies to stabilize the channel bed. Check dams are common mitigation measures worldwide and they used to control channelized debris-flow frequency and volume. In some cases, they used to control shallow slides in the

source area of debris slides. In general, check dams have a high structure cost and thus are constructed where urban areas lie downslope. Debris flows are connected with channel gradients over 25° and take most of their volume by scouring the channel bed. Check dams can be built of reinforced concrete or log cribs.

3) Debris flow retaining walls. These can be built of various kinds of materials. They are designed to prevent the evolution of the debris fall, either by blocking the flow or diverting debris around a prone area.

4) Slope stabilization. The slope can be cut back in a series of terraces rather than a single steep cut. This decreases not only the slope angle but the shear strength by removing overlying material. Additionally, it stops loose materials from rolling to the base. Besides, some drainage techniques like surface and subsurface drains can be used to stabilize the slope.

II. Passive mitigation measures

Passive mitigation measures are usually used to reduce the potential losses. Examples of passive mitigation measures are listed below.

1) Hazard mapping. This is an important issue for disaster prevention and management. The areas prone to debris flow hazards should be taken into account in land-use planning.

2) Land use zoning. Building regulations in hazardous areas to debris flow are useful to reduce damages to constructions. Additionally, in those areas with a high hazard of debris flow, urban development and constructions should be prohibited.

3) Warning system. They are able to detect debris flow and automatically trigger an alarm. Thus they are helpful in protecting urban areas, traffic routes and infrastructure locations. Additionally, they can be used to control and monitor safety measures. The most important characteristics of early warning include data collection, transfer and management, distribution of information, decision hierarchy structure and the response of panning and organization. At present, warning systems are mainly used in associations with traffic routes.

4) Immediate technical assistance. This is a technical measure following a disaster, which includes excavation of buried areas, cleaning of swamped areas and reconstruction of infrastructures.

5) Documentation and control. Established mitigation measures should be controlled regularly. The effectiveness of the existing measures should be evaluated after an event. Thus, weak features in the mitigation measures can be recognized and additional measures can be designed accordingly.

3.5 Karst

3.5.1 Definition of karst

Karst is a special type of landscape that is formed by the dissolution of soluble rocks.

Karst regions contain aquifers that are capable of providing large supplies of water. It is characterized by numerous caves, sinkholes, fissures, and underground streams. Karst topography usually forms in regions of plentiful rainfall where bedrock consists of carbonate-rich rock, such as limestone, gypsum, or dolomite, that is easily dissolved. Surface streams and lakes are usually absent from karst topography. Limestone, with its high calcium carbonate content, is easily dissolved in the acids produced by organic materials. About 10% of the Earth's land surface consists of soluble limestone, which can be easily dissolved by the weak solution of carbonic acid found in underground water.

Karst is closely related to engineering construction. For example, in water conservancy and hydropower construction, reservoir leakage caused by karst is the main engineering geologic problem in water conservancy construction. In the tunnel construction in karst area, once the water pressure of karst pipeline is high, it will cause a lot of water inrush, sometimes accompanied by sediment injection, which will bring serious difficulties to the construction, and even accidents such as an immersed tunnel. In the construction of underground cavern, the construction of high fill is difficult and expensive. In the process of highway construction in the karst area, underground karst water will cause some problems such as subgrade base water absorption, subgrade flooding, subgrade water gushing, tunnel water gushing.

3.5.2 Morphological characteristics of karst

In the region with soluble rocks, the dissolution effect can cause a series of corrosion phenomena on the surface and underground. This is called karst formation characteristics. These features are unique to karst areas, which make the karst landscapes beautiful and fascinating. They are often developed as tourist attractions such as Stone Forest National Park in China, Carlsbad Caverns National Park in the USA, Mole Creek Karst National Park in the north of Tasmania, Australia, Tsingy de Bemaraha National Park in Madagascar, etc. Meanwhile, these features especially caves and underground streams, are sources of engineering geologic problems. Figure 3-18 shows some typical features in the karst area.

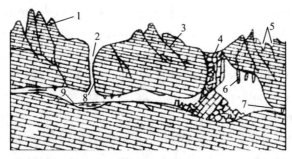

1—Stone forest; 2—Doline; 3—Dissolution fissures; 4—Collapse sink;
5—Solution valley; 6—Stalactites; 7—Cave; 8—Sinkhole; 9—Undergroud stream

Figure 3-18 Schematic of karst morphology

I. Stone forest

The Stone Forest in China has the most unique karst landform in the world due to its long history, large scale, and rich categories. Stone Forest started the formation from about 270 million years ago, extensive deposits of sandstone overlain by limestone accumulated in this basin during the Permian period of geologic time. Uplift of this region occurred subsequent to deposition and exposure to wind and running water shaped these limestone ridges. These formations extend as far as the eye can see, looking like a vast forest of stone, hence the name "the Stone Forest".

II. Doline and sinkhole

The doline is a subcircular bowl or funnel-like depression. The sinkhole is a form which has originated through a gradual or sudden lowering of a portion of the topographical surface. Sinkhole formed when underlying limestone bedrock is dissolved by groundwater. It is considered the most fundamental structure of karst topography. Sinkholes can range in size from a few meters to over 100m deep. A sinkhole can even collapse through the roof of an underground cavern and form as a collapse sinkhole (caving-in), which can become a portal into a deep underground cavern.

III. Solution valley

The collapse of a cavern over a large area can create a feature referred to as a solution valley or basin, sometimes referred to as a karst gulf, which from the air resembles a huge sinkhole. These depressions may reach hundreds of meters across and may contain numerous smaller, local sinkholes.

IV. Disappearing streams

Disappearing streams or sinking streams are streams that terminate abruptly by flowing or seeping into the ground. Groundwater percolating through cracks removes the soluble rock while leaving an enlarged channel for the further flow of water. Since the dissolution often occurs at the surface, there are few continuous surface streams because runoff encounters sinkholes or is otherwise routed underground.

V. Springs

Springs are openings in the ground or rock where underground streams or seeps release water into caves or on the ground. Springs, along with caves, sinkholes, and natural bridges, are all features of karst regions. Slightly acidic groundwater flows through cracks in limestone or dolomite, slowly dissolving the rock. The cracks widen to form cavities and eventually a subterranean drainage system. When a cave is below the water table, it is filled entirely with water. When the cave is above the water table, the cave has air in it and its water flows ever downward. Wherever underground water exits the rock and flows into the open air, it is called a spring. Spring water can discharge from the ground due to gravity or hydrostatic pressure.

VI. Karst cave

Cave entrances are natural openings in the earth large enough to allow a person to enter. Caves may reflect a complex underground drainage system. Inside karst caves, one might find a wide range of speleothems—structures created by the deposition of slowly dripping calcium carbonate solutions. The carbonic acid (H_2CO_3) causes the dissolution of calcium carbonate ($CaCO_3$) in limestone by reacting with the carbonate anion to yield calcium

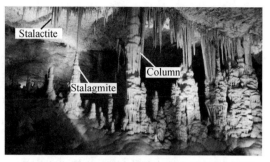

Figure 3-19 Photograph of a karst cave

bicarbonate $Ca(HCO_3)_2$. Dripstones provide the point where slowly dripping water turns into cone-shaped stalactites (those structures which hang from the ceilings of caverns), over thousands of years which drip onto the ground, slowly forming stalagmites. When stalactites and stalagmites meet, they form cohesive columns of rock. Figure 3-19 shows the beautiful displays of stalactites, stalagmites, columns of karst topography.

Ⅶ. Underground streams

Underground streams or rivers, also known as subterranean rivers, flow at least partly beneath the surface of the Earth. Many underground rivers are part of a karst landscape, where eroded limestone often creates caves. Underground rivers may emerge at sinkholes or above ground, as the karst landscape gives way to the soil. In general, a typical karst landscape forms when much of the water falling on the surface interacts with and enters the subsurface through cracks, fractures, and holes that have been dissolved into the bedrock.

3.5.3 Formation conditions of karst

3.5.3.1 Main conditions

Karst topography forms in areas where the underlying bedrock is composed of material that can be slowly dissolved by groundwater. Examples of this type of sedimentary rock include carbonate rocks such as limestone, halite, gypsum, dolomite, and anhydrite. Main conditions for the formation of karst include rock solubility, rock permeability, water corrosion and water flowability.

Ⅰ. Rock solubility

The presence of soluble rock is the basis for karst development. Its composition and structural characteristics can significantly affect the development of karst. Different compositions of the soluble rocks show different extent of solubility. is different, and its solubility is different. According to its composition, soluble rocks are divided into carbonate rocks (limestone, dolomite and gypsum, etc.), sulfate rocks, salt rocks. The solubility of carbonate rocks in the three types of soluble rocks is the least, and the solubility of the salt rock is the largest. However, carbonate rocks are the most widely distributed soluble rocks that contribute to the formation of karst.

II. Rock permeability

The permeability of rock is another necessary condition for karst development. The higher the rock permeability of rock, the stronger is the karst development. The permeability of rocks depends on the fracture, porosity and connection of the rock mass. The development of fractures in the rock tends to control the development of karst. Flow paths in karst are developed preferentially along fault zones, joints, and bedding planes. In addition, due to the increase of weathering fissures, it is beneficial to the movement of groundwater and karst is easy to develop.

III. Water property

The dissolution of rocks and the duration of direct water-rock contact result in variable groundwater quality at discharge points. The mineral components of karst waters depend upon the composition of the rocks through which water is percolating: hydrocarbonate-calcium type of waters is created from the dissolution of calcium carbonate which is a dominant type of water in limestone, while the hydrocarbonate-magnesium type of groundwater is present to a lesser extent, and is regularly connected to dolomitic rocks. Karst water contains carbon dioxide (CO_2) which dissolves carbonate rocks. When the water contains weak acids or chloride (Cl^-) and sulfate ions (SO_4^{2-}), rocks can be more soluble. In addition, with the increase of water temperature, the diffusion rate of CO_2 into the water increases and thus strengthens the karstification.

IV. Water flowability.

The dissolution of carbonate rocks is closely related to the groundwater flow. There are three types of porosity in karst aquifers: inter-granular pores in the rock matrix, common rock discontinuities such as fractures and bedding planes, and solutional enlarged voids such as channels and conduits developed from the initial discontinuities. Whereas groundwater flow in the matrix and small fissures is typically slow and laminar, flow in karst conduits (caves) is often fast and turbulent.

3.5.3.2 Main factors

In karst development areas, various karst forms in space distribution and arrangement are predictable. They are mainly affected by the factors including lithology, geologic structure, crustal movement, topography and climate.

I. Effect of lithology

Lithology and secondary porosity are interconnected. Lithology can be considered an important factor in the development of karst. Karst has a unique hydrology condition. This condition is that karst is composed of soluble lithology and the lithology has a good condition in secondary porosity. The composition of soluble rock is the foundation of karst development and distribution. The calcareous strata with uniform composition and thickness are the most suitable condition for karst development.

II. Effect of geologic structure

The geological structures such as folds, joints and faults control the flow channels of groundwater. Different geologic structures will result in different morphology, location and degree of karst development. The anticlinal axis is developed and the surface water seeps along the joint, forming the vertical karst landform such as doline and sinkhole. It is difficult to predict where sinkholes will form, but many are aligned over joints in the bedrock. In contrast, groundwater flow can down-dip on the syncline limb and is concentrated toward the axis so that karst features are strongly developed along the belt of syncline. In addition, the tensile fracture zone is relatively small, with loose structure and insufficient cementation, which is conducive to groundwater infiltration and karst development.

III. Effect of crustal movement

Groundwater in karst area is controlled by the erosion base level of soluble rock. The change of erosion base level is caused by crustal movement. Therefore, when the earth's crust rises and the base level of erosion is relatively low, the denudation of karst dominates and forms a vertical karst landform. When the crust is relatively stable and the erosion datum time is relatively constant, the groundwater mainly moves horizontally, forming a large horizontal solution hole. Since the crustal fluctuation and stability are interchangeable, the morphology of vertical and horizontal karst is alternated.

IV. Effect of topography

Karst landscapes vary considerably. For example, some karst areas have streams and rivers that will disappear into the ground, only to reappear later as springs on the surface. Some karst regions are sharp jagged hills, while others are soft rolling hills with depressions that used to be sinkholes. In the places where the rock is exposed and steep, the surface water is gathered fast, the flow is fast and seepage is small. In this condition, some landforms like the valley, sinkholes will form. While in the low terrain, the surface runoff is slow, and the infiltration is much lower, doline and caves will form.

V. Effect of climate

Karst development is strongly influenced by climate, both directly (via the moisture balance and temperature regime) and indirectly. For example, climate changes can affect the rate of weathering. Deep water table can form underground karst features. The indirect effects include bio-geomorphic impacts of biota, and base level changes associated with sea-level and river incision or aggradation.

3.5.4 Engineering problems in karst area

Developing and building in karst environments provides many challenges and risks not encountered in other geologic environments. The unpredictability of the subsurface conditions frequently makes planning and construction very challenging.

3.5.4.1 Main engineering problems

The construction of roads, bridges or tunnels in karst areas will often face many difficulties in engineering design or construction, which may result in engineering failure or rebuild if not taken seriously. Unlike most other building environments, subsurface conditions can change after the geotechnical exploration has been completed. Variations in the groundwater levels and construction activities may contribute to the formation of collapse features at a site. For instance, sinkholes can form where water is allowed to enter the subsurface at one location. Because of the connected nature of limestone bedding and fracturing, infiltrating water can lead to sinkhole formation well downstream of the point of entry. Therefore, control of both surface and water at the site is important. The design and implementation of construction controls to avoid excessive water infiltration are vital.

3.5.4.2 Mitigation measures

There are several things we can do to reduce the risks associated with development in karst environments. A preliminary site evaluation, which typically includes a detailed geologic review, site reconnaissance, fracture trace analysis, and traditional subsurface explorations, can help identify areas within a site that may have existing solution features and areas that are at a higher risk. Once a site design has been selected, additional exploration programs, including geophysical explorations and more specific subsurface explorations need to be conducted to provide detailed information.

While karst environments present unique challenges to a construction project, careful evaluation of a site's conditions, proper site and foundation design, and prudent construction operations can reduce the impact of karst environments on site and building performance, and development costs. Karst mitigation is focused on maintaining the quality and quantity of water entering the feature. It includes permanent measures designed to reduce impacts resulting from the project. Mitigation measures include installing peat/sand/gravel filters, vegetative buffers, and lined spill/runoff containment structures, to detain or treat highway runoff prior to discharge.

I. Construction mitigation methods

Karst features are mitigated during construction to maintain water flow or to provide structural integrity for the overlying roadway. The location of the feature and direction of water flow will determine the appropriate mitigation measure. If the location is in the proposed roadway and the water flows through the feature into the underlying karst, it will be capped. If the feature located in the proposed roadway is a spring or seep, the mitigation method will maintain the flow of water from the underlying karst to the surface. The design of roadside ditches and other conveyances must ensure that the surface runoff has received adequate filtration prior to entering karst features located along the roadway.

II. Excavation and plugging

This method is the most common mitigation method used to permanently seal a feature from surface flows. It is used for shallow sinkholes with a depth of 15 feet or less. All soil, rock and debris are removed from within the weak zones. The throat of the soil void is "capped" with concrete, grout, or a rock fill plug and backfilled or compacted to the desired density. Geotextile fabric is placed between the layers of stone and soil of an aggregate cap to prevent fines from entering the feature. A concrete cap is used when the sinkhole will be located under the roadbed to provide a permanent seal.

III. Void-bridging

This method places high-strength geotextile material over potential voids to increase the load-carrying capacity. On embankments, this allows for higher construction and steeper side slopes. It can also be used under lightweight structures to create a barrier through which a top layer of sand and other soils can't pass.

IV. Drainage control

Infiltration of surface water can lead to soil voids, collapse and potential sinkhole formation. The features are protected during and post-construction with lined drainage routes and storm water detention areas. The primary goal is to control entry points of surface runoff and divert subsurface water from known sinkholes. A concrete lined ditch that is cracked would allow water and potential contaminants into a feature and the increased water flow would lead to the collapse of the feature.

V. Spring or seep protection

Springs and seeps require special treatment to reduce impacts to the road and the water resources. The outlet below is designed to capture the flow from a seep located under a roadway and direct it towards its normal flow path. Springs or seeps will continue to flow and cannot be treated by capping. Failure to direct the flow from under the roadway will result in instability of the fill under the roadbed. The spring box is designed to capture flow from a spring or seep located under the fill and directs it toward the bottom of the slope. The construction of bridge piers within karst terrain requires that the pier footing be designed to the terrain.

3.6 Earthquake

Earthquakes are dangerous geologic hazards. They can cause damage to buildings, bridges and other infrastructure; disrupt gas, electric, and telecom service; and sometimes trigger landslides, avalanches, flash floods, fires, and huge, destructive ocean waves (tsunamis).

3.6.1 Basic concepts of earthquake

Earthquake is defined as a sudden, rapid shaking of the ground caused by the passage of seismic waves through Earth's rocks. Seismic waves are produced when some form of energy stored in Earth's crust is suddenly released, usually when masses of rock straining against one another suddenly fracture and "slip". Earthquakes occur most often along geologic faults, narrow zones where rock masses move in relation to one another. The major fault lines of the world are located at the fringes of the huge tectonic plates that make up Earth's crust.

Figure 3-20 shows the schematic of main components of an earthquake, in which the focus is the point where the rocks start to fracture and focus is the origin of the earthquake. The epicenter is the point on land directly above the focus. An isoseismal line

Engineering Geology and Soil Mechanics

is a contour or line on a map bounding points of equal intensity for a particular earthquake. Most parts of the world experience at least occasional **shallow earthquakes**—those that originate within 60km of the Earth's outer surface. In fact, the great majority of earthquake foci are shallow. Of the total energy released in earthquakes, 12% comes from **intermediate earthquakes**—that is, quakes with a focal depth ranging from about 60 to 300km. About 3% of total energy comes from **deeper earthquakes**. The frequency of occurrence falls off rapidly with increasing focal depth in the intermediate range. Below intermediate depth the distribution is fairly uniform until the greatest focal depths, of about 700km, are approached.

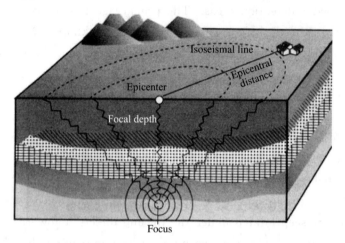

Figure 3-20　Schematic of focus, epicenter and isoseismal line

I. Seismic waves

Earthquakes generate four principal types of seismic waves, which are elastic waves that may travel either along or near the earth's surface called surface waves (including Rayleigh waves and Love waves) or through the Earth's interior called body waves (including Primary waves and Secondary waves).

Primary waves, also known as P-waves or pressure waves, are longitudinal waves travelling at the greatest velocity (about 7~13km/s) through the Earth. They can pass through solids, liquids and gases easily. As they travel through rock, the waves move tiny rock particles back and forth and push them apart and then back together.

Secondary waves, also known as S-waves, shear waves or shaking waves, are transverse waves that travel slower (about 4~7km/s) than P-waves. In this case, particle motion is perpendicular to the direction of wave propagation. S-waves cannot travel through air or water but are more destructive than P-waves because of their larger amplitudes.

While the exact speed of primary waves and secondary waves varies depending on the composition of the material they're traveling through, the ratio between the speeds of the two waves will remain relatively constant in any earthquake. P-waves generally travel 1.7

times faster than S-waves.

Surface waves are similar in nature to water waves and travel just under the Earth's surface. Surface waves are the slowest moving of all waves with an average speed of 3~4km/s, which means they arrive at the last. Surface waves are to blame for most of an earthquake's carnage. They move up and down the surface of the Earth, rocking the foundations of man-made structures. So the most intense shaking usually comes at the end of an earthquake.

Rayleigh waves, also called ground roll, travel as ripples similar to those on the surface of water. The greater part of the shaking felt from an earthquake is because of the Rayleigh wave, which can be considerably bigger than other waves.

Love waves, also known as Q-waves from the German word for horizontal, will cause horizontal shearing of the ground. They usually travel slightly faster than Rayleigh waves. Love waves form on the Earth's surface when S-waves change into a new type of wave. These waves can cause widespread destruction after large earthquakes.

II. Earthquake magnitude

Earthquake magnitude is a measure of the "size" or amplitude, of the seismic waves generated by an earthquake source and recorded by seismographs. A seismograph is the device that scientists use to measure earthquakes. The goal of a seismograph is to accurately record the motion of the ground during a quake. The Richter scale is a standard scale used to compare earthquakes. It is a logarithmic scale, meaning that the numbers on the scale measure factors of 10. So, for example, an earthquake that measures 4.0 on the Richter scale is 10 times larger than one that measures 3.0. On the Richter scale, anything below 2.0 is undetectable to a normal person and is called a microquake. Microquakes occur several hundred times a day worldwide. Moderate earthquakes measure less than 6.0 the Richter scale. Earthquakes measuring more than 6.0 are big earthquakes that can cause significant damage. The major earthquakes which have a magnitude of 7 or larger can occur more than once per month around the world. With regard to the great earthquake, which has a magnitude of 8 and greater, occur about once a year. The biggest quake in the world since 1900 scored a 9.5 on the Richter scale. It rocked Chile on May 22, 1960.

Another way to measure the size of an earthquake is to compute how much energy it released. The amount of energy radiated by an earthquake is a measure of the potential for damage to man-made structures. The relationship between the magnitude and the total energy released in an earthquake is:

$$\lg E = 11.8 + 1.5M \tag{3-2}$$

where, E—energy released, with the unit is J;

M—magnitude of an earthquake.

Table 3-3 shows the value of magnitude and the related energy released. While each whole number increase in magnitude represents a tenfold increase in the measured amplitude, it represents a 32 times more energy release. The numerical value of the magnitude that is the earthquake's size remains constant for a given earthquake.

Table 3-3   Relations between earthquake magnitude and energy released

Earthquake magnitude	Energy/J	Earthquake magnitude	Energy/J
1	2.00×10^{13}	6	6.31×10^{20}
2	6.31×10^{14}	7	2.00×10^{22}
3	2.00×10^{16}	8	6.31×10^{23}
4	6.31×10^{17}	8.5	3.55×10^{24}
5	2.00×10^{19}	8.9	1.41×10^{25}

III. Earthquake intensity

Intensity is the quantum of negative impact of earthquake on surrounding areas. Unlike magnitude, the intensity that is the devastation caused by earthquake varies with the location and is not a single numerical value. The farther an area is from epicenter, the lower is the intensity of earthquake. For calculating intensity, the responses of people in surrounding areas, worsened condition of structures and changes in natural surroundings are noted. Areas near to the epicenter severely feel the shaking intensity and thus are affected critically as compared to those staying farther away.

The Mercalli scale measures the intensity of an earthquake by quantifying the effects of an earthquake on the Earth's surface. Based on human reactions, natural objects and man-made structures, the Mercalli scale rates earthquakes on a scale of 1 to 12, with 1 denoting that nothing was felt and 12 denoting total destruction. There are different intensity scales used in various countries. For example, in China, we use the scale similar to Mercalli scale. 1~5 reflect the earthquakes whether can be felt by people; 6 reflects the buildings are slightly damaged, 7 or more reflects moderate damage to most buildings; 9 or more denotes most buildings will collapse; 11 or more reflects the buildings are generally destroyed. Besides, the Modified Mercalli intensity scale is used in the United States. This scale, composed of increasing levels of intensity that range from imperceptible shaking to catastrophic destruction, is designated by Roman numerals. It does not have a mathematical basis; instead it is an arbitrary ranking based on observed effects. Table 3-4 lists the relations between Modified Mercalli intensity scale and Richter scale. It should be noted that for a given earthquake, the magnitude on Richter scale remains constant, while the intensity varies depending on the location. The farther an area is from epicenter, the lower is the intensity scale. The intensity scale in Table 3-4 reflects the impact of an earthquake on the area around the epicenter.

Table 3-4 Modified Mercalli intensity scale versus Richter scale

Modified Mercalli intensity	Effects	Richter scale (approximate)
I. Instrumental	Not felt	1~2
II. Just perceptible	Felt by only a few people, especially on floors of tall buildings	3
III. Slight	Felt by people lying down, seated on a hard surface, or in the upper stories of tall buildings	3.5
IV. Perceptible	Felt indoors by many, by few outside; dishes and windows rattle	4
V. Rather strong	Generally felt by everyone; sleeping people may be awakened	4.5
VI. Strong	Trees sway, chandeliers swing, bells ring, some damage from falling objects	5
VII. Very strong	General alarm; walls and plaster crack	5.5
VIII. Destructive	Felt in moving vehicles; chimneys collapse; poorly constructed buildings seriously damaged	6
IX. Ruinous	Some houses collapse; pipes break	6.5
X. Disastrous	Obvious ground cracks; railroad tracks bent; some landslides on steep hillsides	7
XI. Very disastrous	Few buildings survive; bridges damaged or destroyed; all services interrupted (electrical, water, sewage, railroad); severe landslides	7.5
XII. Catastrophic	Total destruction; objects thrown into the air; river courses and topography altered	8

The Modified Mercalli intensity value assigned to a specific site after an earthquake has a more meaningful measure of severity than the magnitude because intensity refers to the effects actually experienced at that place. The lower numbers of the intensity scale generally deal with the manner in which the earthquake is felt by people. The higher numbers of the scale are based on observed structural damage. Structural engineers usually contribute information for assigning intensity values of VIII or above.

3.6.2 Types of earthquake

There are two main types of earthquakes: natural and man-made. Naturally occurring earthquakes occur along tectonic plate lines while man-made earthquakes are always related to explosions detonated by man. Natural earthquakes can be divided into tectonic earthquakes, volcanic earthquakes and collapse earthquakes.

I. Tectonic earthquakes

Tectonic earthquakes will occur anywhere there is sufficient stored elastic strain energy to drive fracture propagation along a fault plane. This type of earthquake is directly related to the strong and weak tectonic movement, which is located in the most intense area of tectonic movement since the Cenozoic era. Tectonic earthquakes are the most important type of earthquake, accounting for about 90% of the total earthquakes.

A tectonic earthquake occurs where tectonic plates meet, an area known as the boundary. When two plates push into each other, they form a convergent plate boundary. For example, the oceanic Nazca Plate off the coast of South America along the Peru-Chile trench pushes into and is subducted under the South American Plate. This movement lifts up the South American Plate, creating the Andes mountains. The Nazca Plate breaks into smaller parts that are locked in place for long periods before suddenly shifting to cause earthquakes.

II. Volcanic earthquakes

Earthquakes related to volcanic activity are called volcanic earthquakes. The places where tectonic plates meet give access to the molten mantle. As a result, these fissures are a major source of volcanic activity. Volcanoes are openings in the Earth's crust where pressure buildups are released along with molten materials, preventing the planet from overheating and exploding. They are not always found along fault lines, but frequently mark the edge of a tectonic plate. The Ring of Fire in the Pacific is a prime example of this phenomenon. Volcanic earthquakes are a form of tectonic earthquake where volcanic activity coincides with tectonic forces. There are actually two forms of volcanic earthquake including volcano-tectonic earthquake and long-period earthquakes.

III. Collapse earthquakes

Earthquakes caused by the collapse of ground or subterranean rock are called collapse earthquakes. These earthquakes are of weak magnitude earthquakes happen in the caverns and mines. Sometimes, underground blasts (rock breaking) in the mines become the cause of the collapsing of mines and collapsing of mines produces seismic waves. Although the collapse earthquakes tend to be very low magnitude, they may still result in significant property damage when the collapsed cavities are underneath structures.

3.6.3 Methods of reducing earthquake hazards

Compared with other natural hazards, earthquakes are unique, because there is no warning. Considerable work has been done in seismology to explain the characteristics of the recorded ground motions in earthquakes. Such knowledge is needed to predict ground motions in future earthquakes so that earthquake-resistant structures can be designed. Although earthquakes cause death and destruction through such secondary effects as landslides, tsunamis, fires, and fault rupture, the greatest losses—both of lives and of property—result from the collapse of man-made structures during the violent shaking of the ground. Accordingly, the most effective way to mitigate the damage of earthquakes from an engineering standpoint is to design and construct structures capable of withstanding strong ground motions.

Developing engineered structural designs that are able to resist the forces generated by seismic waves can be achieved either by following building codes based on hazard maps or by appropriate methods of analysis. Many countries reserve theoretical structural analyses for the larger, more costly, or critical buildings to be constructed in the most seismically active regions, while simply requiring that ordinary structures conform to local building codes. An essential part of what goes into engineering decisions on design and into the development and revision of earthquake-resistant design codes is therefore seismological, involving measurement of strong seismic waves, field studies of intensity and damage, and the probability of earthquake occurrence. Earthquake engineering is a subdivision of structural engineering that focuses on designing structures capable of withstanding the immense levels of stress caused by seismic forces. Civil engineers specializing in this field usually work in geographic areas that frequently experience earthquakes, allowing them to test new technologies in real-world earthquake scenarios.

Ⅰ. Lead-rubber bearings

To withstand an earthquake, buildings need to be designed with seismic control—especially taller buildings, as their collapse could cause significant damage. One inexpensive method of achieving seismic control is base isolation. This passive method isolates the base of a structure from its foundation using a set of lead-rubber bearings within the structure's foundation that can effectively deflect or absorb the vibrations caused by seismic waves. Lead-rubber bearings are comprised of a lead core set within a rubber housing, which is then encased between two thick steel plates and fixed at the base of a building's foundation. The flexibility of this design aids in deflecting seismic waves, while the plasticity of the rubber components absorbs energy from vibrations that would otherwise cause significant damage. Finally, the solid lead core dissipates residual energy that has not already been absorbed or deflected by the outer layers.

Ⅱ. Steel plate shear walls

Steel plate shear wall systems have been used to reinforce buildings since the 1970s. They are considered as a promising alternative to conventional earthquake-resistant systems that are being used in many high-risk seismic regions. These walls are designed to limit lateral force in buildings by using steel shear walls that absorb stress and bend but do not entirely buckle under pressure. The walls are also significantly thinner than concrete shear walls, offering similar levels of resistance and stability, reducing construction costs and lowering total building weight—all without compromising public safety. In addition, steel walls don't need to be cured, allowing a much faster and fluid erection process.

Ⅲ. Controlled rocking

Controlled rocking systems prevent damage by minimizing the drifts that occur in a structure during an earthquake. These high-performance systems utilize braced steel frames that have elastic properties; this allows the steel frames to rock upon the foundation. The elastic element creates a self-centering, restoring force that dissipates

Engineering Geology and Soil Mechanics

seismic vibrations throughout the structure and allows the frame of a structure to rock in a controlled fashion within a gap that has been intentionally placed in the foundation. Another component of engineering effective controlled rocking systems is the application of replaceable energy-dissipating devices that produce high initial system stiffness.

Ⅳ. Tuned mass dampers

Traditional mass dampers are designed to control the movement of high-rise buildings. To create a mass damper, civil engineers suspend large metal pendulums attached to cables at the top of a tall building; these pendulums act as an inertial counterweight that keeps the building as centered as possible. These dampers effectively lower the speed at which a building is allowed to oscillate, as well as the total distance of each oscillation. In circumstances where the use of a traditional mass damper has been deemed unsafe or unreasonable because of excessive amounts of sway, tuned mass dampers may be utilized instead. Tuned mass dampers work similarly to traditional mass dampers but include the use of an additional control system, such as an electromagnet, to limit the motion of the damper's pendulum element.

Ⅴ. Seismic cloaking

Seismic cloaks are currently being tested as a means of creating protective barriers capable of rerouting seismic energy away from aboveground structures. Seismic cloaking involves the modification of soil and other ground materials surrounding a building to deflect or redirect the force created by an earthquake. This innovation revolves around the theory that seismic waves pass energy between the potential energy stored in the planet's crust and the kinetic energy within the seismic wave itself. Armed with this theoretical knowledge, earthquake engineers are tasked with creating a cloaking structure that can control these destructive seismic waves.

Knowledge expansion

About Geotechnical Investigation

Geotechnical investigation is a process in which the physical properties of a site are assessed for the purpose of determining which uses of the site will be safe. Before land can be developed or redeveloped, geotechnical investigation is often required. This process is also required or recommended in the wake of incidents like earthquakes, the emergence of foundation cracks on land which was thought to be solid, and so forth. The goal of such investigation is to confirm that the land is safe to build on. A number of things can be a concern when evaluating a site which has been proposed for development. They may be concerned that the soil cannot safely support a structure, or that structures above a certain size could be dangerous. Geotechnical investigation can also reveal issues which could be problematic in an earthquake, such as soils which are subject to liquefaction. The investigation could also be used to find a formation which would be capable of supporting development, such as bedrock which can be used as an anchor to reduce the risk of damage in an earthquake.

Site investigation or sub-soil explorations are done for obtaining the information about subsurface conditions at the site proposed for construction. Objectives of site investigation or subsurface exploration are: ① to know about the order of occurrence of soil and rock strata; ② to know about the location of the groundwater table level and its variations; ③ to determine engineering properties of soil; ④ to select a suitable type of foundation; ⑤ to estimate the probable and maximum differential settlements; ⑥ to find the bearing capacity of the soil; to predict the lateral earth pressure against retaining walls and abutments; ⑦ to select suitable soil improvement techniques; ⑧ to select suitable construction equipment; ⑨ to forecast problems occurring in foundations and solutions. When test holes are required, it is important that they are conducted by qualified personnel and that the needed data is obtained from the investigation. Site investigation or sub-soil exploration is carried out stage-wise as given below.

1) Site reconnaissance

Site reconnaissance is the first stage, in which the visual inspection of the site is done and information about topographical and geological features of the site are collected. The general observations made in this stage are: ① presence of drainage ditches and dumping yards etc.; ② location of groundwater table by observing well in that site; ③ presence of springs, swamps, etc.; ④ high flood level marks on the bridges, high rise buildings, etc.; ⑤ presence of vegetation and nature of the soil; ⑥ past records of landslides, floods, shrinkage cracks, etc. of that region; ⑦ study of aerial photographs of the site, blueprints of present buildings, geological maps, etc.; ⑧ observation of deep cuts to know about the stratification of soils; ⑨ Observation of settlement cracks of existing structures.

2) Preliminary site exploration

Preliminary site exploration is carried out for small projects, light structures, highways, airfields, etc. The main objective of preliminary exploration is to obtain an approximate picture of sub-soil conditions at low cost. It is also called general site exploration. The soil sample is collected from experimental borings and shallow test pits and simple laboratory tests such as water content test, density, unconfined compressive strength test, etc. are conducted. Simple field tests such as penetration methods, sounding methods, geophysical methods are performed to get the relative density of soils, strength properties, etc. The data collected about subsoil should be sufficient enough to design and build light structures. Some of the general information obtained through primary site exploration are: ① approximates values of soil's compressive strength; ② position of the groundwater table; ③ depth and extent of soil strata; ④ soil composition; ⑤ depth of hard stratum from ground level; ⑥ engineering properties of soil.

3) Detailed site exploration

Detailed exploration is preferred for complex projects, major engineering works, heavy structures like dams, bridges, high rise buildings, etc. A huge amount of capital is required for a detailed site exploration hence, it is not recommended for minor engineering works where the budget is limited. For such type of works, data collected through preliminary site exploration is enough. In this stage, numerous field tests such as in-situ vane shear test, plate load test, etc. and laboratory tests such as permeability tests, compressive strength test on undistracted soil samples are conducted to get exact values of soil properties.

4) Preparation of report of sub-soil exploration

After performing preliminary or detailed site exploration methods a report should be prepared. A sub-soil investigation or exploration report generally include: ① introduction; ② scope of site investigation; ③ description of the proposed structure, purpose of site investigation; ④ site reconnaissance details; ⑤ site exploration details such as number, location and depth of boreholes, sampling details etc.; ⑥ methods performed in site exploration and their results; ⑦ laboratory tests performed and their results; ⑧ details of groundwater table level and position; ⑨ recommended improvement methods if needed and recommended types of foundations, structural details, etc.; ⑩ conclusion.

Exercises

[3-1] What is a geologic hazard? Introduce some common geologic hazards.

[3-2] What is a land subsidence? What are the main types of subsidence? What are the main reasons causing a land subsidence?

[3-3] What are groundwater types? How to control the groundwater?

[3-4] What is a landslide? What are the characteristics of a landslide?

[3-5] What are the conditions for the formation of landslides? What are the factors that affect landslides?

[3-6] What are the prevention and mitigation methods of landslides?

[3-7] What is a debris flow? What are the conditions for the formation of debris flow? What are its developmental characteristics?

[3-8] What are the main types of debris flow? How to prevent and mitigate debris flow hazards?

[3-9] What is a karst? What are the main morphological characteristics forms of a karst landscape?

[3-10] What are the main engineering geologic problems in karst area? What are the

Chapter 3 Geologic Hazards

common measures to deal with the problems?

[3-11] What is an earthquake? What are the types of earthquakes?

[3-12] What is a seismic wave? What are the types of seismic waves?

[3-13] What is the earthquake magnitude? What is the earthquake intensity?

[3-14] What are the relations between the magnitude scale and intensity scale?

[3-15] What are the main methods to reduce the earthquake hazards in civil engineering practice?

Chapter 4 Engineering Properties of Soil

4.1 Introductory case

Soil is an important engineering material. The geological conditions and basic soil properties are extremely significant for engineering facilities. Soil comprises materials that are found in the surface layer of the Earth's crust. Soil mass is generally a three-phase system as it contains soil solid with water and air. The space covered by water and air in the soil mass is called voids. These voids are sometimes filled completely with water and sometime by water and air. Since water and air have various contents in different conditions, soil is normally characterized by specific nature.

The behavior of foundation depends on the engineering properties of the underlying deposits of the soil or rock. In some cases, there is no need to stabilize the soil foundation since the soil properties are good enough to hold the load of the building. However, most soils cannot be directly used as natural foundations. For example, collapsible loess is widely distributed in China, which is not suitable as natural foundation because it will cause structural damage due to water invasion. Therefore, soil replacement is often conducted for foundation construction for collapsible loess. Upon backfilling of excavation, plain soil or lime soil is used and compacted to make foundations good for construction. Besides, borrow pit problems often occur when the source material (soil) at one location is removed for use as fill in the construction of road embankments at another location. As we know the total volume of soil needed in its compacted state for use as fill, we need to calculate the volume that has to be excavated from the borrow pit since the soil volume will reduce after the compaction process. This phenomenon shows that soil is not entirely composed of solid particles. There are voids in the soil, which are filled with water and/or air. Thus, we can say that the soil solid particles, water and air form a three-phase system of soil. We need to study the phase relationship in soil.

4.2 Soil formation and structure

The term soil has various meanings, depending upon the general field in which it is being considered. To a pedologist, soil is the substance existing on the Earth's surface, which grows and develops plant life. To a geologist, soil is the material in the relative thin surface zone within which roots occur, and all the rest of the crust is grouped under the term rock irrespective of its hardness. To an engineer, soil is the un-aggregated or un-cemented deposits of mineral and/or organic particles or fragments covering large portion of the Earth's crust.

Soil Mechanics is one of the youngest disciplines of Civil Engineering involving the study of soil, its behavior and application as an engineering material. According to Karl Terzaghi: "Soil Mechanics is the application of laws of mechanics and hydraulics to engineering problems dealing with sediments and other unconsolidated accumulations of solid particles produced by the mechanical and chemical disintegration of rocks regardless of whether or not they contain an admixture of organic constituent."

4.2.1 Soil formation

Soil material is the product of rock, which is termed the "parent material". The geologic process that produces soil is weathering. The parent material may be directly below the soil, or great distances away if wind, water or glaciers have transported the soil. In addition to the soil parent material, soil formation is also dependent upon other prevailing processes affecting soil formation. The soil formation process is termed "pedogenesis". Climatic conditions are important factors affecting both the form and rate of physical and chemical weathering of the parent material. The formation of soils can be seen as a combination of the products of weathering, of structural development of the soil, of differentiation of that structure into horizons or layers, and lastly of its movement or translocation. In fact there are many ways in which soil may be transported away from the location where it was first formed.

Based on the method of formation, soil may be categorized as residual soil and transported soil. Residual soils are formed from the weathering of rocks and practically remain at the location of origin with little or no movement of individual soil particles. Residual soil generally comprises a wide range of particle sizes, shapes, and composition. Transported soils are those that have formed at one location (like residual soils) but are transported and deposited at another location. Weathered materials have been moved from their original location to new locations by one or more of the transportation agencies, viz., water, glacier, wind, and gravity, and deposited to form transported soil. Soils that are carried and deposited by rivers are called alluvial deposits. If coarse and fine-grained deposits are formed in seawater areas, then they are called marine deposits. Glacial soils transported by rivers from melting glacial water create deposits of stratified glacial drift and are referred to as glacio fluvial deposit or stratified drift. Soils carried by wind are subsequently deposited as aeolian deposits. Wind-blown silts and clays deposited with some cementing minerals in a loose, stable condition are classified as loess. Gravity can transport materials only for a short distance. Gravity deposits are termed talus. They include the material at the base of cliff and landslide deposits. Besides, in water-stagnated areas where the water table is fluctuating, and vegetational growth is possible, swamp and marsh deposits develop. Soils transported and deposited under this environment are soft, high in organic content, and unpleasant in odor. Accumulation of partially or fully decomposed aquatic plants in swamps or marshes is termed muck or peat. Muck is a fully decomposed material, spongy, light in weight, highly compressible, and not suitable for construction purposes.

4.2.1.1 Solid phase

The solid phase of soils consists of both inorganic and organic components. Inorganic components range in size from tiny colloids to large gravels and rocks, and include many soil minerals, both primary and secondary. Inorganic components usually control soil properties and its suitability as a plant growth medium.

I. Composition

The mineral composition of soil particle is mainly depends on the mineral composition of the parent rock and the weathering. The inorganic mineral in soil is the product of rock weathering, which is the main part of mineral in soil. There are two main types of minerals as primary minerals and secondary minerals. The key difference between primary and secondary minerals is that primary minerals form from igneous primary rocks whereas secondary minerals from form weathering of primary rocks. In addition, solid phase also include soluble salts and organic matter.

1) Primary minerals. Primary minerals are substances that are formed from primary igneous rocks via original crystallization. That means primary minerals form from solidification processes. The category of primary minerals includes essential minerals (that are used to assign a classification name to the rock) and accessory minerals (that are less abundant). Moreover, the dominant form of primary minerals is silicate mineral. Some examples of primary minerals include quartz, feldspar, mica, muscovite, granite, etc.

2) Secondary minerals. Secondary minerals are substances that are formed from the alteration of primary minerals. That means secondary minerals form when primary minerals undergo chemical and geological alterations such as weathering and hydrothermal alteration. During weathering, water accompanied by CO_2 from the atmosphere plays an important role in processes, like hydrolysis, hydration and solution. As a result the primary minerals are altered or decomposed. The most commonly formed secondary minerals are clay minerals (e.g. kaolinite, illite and montmorillonite) and iron and aluminum oxides.

3) Soluble salts. Soluble salts are minerals dissolved in water. According to their solubility in water, the soluble salts in soil can be divided into three types: soluble salts, moderately soluble salts and insoluble salts. The dissolution and crystallization of salt in soil will affect the engineering properties of soil, and sulfate also has certain corrosion effect on metal and concrete. Therefore, there are certain restrictions on the content of soluble salt and medium soluble salt in the project. Due to the existence of salt in soil, the

concentration and type of ions in pore water will be affected, which is one of the factors affecting soil physical properties.

4) Organic matter. Soil organic matter is the fraction of the soil that consists of plant or animal tissue in various stages of breakdown (decomposition). Organic matter is usually concentrated in local soil layer, and the organic matter in soil has a significant impact on soil properties. The soil with organic matter content less than 5% is called inorganic soil. The chemical composition of organic matter can be divided into carbohydrate, protein, fat, hydrocarbon and carbon. Thoroughly decomposed soil organic matter is called humus. The particles of humus are very fine, mostly gelatinous, with charge, and have strong adsorption.

II. Clay mineral

Clay minerals are the most abundant secondary minerals in soil, and they are the most important factors affecting the properties of clayey soil. There are many kinds of clay minerals, which exist in crystalline form. The basic structural units of most clay minerals are a silicon-oxygen tetrahedron and an aluminum-hydroxyl octahedron, as shown in Figure 4-1a, b. The silica unit consists of a silicon atom surrounded by four oxygen atoms, equidistant from the silicon, at the corners of the tetrahedron with the silicon at the center of the tetrahedron. Thus, the silica tetrahedral unit has four (tetra=four) triangular faces with three faces on the sides and one at the bottom. The O-O distance is 2.55Å (Å = Angstrom = 10^{-10} m), with the silicon of radius 0.5Å, fitting into the central space of radius 0.55Å between the oxygen atoms. An aluminum octahedral unit has six OH ions (known as hydroxyl ions) at each of the six corners of an octahedron, with the aluminum atom at the center of the octahedron. The octahedral unit has eight triangular faces, six on the sides and one each at the top and bottom of the octahedron.

There are valency imbalances in both units, resulting in net negative charges. The basic units, therefore, do not exist in isolation but combine to form sheet structures. The tetrahedral units combine by the sharing of oxygen ions to form a silica sheet. The octahedral units combine through shared hydroxyl ions to form a gibbsite sheet. The silica sheet and gibbsite sheet structures are represented symbolically in Figure 4-1c, d. The silica sheet retains a net negative charge but the gibbsite sheet is electrically neutral. Silicon and aluminum may be partially replaced by other elements, this being known as isomorphous substitution, resulting in further charge imbalance. The type and amount of isomorphous substitution within the crystal structure brings about the differences among minerals within the clay mineral groups. Layer structures then form by the bonding of a silica sheet with either one or two gibbsite sheets. Clay mineral particles consist of stacks of these layers, with different forms of bonding between the layers. Figure 4-2 shows the the structures of three principal clay minerals.

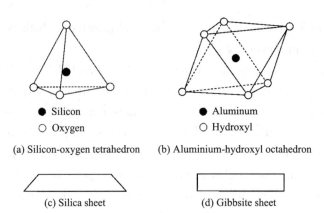

Figure 4-1 Basic units of clay minerals

Kaolinite consists of a structure based on a single sheet of silica combined with a single sheet of gibbsite. There is very limited isomorphous substitution. The combined silica-gibbsite sheets are held together relatively strongly by hydrogen bonding. A kaolinite particle may consist of over 100 stacks. Illite has a basic structure consisting of a sheet of gibbsite between and combined with two sheets of silica. In the silica sheet there is partial substitution of silicon by aluminium. The combined sheets are linked together by relatively weak bonding due to non-exchangeable potassium ions held between them. Montmorillonite has the same basic structure as illite. In the gibbsite sheet there is partial substitution of aluminium by magnesium and iron, and in the silica sheet there is again partial substitution of silicon by aluminium. The space between the combined sheets is occupied by water molecules and exchangeable cations other than potassium, resulting in a very weak bond. Considerable swelling of montmorillonite can occur due to additional water being adsorbed between the combined sheets.

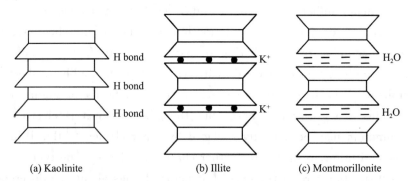

Figure 4-2 Three principal clay minerals

III. Identification of mineralogy

Mineralogy and microfabric of the clay structure can be evaluated by methods of differential thermal analysis, X-ray diffraction, chemical analysis, scanning electron microscope, infrared spectroscopy, etc. Differential thermal analysis (DTA) is a technique in which the difference in temperature between the sample and a reference material is monitored against time or temperature while the temperature of the sample, in a specified atmosphere, is programmed. DTA is the most commonly method used to identify clay minerals. X-ray diffraction (XRD) is a powerful nondestructive technique for characterizing crystalline materials. X-ray is a kind of light wave with short wavelength and strong penetrability. When X-ray is injected into the crystal lattice of clay minerals, due to the phase difference between reflected and refracted light, diffraction phenomena will occur. The process of chemical analysis is to use the method of total chemical analysis to measure the relative content of main elements in clay minerals, which is usually expressed by the content of oxides. Then, according to the molecular formula of different minerals, the type and content of clay minerals contained in them are calculated by the optimal fitting method. Scanning electron microscope (SEM) is a type of electron microscope designed for directly studying the surfaces of solid objects, that utilizes a beam of focused electrons of relatively low energy as an electron probe that is scanned in a regular manner over the specimen.

IV. Grain size and gradation

Soils can behave quite differently depending on their geotechnical characteristics. In coarse grained soils, where the grains are larger than 0.075mm, the engineering behavior is influenced mainly by the relative proportions of the different sizes present, the shapes of the soil grains, and the density of packing. These soils are also called granular soils. In fine-grained soils, where the grains are smaller than 0.075mm, the mineralogy of the soil grains, water content, etc. have greater influence than the grain sizes, on the engineering behavior. The borderline between coarse and fine grained soils is 0.075mm, which is the smallest grain size one can distinguish with naked eye.

The grain size distribution of a coarse grained soil is generally determined through sieve analysis, where the soil sample is passed through a stack of sieves and the percentages passing different sizes of sieves are noted. The grain size distribution of the fines are determined through hydrometer analysis, where the fines are mixed with distilled water to make 1000mL of suspension and a hydrometer is used to measure the density of the soil-water suspension at different times. The time-density data, recorded over a few days, is translated into grain size and percentage finer than that size. Very often, soils contain both coarse and fine grains and it is necessary to do sieve and hydrometer analyses to obtain the complete grain size distribution data. Here, sieve analysis is carried out first, and on the soil fraction passing 75μm sieve, a hydrometer analysis is carried out. Hydrometer analysis is based on Stokes law. According to this law, the velocity at which grains settles out of suspension, all other factors being equal, is dependent upon the shape, weight and size of the grain.

The grain size data thus obtained from sieve and hydrometer analyses are generally presented graphically. Logarithmic scale is used for the grain sizes since they vary in a wide range. The percentage passing is generally cumulative. The grain size distribution curve gives a complete and quantitative picture of the relative proportions of the different grain sizes within the soil mass. Figure 4-3 shows the grain size distribution curves of course-grained and fine-grained soils. The d_{30} is a size such that 30% of the soil grains are smaller than this size. The d_{10}, d_{50}, d_{60}, etc. can be defined in similar manner. The d_{10} is called the effective grain size, which gives a good indication of the permeability characteristics of a coarse grained soil. The shape of the grain size distribution curve can be described through two simple parameters, namely, coefficient of uniformity ($C_u = d_{60}/d_{10}$) and coefficient of curvature ($C_c = d_{30}^2/d_{10}d_{60}$).

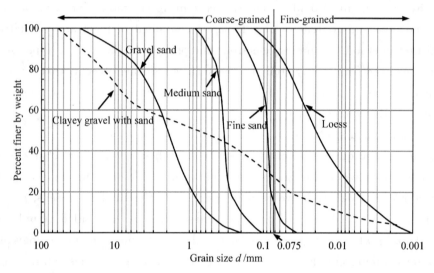

Figure 4-3　Distribution curves of course-grained and fine-grained soils

A coarse-grained soil is said to be well graded if there is a good distribution of sizes in a wide range, where smaller grains fill the voids created by the larger grains thus producing a dense packing. The grain size distribution curves for such soils would generally be smooth and concave as shown in Figure 4-4(soil A). A sand or gravel is well graded if $C_u >$ 5 and $C_c = 1 \sim 3$ as per Chinese code of soil classification. A soil that is not well graded is poorly graded. Uniform soils and gap-graded soils are special cases of poorly graded soils. In uniform soils, the grains are about the same size. When there are smaller and larger grains, but none in an intermediate size range, the soil is known as a gap-graded soil. Typical grain size distribution curves of well graded (soil A), gap graded (soil B) and uniform (soil C) soils are shown in Figure 4-4, respectively.

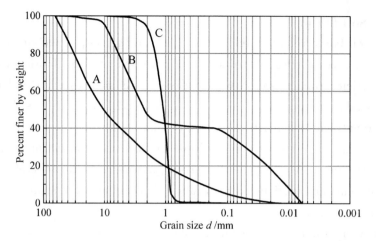

Figure 4-4 Different types of grain size distribution curves

4.2.1.2 Liquid phase

The liquid phase of soil is mainly consisted of soil water. Soil water is the term for water found in naturally occurring soil. There are three main types of soil water: gravitational water, capillary water, and hygroscopic water. Gravitational water is free water moving through soil by the force of gravity. It is largely found in the macropores of soil and very little gravitational water is available to plants as it drains rapidly down the water table in all except the most compact of soils. Capillary water is water held in the micropores of the soil, and is the water that composes the soil solution. Capillary water is held in the soil because the surface tension properties (cohesion and adhesion) of the soil micropores are stronger than the force of gravity. However, as the soil dries out, the pore size increases and gravity starts to turn capillary water into gravitational water and it moves down. Hygroscopic water forms as a very thin film surrounding soil particles and is generally not available to the plant. This type of soil water is bound so tightly to the soil by adhesion properties that very little of it can be taken up by plant roots. Since hygroscopic water is found on the soil particles and not in the pores, certain types of soils with few pores (clays for example) will contain a higher percentage of it.

4.2.1.3 Gaseous phase

In addition to water, there is gas in the pores of unsaturated soil. The gas in soil is mainly air, and sometimes carbon dioxide, methane and hydrogen sulfide may also exist. The gaseous composition of soil air is quite similar to the atmospheric air. It varies only in terms of carbon dioxide, which is high than that in the atmospheric air. The amount of oxygen in the soil air is much less than that of the atmospheric air.

4.2.2 Soil structure

The physical arrangement of particles and particle groups in different patterns is known as soil fabric. Soil structure includes soil fabric as well as the intra-and inter-particle forces of attraction and repulsion.

4.2.2.1 Soil fabric

The term soil fabric refers to the geometrical arrangement of individual particles in a soil, including the geometrical distribution of pore spaces. Soil fabric may be used to describe the particle arrangement in both cohesive soils such as clays and granular soils such as silts, sands, and gravels. However, the term packing is also used to describe the geometrical arrangement of particles in pure granular soils where clays are not present.

In general, clay particles occur in multiple groups called fabric units, rather than as individual discrete units. A comprehensive description of soil fabric therefore involves the arrangement of individual particles in each fabric unit and also the arrangement of various fabric units and pore spaces or voids.

Fabric units in clays that are visible to naked eye are known as peds. Fabric units that can be distinguished using an optical microscope are called clusters. Several clusters together form peds. Soil fabric for clays is different from that of sands or other coarse-grained particles because of their shape. As evident from the mineral structure, clay particles are formed by combinations of silica and alumina sheets. Thus, the individual clay particles are in the shape of a sheet or plate with thickness considerably less than the other dimensions. This is in contrast to the sand particles that are more or less equi-dimensional in all directions, though they are not truly spherical.

4.2.2.2 Attractive and repulsive forces in clays

By considering the attractive and repulsive forces existing at the inter-particle and particle group level along with the soil fabric, soil structure helps to understand the engineering behavior of clays under structural loads. Clay particles are not electrically neutral but carry some charge due to: ① broken bonds at edges leading to positive charge at the particle edges as in kaolinite; ② isomorphous substitution resulting in negative charge on particle surface. The negatively charged particle surface attracts positively charged ions called cations (pronounced as cat-eye-ons) available in the surrounding pore water to balance the negative charge for equilibrium. The magnitude of this charge depends on cation exchange capacity (CEC) and specific surface, both of which are high in montmorillonite, medium in illite, and the least in kaolinite. CEC is a measure of the soil's ability to hold positively charged ions (cations). It is a very important soil property influencing soil structure stability. A cation can be either basic or acidic. Specific surface is the surface area of the given clay mineral per unit weight. The higher the specific surface, the more is the magnitude of the surface forces of attraction and repulsion of electrical nature per unit weight of soil and their influence over the gravity forces.

Attractive forces between particles occur due to: ① attraction between positive edge and negative surface when there is edge-to-face particle association; ② short range London-van der Waals forces when clay particles are at atomic distances apart; ③ presence of cementing or bonding materials such as iron oxide, aluminum oxide, carbonates, etc. ④ organic matter present in the soil. Repulsive forces occur in clays mainly due to repulsion between diffuse double layers of adjacent clay particles. The magnitude of the repulsive forces, relative to the gravity forces due to weight of the clay particles, increases with increase in the thickness of diffuse double layers, decrease in particle thickness and size.

The negatively charged particle surface attracts positively charged ions (cations) present in the pore water. The cations, attracted to the particle surface, are not a fixed part of the clay particle, but are "exchangeable" by other cations of higher valence or size, whenever there is a change in the soil chemistry due to natural causes or environmental pollution. The number of cations on a particle surface depends on the ion valence and the CEC of the mineral. The lower the valence of the cations, the more is the number of cations on the particle surface. The higher the CEC of the soil, the more is the number of exchangeable cations on the particle surface. When the soil is in dry state, the exchangeable cations cling or adhere to the particle surface. When the water content of soil increases due to rainfall or the flow of groundwater, water acts as a shield and it penetrates the space between the particle surface and the cations, decreasing the attractive force between the two.

The cations thus spread over a distance from the particle surface. As the influence of the negative charge of the particles is maximum near the surface and decreases with the increase in distance, the concentration of cations is more at the surface and decreases with the increase in distance from the particle surface. Further, the cations, because of their positive charge, attract negatively charged ions (anions) in the pore water. The concentration of anions is minimum near the particles surface because of repulsion from the negatively charged surface, and it increases with the increase in distance from the particle surface.

The pore water, in the vicinity of the particle surface, is not free, but it is under the influence of attractive force between the cations and the negatively charged particle surface and anions and also the repulsive force between the particle surface and the anions. This layer of water that is under the influence of electrical forces of attraction and repulsion and in which the cations and anions diffuse, up to some distance from the particle surface, is known as diffuse double layer. The interaction between two particles or fabrics units, which are in proximity to each other, is through the inter-action of their diffuse double layers. The thickness of diffuse double layer significantly influences the soil properties in clays such as swelling, shrinkage, compressibility, as well as the permeability and shear strength.

4.2.2.3 Soil structure types

I. Single grained soil structure

Single grained structures are present in cohesionless soils like gravel and sand. The grains of cohesionless soils have less surface force and more gravitational force. So, when we pour some amount of sand or gravel on the ground, the grains will settle using gravitational force rather than surface force.

Ⅱ. Honeycomb soil structure

Honeycomb structure found in soil contains particle of size 0.02mm to 0.002mm which are generally fine sands or silts. When this type of soils is allowed to settle on the ground, the particles will attract each other and joins one with another and forms a bridge of particles. A large void is also formed between those bridges which makes the soil very loose in nature. The attraction of particles is due to cohesion between them. This cohesion is just because of their size but these soils are not plastic in nature. In fine sands, when water is added to dry fine sand bulking of sand occurs which is nothing but a structure of honeycomb. Honeycomb structured soil is limited for static load condition. They cannot resist vibrations and shocks under building and may cause large deformations to the structure.

Ⅲ. Flocculated soil structure

A flocculated or flocculent structure in clays consists of clay particles with edge-to-face association. In this arrangement, the edge of one particle is in contact with the face of the other particle. Flocculated structure is present in clay particles which contains larger surface area. These are charged particles which have positive charge on the edges and negative charge on the face of the particle. When there is net attractive force between the particles, then positive charged particles attracted towards negatively charged faces which results in the formation of flocculated structure. Clay in the marine area is the best example for flocculated structure. Salt in the marine water acts as electrolyte and reduces the repulsive force between the particles which leads to the formation of flocculent structure. This type of soils has high shear strength. Because of edge to face orientation void ratio is high in this type soil and water content is also high but they are light in weight. The compressibility is low for this type of soils.

Ⅳ. Dispersed soil structure

Dispersed structure in clays consists of face-to-face association of clay particles or fabric units. Because of the face-to-face association, the repulsive forces between the particles or between the fabric units are more in dispersed structure. Dispersed structure also occurs in clay particles when the clay is remolded. Remolding reduces the shear strength of soil which reduces the net attractive forces between the particles. Hence, due to repulsion between them, the edge-to-face orientation turns into face-to-face orientation. Finally, dispersed structure of clay will form. This type of soil is highly compressible and less permeable. The loss of strength during remolding is slowly achieved by the soil with the time. The process of regaining its strength after remolding is called thixotropy.

Ⅴ. Coarse-grained skeleton soil structure

Coarse-grained skeleton is a structure of soil which is present in composite soils containing both fine and coarse-grained particles. But it is formed when there is large amount of coarse-grained particles than the fine-grained particles. The coarse-grained particle forms a skeleton structure and voids between them are filled by fine-grained or clayey particles. If it is undisturbed, it will give good results against heavy loads. If it is disturbed, the strength will extensively reduced.

Ⅶ. Clay matrix soil structure

Clay matrix structure, which also occurs in composite soils, but the amount of clay particles or fine-grained particle is more compared to coarse-grained particles. Coarse particles in this soil are separated with each other as they are less in quantity. This type of soil is very stable in nature and it has same properties of an ordinary clay deposit.

4.3 Physical and index properties

The physical properties of soil are used to describe the percentage of solid particles by volume or mass (weight) per sample size, and water/air per unit pores. They also reflect the mechanical properties of soil to some extent. The solid particles (solid phase) in the soil are generally assumed to be constant in weight, mass, and volume.

There are two methods to determine the physical properties of soil. One of which is through direct measurement, such as water content, density and specific gravity, known as direct measurement parameters. Another is through calculation based on direct parameters, e.g. void ratio, porosity, degree of saturation, etc. To explain the definition of the physical properties and their relationship, a three-phase diagram is often used to represent the relative content of each phase of a soil sample (Figure 4-5). Three-phase refers to solid phase, water phase and air phase in the soil. Generally, "soil" refers to the macro scale unit containing of three phases (sometimes two phases without air). "Soil particle" refers to the solid phase in soil.

The three-phase diagram (Figure 4-5) shows the three phases of a soil mass separately. All the soil particles are assumed to be collected together to form a single soil solid layer at the bottom, above which is the middle water layer and top air layer. It can be seen from Figure 4-5 that two units (mass and volume) are used to measure the three phases of a soil, where m indicates the mass, V represents the volume, and the subscript s, w, a, and v represent soil solids, water, air, and void, respectively. Since the density of air ($\rho_a \approx 1.29 \text{kg/m}^3$) is much smaller than density of water ($\rho_w \approx 1000 \text{kg/m}^3$), we

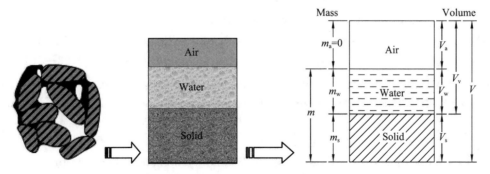

Figure 4-5 Three-phase diagram of a soil mass

assume that the mass of air, $m_a = 0$. Therefore, the mass of a given soil (m) is the sum of water mass (m_w) and soil solids mass (m_s). In contrast, the volume of a given soil (V) is the sum of air volume (V_a), water volume (V_w) and soil solids volume (V_s). The air volume (V_a) and water volume (V_w) together consist of the void volume V_v.

4.3.1 Direct measured physical properties

4.3.1.1 Density and bulk density of soil

The soil density (ρ) is defined as the ratio of the mass to the volume of a soil sample, with the unit of g/cm³ (or kg/m³).

$$\rho = \frac{m}{V} = \frac{m_s + m_w}{V_s + V_w + V_a} \tag{4-1}$$

Standard test methods for laboratory determination of density (unit weight) of soil specimens can be found following the referenced standards from different countries, e.g., the Chinese standard for soil test method (GB/T 50123—2019), the American standard (ASTM D7263), and the British standard (BS1377 series).

Bulk density (or unit weight) of soil (symbolized γ) is the measure of the bulk of the soil and is defined as the weight per volume that can be written as

$$\gamma = \frac{W}{V} = \frac{m \times g}{V} = \rho g \tag{4-2}$$

where, W—weight of soil, kN;

g—gravitational acceleration, about 9.8 m/s².

4.3.1.2 The specific gravity of soil solid

The specific gravity of soil solid (symbolized G_s) is defined as the density (or unit weight) of the soil solids to the density (or unit weight) of distilled water at 4℃.

$$G_s = \frac{m_s}{V_s \times (\rho_w)_{4℃}} = \frac{\rho_s}{(\rho_w)_{4℃}} \tag{4-3}$$

or
$$G_s = \frac{W_s}{V_s \cdot (\gamma_w)_{4℃}} = \frac{\gamma_s}{(\gamma_w)_{4℃}} \tag{4-4}$$

where, ρ_s—density of the dry soil solid;

γ_s—unit weight of the soil solid;

$(\rho_w)_{4℃}$—density of distilled water at 4℃;

$(\gamma_w)_{4℃}$—unit weight of distilled water at 4℃.

We can measure the specific gravity of soil solid following the related method in soil testing standard. For example, as per Chinese standard (GB/T 50123—2019), the specific gravity of soil solids can be determined in Eq. (4-5) by means of a water pycnometer. It is measured by adding distilled water to the bottle with dry soil solids.

$$G_s = \frac{m_s}{m_1 + m_s - m_2} \tag{4-5}$$

where, m_1—total mass of water, dry soil solid and pycnometer when it is full;

m_2—total mass of water and pycnometer when it is full;

m_s—mass of dry soil solid.

The specific gravity of soil solids depends on soil mineral composition. The typical values of specific gravity lie in 2.65~2.80. For example, the specific gravity of sand usually falls within 2.65~2.68 and that of clay is about 2.70~2.75. This is a unitless parameter. The smaller values stand for soils of more coarse-grained particles. The specific gravity will be smaller if the soil contains organic matter.

4.3.1.3 Water content of soil

The water (moisture) content of soil (symbolized w) is defined as the ratio of the mass (weight) of water to the mass (weight) of soil solid, expressed as a percentage.

$$w = \frac{m_w}{m_s} \times 100\% = \frac{W_w}{W_s} \times 100\% \tag{4-6}$$

We can determine the water content of soil following the related method in soil testing standard. For example, as per Chinese standard (GB/T 50123—2019), an oven at the standard drying temperature (105~110℃) is used to dry the wet soil to get the water content. The water content is one of the most significant properties used in establishing a correlation between soil behavior and its index properties. The water content can be used in expressing the phase relationships of air, water, and solids in a given volume of soil.

4.3.2 Other index properties

The above mentioned three physical properties (i.e., ρ or γ, G_s, and w) can be directly measured by laboratory tests. The following index properties can be obtained based on the three-phase diagram of soil (Figure 4-5).

I. Void ratio (e)

The void ratio of soil is defined as the ratio of the volume of voids (V_v) in soil to the volume of soil solids (V_s), which can be expressed as a decimal or a fraction.

$$e = \frac{V_v}{V_s} \tag{4-7}$$

Since voids and solid particles are of irregular shapes, there is no direct way to measure their volumes. It can be determined through the phase relationships.

II. Porosity (n)

The porosity of soil is defined as the volume of voids (V_v) in soil per the total volume (V) of soil, expressed as a percentage.

$$n = \frac{V_v}{V} \times 100\% \tag{4-8}$$

The void ratio and porosity differ only in the denominator. For a given soil, larger void ratio or porosity indicates the soil is looser and is more compactable.

III. Saturation degree (S_r)

The degree of saturation is defined as the ratio of the water volume (V_w) to the void volume (V_v).

$$S_r = \frac{V_w}{V_v} \times 100\% \tag{4-9}$$

Saturation reflects the extent to which soil void are filled with water. In practical, degree of saturation varies in the range of 0~100%, where 0 means soil is in dry condition and 100% is in fully staturated condition.

IV. Dry density (ρ_d) and dry unit weight (γ_d)

The dry density of soil is the mass of soil solid (m_s) divided by the total volume (V).

$$\rho_d = \frac{m_s}{V} \tag{4-10}$$

The dry unit weight of soil is ratio of the weight of soil solid (W_s) to the total volume.

$$\gamma_d = \frac{W_s}{V} = \frac{m_s \times g}{V} = \rho_d \times g \tag{4-11}$$

The total volume will be reduced upon drying, but the mass of soil solid is unchanged. Therefore, the dry density (ρ_d) of soil is not the same as the density of dry soil solid (ρ_s). The dry density or dry unit weight of soil is also an indicator to evaluate the compaction behavior of soil. The greater the dry density or dry unit weight, the more the soil is compacted. Dry density is often used as an indicator to control the compaction effect in the construction of backfill project.

V. Saturated density (ρ_{sat}) and saturated unit weight (γ_{sat})

The saturated density of soil is defined as the mass of saturated soil per total volume when the soil void is completely filled with water (in fully saturated state),

$$\rho_{sat} = \frac{m_s + V_v \times \rho_w}{V} \tag{4-12}$$

In fully saturated state, the saturated weight of soil per total volume is defined as the saturated unit weight, which can be expressed as

$$\gamma_{sat} = \frac{W_s + V_v \times \gamma_w}{V} = \frac{m_s \times g + V_v \times \rho_w \times g}{V} = \rho_{sat} g \tag{4-13}$$

VI. Buoyant density (ρ') and buoyant unit weight (γ')

Below the groundwater table, the effective weight of soil is reduced by the buoyant force of water. So the concept of buoyant unit weight is proposed, which is also called effective (submerged) unit weight. The expression is shown below

$$\gamma' = \frac{W_s - V_s \times \gamma_w}{V} = \frac{m_s \times g + V_v \times \rho_w \times g - V \times \rho_w \times g}{V} = \gamma_{sat} - \gamma_w \tag{4-14}$$

Therefore, the buoyant density can be obtained by

$$\rho' = \frac{m_s - V_s \times \rho_w}{W} = \frac{m_s + V_v \times \rho_w - V \times \rho_w}{V} = \rho_{sat} - \rho_w = \frac{\gamma'}{g} \tag{4-15}$$

From the definitions of the four unit weights of soil, we can get the relations: $\gamma_{sat} \geqslant \gamma \geqslant \gamma_d > \gamma'$. For four types of density, we have: $\rho_{sat} \geqslant \rho \geqslant \rho_d > \rho'$.

4.3.3 Basic relations among physical and index properties

I. Relationship between void ratio and porosity

Based on Eq. (4-7), if we assume that the volume of soil solid $V_s = 1$, then we can get the volume of voids $V_v = e$ and the total volume $V = 1 + e$, hence we get,

$$n = \frac{V_v}{V} = \frac{e}{1+e} \tag{4-16}$$

$$e = \frac{n}{1-n} \tag{4-17}$$

II. Relationship among dry density, density and water content

According to Eq. (4-10), is we assume that the total volume $V=1$, then the mass of soil solid $m_s = \rho_d$. According to the Eq. (4-6), the mass of water $m_w = w\rho_d$. Thus, we can obtain

$$\rho = \frac{m}{V} = \frac{\rho_d + w\rho_d}{1} = \rho_d(1+w) \tag{4-18}$$

or

$$\rho_d = \frac{\rho}{1+w} \tag{4-19}$$

III. Relationship among void ratio, specific gravity and dry density

According to Eq. (4-7), if we assume that the volume of soil solid $V_s = 1$, then the voids volume $V_v = e$, and the mass of soil solid $m_s = \rho_s$. Thus, we can get

$$\rho_d = \frac{m_s}{V} = \frac{\rho_s}{1+e} \tag{4-20}$$

Combined Eq. (4-20) and Eq. (4-3), we have

$$e = \frac{G_s \rho_w}{\rho_d} - 1 \tag{4-21}$$

IV. Relationship among saturation degree, water content, specific gravity and void ratio

According to Eq. (4-7), if we assume that the volume of soil solid $V_s = 1$, then the volume of voids $V_v = e$ and the mass of soil solid $m_s = \rho_s$. According to Eq. (4-6), the mass of water $m_w = w\rho_s$, then $V_w = w\rho_s/\rho_w$. Therefore, we can get

$$S_r = \frac{V_w}{V_v} = \frac{\frac{w\rho_s}{\rho_w}}{e} = \frac{wG_s}{e} \tag{4-22}$$

When the soil is fully saturated, $S_r = 100\%$, then

$$e = w_{sat} G_s \tag{4-23}$$

where, w_{sat}—saturated water content.

V. Relationship among buoyant unit weight, specific gravity and void ratio

If the volume of soil solid $V_s = 1$, the volume of voids $V_v = e$ from Eq. (4-7), and the mass of soil solid $m_s = \rho_s$. Thus, following equation can be derived from Eq. (4-14)

$$\gamma' = \frac{m_s g - V_s \rho_w g}{V} = \frac{\rho_s - \rho_w}{1+e} g = \frac{(G_s - 1)\gamma_w}{1+e} \tag{4-24}$$

Table 4-1 Basic equations among commonly used indexes of soil

Main index obtained by lab tests	Commonly used indexes		
	Index name	Expression	Relations with other indexes
w—water content ρ—density of soil G_s—specific gravity of soil solid	Void ratio	$e = \dfrac{V_v}{V_s}$	$e = \dfrac{\rho_s}{\rho_d} - 1 = \dfrac{\rho_s(1+w)}{\rho} - 1$, $e = \dfrac{n}{1-n}$ $e = \dfrac{wG_s}{S_r}$

Continued

Main index obtained by lab tests	Commonly used indexes		
	Index name	Expression	Relations with other indexes
w—water content ρ—density of soil G_s—specific gravity of soil solid	Porosity	$n = \dfrac{V_v}{V} \times 100$	$n = 1 - \dfrac{\rho_d}{\rho_s} = 1 - \dfrac{\rho}{\rho_s(1+w)}$, $n = \dfrac{e}{1+e}$
	Saturation degree	$S_r = \dfrac{V_w}{V_v}$	$S_r = \dfrac{wG_s}{e}$
	Dry density	$\rho_d = \dfrac{m_s}{V}$	$\rho_d = \dfrac{G_s \rho_w}{1+e}$, $\rho_d = \dfrac{\rho}{1+w}$, $\rho_d = \dfrac{nS_r}{w}\rho_w$
	Saturated density	$\rho_{sat} = \dfrac{m_s + V_v \rho_w}{V}$	$\rho_{sat} = \dfrac{G_s + e}{1+e}\rho_w$, $\rho_{sat} = \rho' + \rho_w$ $\rho_{sat} = G_s \rho_w (1-n) + n\rho_w$
	Buoyant unit weight	$\gamma' = \dfrac{m_s g - V_s \gamma_w}{V}$	$\gamma' = \dfrac{G_s - 1}{1+e}\gamma_w$, $\gamma' = \gamma_{sat} - \gamma_w$ $\gamma' = (G_s - 1)(1-n)\gamma_w$

Each soil index is recommended to be remembered and understood. The basic relations between the commonly used indexes are summarized in Table 4-1 for simple calculation in engineering applications.

【Example 4-1】 A specimen in the natural state has a volume $V = 60 \text{cm}^3$, the total mass $m = 108\text{g}$, the dry mass $m_s = 96.43\text{g}$, and a specific gravity of soil solid $G_s = 2.7$. Determine the soil density ρ, water content w, void ratio e, porosity n and saturation degree S_r.

【Solution】

(1) Known $V = 60 \text{cm}^3$, $m = 108\text{g}$, from Eq. (4-1),

$$\rho = \frac{m}{V} = \frac{108}{60} = 1.80 \text{g/cm}^3$$

(2) $m_s = 96.43$, $m_w = m - m_s = 108 - 96.43 = 11.57\text{g}$.

By Eq. (4-6)
$$w = \frac{m_w}{m_s} = \frac{11.57}{96.43} = 12.0\%$$

(3) Known $G_s = 2.7$,

$$V_s = \frac{m_s}{\rho_s} = \frac{96.43}{2.7} = 35.7 \text{cm}^3$$

$$V_v = V - V_s = 60 - 35.7 = 24.3 \text{cm}^3$$

By Eq. (4-7)
$$e = \frac{V_v}{V_s} = \frac{24.3}{35.7} = 0.681$$

(4) By Eq. (4-8),
$$n = \frac{V_v}{V} = \frac{24.3}{60} = 40.5\%$$

(5) $V_w = \dfrac{m_w}{\rho_w} = \dfrac{11.57}{1} = 11.57 \text{cm}^3$

By Eq. (4-9),
$$S_r = \frac{V_w}{V_v} = \frac{11.57}{24.3} = 47.6\%$$

【Example 4-2】 A soil sample with water content $w_1 = 12\%$ and unit weight $\gamma = $

$19kN/m^3$, if the void ratio e remains unchanged, when the water content increased to $w_2 = 22\%$, how much water in weight need to be added to $1m^3$ soil?

【Solution】

For the same type of soil with two conditions,

$$e_1 = \frac{G_{s1}\gamma_w(1+w_1)}{\gamma_1} - 1, \quad e_2 = \frac{G_{s2}\gamma_w(1+w_2)}{\gamma_2} - 1$$

since $G_{s1} = G_{s2}$, $e_1 = e_2$,

$$\gamma_2 = \gamma_1 \frac{1+w_2}{1+w_1} = 19 \times \frac{1+0.22}{1+0.12} = 20.7 kN/m^3$$

$1m^3$ soil needs $\Delta W_w = 20.7 - 19 = 1.7 kN$.

【Other solution】 By Eq. (4-19), $\rho_d = \frac{\rho}{1+w}$

$$\gamma_d = \frac{\gamma}{1+w}$$

When soil water content $w_1 = 12\%$, $\gamma_d = \frac{19}{1+0.12} = 16.96 kN/m^3$.

For $1m^3$ soil: $\quad W_{w1} = 19 - 16.96 = 2.04 kN$

$$W_{w1} = G_{s1} \cdot \gamma_w \cdot w_1$$
$$W_{w2} = G_{s2} \cdot \gamma_w \cdot w_2$$

For the same type of soil, $G_{s1} = G_{s2}$, thus

$$W_{w2} = W_{w1} \cdot \frac{w_2}{w_1} = 2.04 \times \frac{0.22}{0.12} = 3.74 kN$$

$$\Delta W_w = 3.74 - 2.04 = 1.7 kN$$

4.4 Consistency of cohesive soil

In geotechnical engineering, soil can be divided by coarse-grained soil and fine-grained soil. For fine-grained soil, their engineering behaviors would be greatly influenced by the consistency. Consistency of soil is the physical state of soil with respect to water content present at that time. Consistency reflects the relative ease with which soil can be deformed.

4.4.1 Soil cohesion

Soil cohesion refers to the cohesive properties of soil particles. Soil cohesion can be reflected by the determination of cohesive strength. Soils that have no cohesive strength and exhibit cohesionless features are called non-cohesive soils.

The cohesive strength of soil is comes from the following sources:

1) Water absorption. When a saturated soil is under loading or evaporation, the water content reduces and the thickness of adsorbed water around soil particles gets thinner, and the soil spacing decreases. The adsorbed water film gradually enhances the cohesion of soil particles, and the cohesive strength will increase as a result. Once the cohesive soil is disturbed, the cohesive strength that caused by water absorption would recover quickly.

2) Cementation. Soil particles can be combined by some chemicals, such as the

crystallization of soluble salts (chlorides, carbonates, sulfates), and the cementation of the oxides and humic substances in solutions, which can bond soil particles together. Moreover, if soils under long-term pressure, minerals that have contact with the soil particles may recrystallize and bond the soil particles together. Once the cohesive soil got disturbed, the cohesive strength caused by cementation cannot be recovered immediately and it takes a long time to recover.

For example, the water content of collapsible loess is low and the strength is high under nature conditions. However, if loess is wetted, the strength will decrease rapidly and the collapsible deformation will occur. This goes through a complicated physical and chemical process. It is generally believed that the invasion of water causes the adsorbed water film become thicken and dissolves the cemented materials, resulting in the reduction or disappearance of cohesion for loess.

3) Capillary force. Capillary force of water or ice can also enhance the cohesion of soil. Once after wetting, the capillary force is no longer exist. In addition, ice would melt as a result of increased temperature, which will soon lead to the disappearance of capillary cohesion.

4.4.2 Plasticity

Plasticity is the property of a soil to undergo large deformations when stressed, without cracking or crumbling. Upon the removal of external force, the shape can be maintained. Soil of this nature is called plastic soil, also known as plasticity. Plasticity depends on the water content of a clay.

The plasticity of cohesive soil can be attributed to water absorption. Clay particles are negatively charged, which could attract the hydrated cations in pores and polar water molecules to have adsorbed water films, thus to form soil aggregate by negatively charged particles and water films. When the relative positions of the soil particles change, the adsorbed water film remains the same. Therefore, it is necessary to maintain a certain thickness of the adsorbed water film between the soil particles. This requires the soil to maintain a certain water content.

The factors that affect plasticity of cohesive soils are those affecting the thickness of adsorbed water film. With respect to the nature of plastic soil, they are mineral composition, particle size and shape, cation exchange capacity; with respect to the water properties, they are ion concentrations, ion types, ion valence, pH, etc.

4.4.3 Atterberg limits

4.4.3.1 Concept of Atterberg limits

Cohesive soils can be in different states with different water contents. Atterberg limits is an index describing the degree of plasticity of cohesive soil and the physical characteristics of soil at a certain water content.

Cohesive soils can exhibit plasticity only in a certain range of water content. The shape of the cohesive soil can be changed continuously without cracking under the action of external force, and the shape will remain after the external force disappears. When the water content exceeds a certain value, it will lose its plasticity and be in the flow state. In contrast, when the water content is lower than a given value, it will not show plasticity and show properties similar to a solid.

As shown in Figure 4-6, as the water content increases, the volume of soil will increase. In contrast, the volume of soil will decrease as the water content decreases. When the water content of cohesive soil reduced to a certain limit value, the soil enters the solid state. Even if the water content of soil is further reduced, the volume of soil will not shrink. At this time, air enters into the soil, and the color of soil will become lighter. There is a need to introduce the three

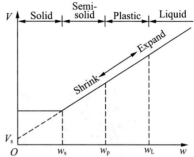

Figure 4-6 Schematic of state transition for cohesive soil

boundaries of water contents when discuss the soil consistency, commonly known as the consistency limits or the Atterberg limits. The boundary water contents are important indexes used to describe the state of cohesive soils. They are defined as follows:

Liquid limit(w_L) refers to the transition point from the liquid to the plastic state of soil, that is, the lower limit of viscous flow. Plastic limit (w_p) refers to the transition point from the plastic state to the semi-solid state, where the water content is the minimum at the plastic state. Shrinkage limit (w_s) refers to the water content of soil transferred from the semi-solid state to the solid state.

It should be pointed out that the transition of cohesive soil from one state to another is gradual and there is no clear boundary. At present, it can be measured by remolded or reconstituted soil samples following the procedures in the relevant codes or standards.

4.4.3.2 Determination of Atterberg limits

The methods of measuring liquid limit include cone penetration and Casagrande method. They follow the same steps for preparing the soil specimens. At least three samples are needed with different target water contents. Each sample mass should be larger than 200g and put into the bowl. Add distilled water into the bowel to prepare the target three samples with water content that one sample around or larger than the liquid limit, one sample around or less than plastic limit and one sample between the two limits. The cone penetration method is used following the Chinese code (GB/T 50123—2019) and the combined testing device (Figure 4-7) is used to run the test on three samples. Draw the relationship curve of penetration depth after 5 seconds and water content in the log-log scale (Figure 4-8). Connect the three dispersed points in the coordinate, and the curve will be a straight line. In Figure 4-8, read the penetration depth 17mm or 10mm, and record the corresponding water content as liquid limit w_L(%) at 17 mm or 10mm respectively; read the penetration depth 2mm, and record the corresponding water content as plastic limit w_p(%), with an accuracy to 0.1%. It should be noted that the liquid limit at 17 mm cone penetration depth is comparable to the liquid limit obtained using the Casagrande method in Chinese code (GB/T 50123—2019) or American code ASTM D4318. The liquid limit at 10mm depth has been used in China and Russia.

Figure 4-7 Photograph of combined testing device for liquid and plastic limit

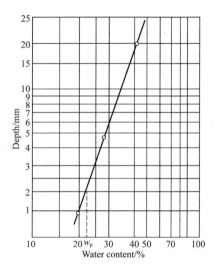

Figure 4-8 Water content versus cone penetration depth

When we use the Casagrande method to measure the liquid limit of cohesive soil, the Casagrande device is used as shown in Figure 4-9. It consists of a brass cup which is connected so that it is in the inclined position when resting on the rubber base. Brass cup when rotated by its handle lifts to the height of 1cm and drops freely on the rubber base. The Chinese code (GB/T 50123—2019) or American code ASTM D4318 show similar testing procedures. During the test, a portion of soil paste is placed in the cup of Casagrande apparatus and the surface is smoothened and leveled. Then a sharp groove is cut symmetrically through the sample using a standard grooving tool, as shown in the brass cup in Figure 4-9. After the soil pat has been cut the handle of the apparatus is turned at a rate of 2 revolutions per second, which applies blows to the grooved soil pat. And due to which two halves of the soil pat come into contact. When the contact length of 13 mm is achieved we stop turning the handle. The number of blows required to close the groove for the length of 12mm or 13mm is recorded. About 10g of soil near the closed groove is taken for the water content determination. Using the samples with different water content to repeat the test and record the number of blows required to close the groove in the range of 15 to 35. Plot the number of blows on a logarithmic scale and water content on an arithmetical scale and join them by a straight line (Figure 4-10). Read the water content corresponding to 25 blows from the curve and report it as the liquid limit (w_L).

 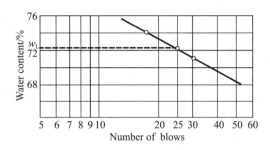

| Figure 4-9 Photograph of Casagrande device | Figure 4-10 Water content versus number of blows |

The plastic limit is determined by the hand-rolling method as per Chinese code (GB/T 50123—2019) or American code ASTM D4318. The Plastic limit test is performed by repeated rolling of an ellipsoidal-sized soil mass by hand on a non-porous surface. Casagrande defined the plastic limit as the water content at which a thread of soil just crumbles when it is carefully rolled out to a diameter of 3mm. The soil mass is rolled between the palm or fingers and the ground-glass plate with just sufficient pressure to roll the mass into a thread of uniform diameter throughout its length. Gather the portions of the crumbled thread to measure the water content, which is the plastic limit. Additionally, the plastic limit can also be obtained using cone penetration method, as shown in Figure 4-8. Read the water content at the penetration depth of 2mm is the plastic limit. The plastic limit obtained using cone method is comparable to that measured by hand-rolling method.

The shrinkage limit is determined by shrinkage plate method, where soil is filled in layered container with scraping surface. Then drying the specimen and weighing to measure dried sample with an accuracy to 0.1g, the shrinkage limit is calculated by following equation.

$$w_s = w - \frac{V_1 - V_2}{m_s} \rho_w \times 100\% \tag{4-25}$$

where, w_s—shrinkage limit of soil, %;

w—water content before drying, %;

V_1—volume of a wet sample (the volume of a shrink dish), cm³;

V_2—volume of dry sample, cm³.

4.4.3.3 Plasticity index and liquidity index

Plasticity index is an index to describe the range of water content over which a soil behaves plastically. The liquidity index is used for scaling the nature water content of a soil sample to the limits.

I. Plasticity index

The difference between the liquid limit and the plastic limit is called the plasticity index, represented by I_p.

$$I_p = w_L - w_p \tag{4-26}$$

The plasticity index can describes the range of water content of clay. The plasticity index depends on the amount of water that soil absorb, soil particle size, mineral composition, cation exchange capacity, etc. Therefore, the plasticity index is an index which can reflect the soil properties. Generally, the higher the plasticity index, the higher is the clay content of a soil, or the higher is the content of montmorillonite in clay. Plasticity index is often used as a classification index of cohesive soil.

II. Liquidity index

Water content has a great influence on the state of cohesive soil. But for different soils, they may not be in the same state, even with the same water content. The natural state of a cohesive soil can be scaled by the liquid index.

$$I_L = \frac{w - w_p}{w_L - w_p} = \frac{w - w_p}{I_p} \qquad (4\text{-}27)$$

where, I_L—liquidity index, expressed in decimal;

w—natural water content of soil;

I_p—plasticity index.

The liquidity index represents the relative relationship between the natural water content and the limit water contents of the soil.

When $w \leqslant w_p$, $I_L \leqslant 0$, the soil is in a brittle state.

When $w_p < w \leqslant w_L$, $0 < I_L \leqslant 1.0$, the soil is in a plastic state.

When $w_L < w$, $I_L > 1.0$, the soil is in a liquid state.

The Chinese code GB 50007—2011 "Code for design of building foundation" shows the criteria for classifying the state of cohesive soil by liquid index, as shown in Table 4-2.

Table 4-2 Classification of state of cohesive soil

State	Brittle	Semi-solid	Plastic	Semi-liquid	Liquid
Liquidity index	$I_L \leqslant 0$	$0 < I_L \leqslant 0.25$	$0.25 < I_L \leqslant 0.75$	$0.75 < I_L \leqslant 1.0$	$I_L > 1.0$

[Example 4-3] The natural water content of a soil sample is 38.8%. The liquid limit is 49% and the plastic limit is 24%. Determine the plastic index and the state of soil.

[Solution]

$w = 38.8\%, w_L = 49\%, w_p = 24\%$.

By Eq. (4-26), $I_p = 49 - 24 = 25$

By Eq. (4-27), $I_L = \dfrac{w - w_p}{w_L - w_p} = \dfrac{38.8 - 24}{49 - 24} = 0.59$

So the soil is in a plastic state.

4.5 Relative density of non-cohesive soil

For non-cohesive or cohesionless soil, the degree of compaction has a great influence on the mechanical engineering properties of soil. The index of density for non-cohesive soil include dry density and void ratio. The void ratio, to some extent, reflects the compactness of non-cohesive soil. The range of void ratio is influenced greatly by particle size, soil shape and gradation. Therefore, even if two non-cohesive soils have the same void ratio, they may not essentially have the same compaction state.

Relative density (D_r) is commonly used in engineering practice to measure the compactness of non-cohesive soil, which is defined as

$$D_r = \frac{e_{max} - e_0}{e_{max} - e_{min}} \tag{4-28}$$

where, D_r—relative density;

e_{max}—void ratio of non-cohesive soil at its loosest state;

e_{min}—void ratio of non-cohesive soil at its densest state;

e_0—in situ void ratio of the soil.

According to Eq. (4-28), a practical expression of relative density can be obtained.

$$D_r = \frac{(\rho_d - \rho_{dmin})\rho_{dmax}}{(\rho_{dmax} - \rho_{dmin})\rho_d} \tag{4-29}$$

where, ρ_{dmax}—maximum dry density of non-cohesive soil;

ρ_{dmin}—minimum dry density of non-cohesive soil;

ρ_d—dry density of non-cohesive soil.

The maximum and minimum dry density of non-cohesive soil can be determined directly by test. The air dried non-cohesive soil was measured by funnel method, and its maximum dry density was measured by rapping method. The specific test procedure of relative density test can be found in Chinese code GB/T 50123—2019 or related codes in other countries.

It should be noted that the relative density is determined by laboratory test with remolded soil. However, the dry density of nature soil may be greater than the measured maximum dry density.

From Eq. (4-28), we can see that when $D_r = 1$, $e_0 = e_{min}$ indicating that the soil was in the densest state. And, when $D_r = 0$, $e_0 = e_{max}$, the soil was in loosest state. In engineering, the compaction state of non-cohesive soil is classified as following:

Loose $0 < D_r \leq 1/3$

Medium $1/3 < D_r \leq 2/3$

Dense $2/3 < D_r \leq 1$

[Example 4-4] A sample of sandy soil was tested. The water content is $w = 11\%$, the density is $\rho = 1.77 \text{g/cm}^3$, the minimum dry density is 1.41g/cm^3, and the maximum dry density is 1.75g/cm^3. Determine the relative density of soil sample.

[Solution] $\rho = 1.70 \text{g/cm}^3$, $w = 10\%$, According to the Eq. (4-19), the dry density of the sand can be calculated as

$$\rho_d = \frac{\rho}{1+w} = \frac{1.77}{1+0.11} = 1.59 \text{g/cm}^3$$

$\rho_{dmin} = 1.41 \text{g/cm}^3$, $\rho_{dmax} = 1.75 \text{g/cm}^3$, from Eq. (4-29),

$$D_r = \frac{(\rho_d - \rho_{dmin})\rho_{dmax}}{(\rho_{dmax} - \rho_{dmin})\rho_d} = \frac{(1.59 - 1.41) \times 1.75}{(1.75 - 1.41) \times 1.59} = 0.58$$

$1/3 < D_r < 2/3$, the sandy soil layer is in the medium state.

4.6 Compaction characteristics of soil

Compaction of soil is required for the construction of earth dams, highways, embankments, retaining structures, and many other engineering applications. Compaction of soil is the process of increasing the soil density (unit weight) by forcing the soil solids into a dense state and reducing the air voids of soil. Compaction leads to increase in shear strength and helps improve the stability and bearing capacity of soil. It also reduces the compressibility and permeability. If performed improperly, settlement of the soil could occur and result in unnecessary maintenance costs or structural failure. It is necessary to understand the compaction characteristics of soil.

4.6.1 Soil compaction tests

Compaction of soil is the application of mechanical energy to a soil to rearrange the particles and reduce the void ratio and increase soil density. The compaction of soil can be conducted by two methods: field test and laboratory proctor test. The determination of the field compaction is essential to acquiring the relative compaction of soil. Field compaction of soils is mainly done with various types of rollers (Figure 4-11): ① sheepsfoot rollers, used mainly for clayey and silty soils; ② smooth-drum rollers, used primarily for non-cohesive soils; ③ vibratory rollers, used primarily for granular soils.

The laboratory proctor test is used to determine the compaction of different types of soil and the properties of soil with a change in water content. Compaction is a type of mechanical stabilization where the soil mass is densified with the application of mechanical energy also known as compactive effort. During compaction, the soil particles are relocated, and the air volume is reduced. Compaction should not be confused with consolidation, where the density of saturated soils is increased due to a reduction in the volume of voids brought about by the expulsion of water under the application of static load.

(a) Sheepsfoot roller (b) Smooth-drum roller

Figure 4-11 Rollers for field compaction

The laboratory compaction tests can be performed following the related codes in different countries. For example, in Chinese code (GB/T 50123—2019) "Standard for soil test method", the compaction tests include light and heavy compaction tests aim at determining the relationship between molding water content and dry density (or unit weight) of soils (compaction curve). For light compaction test, soil was

Figure 4-12 Two molds for compaction tests

compacted in a 102mm diameter mold with a 2.5 kg rammer dropped from a height of 305mm producing a compactive effort of 592.2kJ/m³. For heavy compaction test, soil was in a 152mm diameter mold with a 4.5kg rammer dropped from a height of 457mm with compactive effort of 2684.9kJ/m³. Figure 4-12 shows the photograph of the two molds. The 102mm diameter mold and 2.5kg rammer are shown on the left-hand side. Besides, similar methods can be found in American codes, such as the American Association of State Highway and Transportation Officials (AASHTO) test designation T99, or American Society of Testing Materials (ASTM) test designation D698 (with a standard effort of 600kJ/m³). This test is called standard proctor test since it was developed by R. R. Proctor in 1933. There is also a modified proctor compaction test as per ASTM D1557 with a modified effort of 2700kJ/m³).

4.6.2 Factors affecting soil compaction

I. Water content

As water is added to a soil (at low water content), it becomes easier for the particles to move past one another during the application of the compacting forces. As the soil compacts, the voids are reduced and this causes the dry density (or dry unit weight) to increase. Then, as the water content increases so does the dry density. However, the increase cannot occur indefinitely because the soil state approaches the zero air voids line which gives the maximum dry unit weight for a given water content. Thus, as the state approaches the zero air voids line further water content increase must result in a reduction in dry density. The relation between dry density (ρ_d) and water content (w) can be obtained shown in Figure 4-13, which is called the compaction curve. The compaction curve shows that there is a water content by which makes the dry density to reach the maximum and to have the best compaction effect. This water content is called the optimum water content, which is expressed as w_{op}. The dry density obtained at the optimum water content is called the maximum dry density of the soil, expressed as ρ_{dmax}. When the water

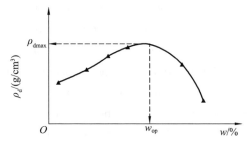

Figure 4-13 Compaction curve of cohesive soil

content is lower than the optimum water content (called partial dry state), dry density of soil increases with the increase of water content. When the water content is higher than the optimum water content (called wet state), the dry density of soil decreases with the increase of water content.

II. Compactive effort

Increased compactive effort (number of blows) enables greater dry density (or dry unit weight) to be achieved (Figure 4-14). Because of the shape of the zero air voids line, the dry unit weights must occur at lower optimum water contents. The compaction curves obtained by compacting the same soil with different compaction efforts are different. The maximum dry density and the optimum water content varies

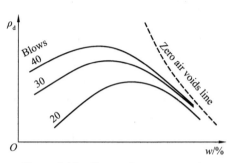

Figure 4-14 Compaction curves under different number of blows

under different compaction efforts; ρ_{dmax} increases with the increasing number of blows, while w_{opt} decreases gradually with the increasing number of blows. At low water content, the effect of compative effort is more obvious; at high water content, the compaction curves are close to the zero air voids line, as shown in Figure 4-14, where the effect of compactive effort is not obvious.

III. Soil type

Type of soil has a great influence on its compaction characteristics. Normally, cohesive soils such as heavy clay, clay or silt can offer higher resistance to compaction whereas non-cohesive soils like sandy soils and coarse-grained or gravelly soils can be easily compacted. The coarse-grained soils yield higher densities in comparison to clay.

The optimum water content of cohesive soil is generally near its plastic limit, and is about 0.55~0.65 times the liquid limit. At the optimum water content, the force between soil particles are relatively weak since the thickness of absorbed water on soil particles is moderate. Therefore, it is easy to be compacted.

Non-cohesive soil is different from cohesive soil. As shown in Figure 4-15, the compaction of non-cohesive soil is also related to water content yet there is no optimum water content. Generally, it is easy to obtain a higher dry density in a completely dry or fully saturated conditions. In wet condition, because of the weak capillary force, the displacement between soil particles encounters more resistance, thus it is difficult to be compacted and the dry density is small. The

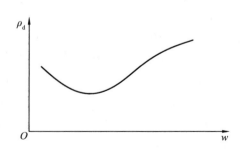

Figure 4-15 Compaction curve of non-cohesive soil

compaction of non-cohesive soil is usually measured by relative density D_r. It generally requires coarse grained soil to be compacted to $D_r > 2/3$, that is, to reach a dense state.

Besides, the existence of organic matter has a negative effect on soil compaction. Because of the strong hydrophilicity of the organic matter, it is not easy for soil to be compacted reaching a greater dry density.

IV. Soil gradation

For a given soil, a well-graded soil has a higher maximum dry unit weight (or dry density) and lower optimum water content than a poorly graded soil, as shown in Figure 4-16. This is because a well-graded soil contains particles of all sizes and the finer size particles fill the void space between the coarser particles resulting in lower air voids and higher unit weight. For coarse-grained soil, the addition of a small amount of fines to it will increase its maximum dry unit weight. However, when the amount of fines added is more than that needed to fill the voids, the maximum dry unit weight again decreases.

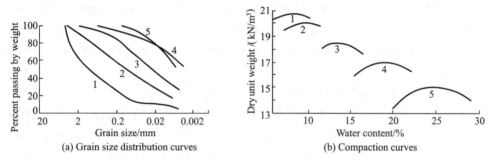

Figure 4-16 Effect of soil gradation on soil compaction

V. Compaction method

Compaction of soils in the field can be done by a variety of compaction equipment. For example, smooth-drum rollers can be used for static compaction; sheepsfoot rollers can be used for kneading compaction; vibratory rollers can be used for vibration compaction; tampers can be used for tamping. A higher contact pressure between the soil and wheels of the equipment increases the dry density and lowers the optimum water content. The slower the speed of roller travel, the more vibrations at a given point and the lesser number of pass required to attain a given density.

4.7 Classification of soil

4.7.1 Soil classification for engineering purpose

For natural soil, the mineral composition and structure change constantly, thus the engineering properties are different. To distinguish the engineering characteristics of soil and to evaluate its suitability as foundation soil or building materials, it is necessary to systematically classify the soil. Taking soil as the object, considering its engineering purpose on the state of soil and its characteristics, we can proceed the engineering classification of soil. The main principle of soil classification is to classify the soil of similar nature into a system and to make reasonable evaluation of the soil.

The classification of soil has two categories: general classification and special classification. Special classification is to meet certain construction purposes (industrial and civil buildings, road engineering, water conservancy engineering etc.), according to a single or a few index properties. This classification generally combines soil properties and engineering needs closely, where the engineering characteristics is obvious. For example, sands can be categorized by density and cone penetration resistance; loess can be categorized by collapsibility index; clays can be categorized by compression index. In addition, the classification is closely related to the needs of a certain type of construction. For example, the AASHTO soil classification system is for road engineering in the U.S. and soil classification method in GB 50007—2011 "Code for design of building foundation" in China.

General classification applies to all kinds of engineering constructions, involving most types of nature soil. Regional soil classification is used by a certain region, which belongs to the general classification. General classification is the basis of special classifications. The special classification has more details and is the supplement to the general classification.

Soil classification should meet the following requirements:

1) Determine the category of the soil. The soil classification in engineering application originates from the stratigraphic division in geology, and it is used for drawing geological engineering plans, sections and diagrams. The category of soil should be comprehensive and practical, considering the nature of special soils.

2) The category of soil is correlated with the engineering properties of soil. Thus the suitability of soils as foundations or building materials can be preliminarily evaluated.

3) The primary contents and testing items for different types of soil can be clearly defined, and applicable testing methods can be chosen.

Soil classification systems worldwide capture great physical insight and enable geotechnical engineers to anticipate the properties and behavior of soils by grouping them into similar response categories based on their index properties. Therefore, number of classification systems have been developed depending on the intended purpose of the system. Different countries may use different soil classification systems. Table 4-3 lists the soil classification codes for engineering purposes in some countries.

Table 4-3 Soil classification codes in some countries

Country	Soil classification code	Brief description
China	GB/T 50145—2007	This code issued in 2007 is the Chinese code "Standard for engineering classification of soil". It replaced the previous code GBJ 145—90 that issued in 1990.
USA	ASTM D2487—17	This is the latest version of American code "Standard Practice for Classification of Soils for Engineering Purposes (Unified Soil Classification System)" issued by the American Society for Testing and Materials (ASTM).

Continued

Country	Soil classification code	Brief description
UK	BS 5930—2015	This is the latest version of British standard "Code of practice for ground investigations", which introduces the British Soil Classification System (BSCS).
Germany	DIN 18196—2011	This is the latest version of German code "Soil Classification For Civil Engineering Purposes". It was issued by German Institute for Standardization.
Japan	JGS 0051—2009	This is the latest version of Japanese code "Method of Classification of Geomaterials for Engineering Purposes". It was issued by the Japanese Geotechnical Society (JGS).

It should be noted that the Unified Soil Classification System (USCS) is the foundation for classification systems worldwide. The USCS places emphasis on particle size and uses the percentage retained on the No. 200 sieve (0.075mm) to separate coarse-grained soils (more than 50% retained) from fine-grained soils (more than 50% passing). This USCS system was originally developed by A. Casagrande (1948) for the purpose of airfield construction during World War II. Afterwards, it was modified by Professor Casagrande, the U.S. Bureau of Reclamation, and the U.S. Army Corps of Engineers to enable the system to be applicable to dams, foundations, and other construction. This classification system is based on particle-size characteristics, liquid limit, and plasticity index. According to ASTM D2487-17, there are three major soil divisions: coarse-grained soils, fine-grained soils, and highly organic soils. These are subdivided further into 15 basic groups, each with its own group symbol.

In the U.S., there is a special soil classification system called AASHTO system for road engineering applications. AASHTO refers to the American Association of State Highway and Transportation Officials. This system was originally developed in 1920 by the U.S. Bureau of Public Roads for the classification of soil for highway subgrade use. This system is developed based on particle size and plasticity characteristics of soil mass. After some revision, this system was adopted by the AASHTO in 1945. The AASHTO Soil Classification System classifies soils into seven primary groups, named A-1 through A-7, based on their relative expected quality for road embankments, sub-grades, sub-bases, and bases. Some of the groups are in turn divided into subgroups, such as A-1-a and A-1-b. Furthermore, a Group Index may be calculated to quantify a soil's expected performance within a group. To determine a soil's classification in the AASHTO system, one first determines the relative proportions of gravel, coarse sand, fine sand, and silt-clay.

Besides, European Soil Classification System (ESCS) for engineering purposes uses soil description and symbols per European standard EN ISO 14688-1, and 34 European countries are bound to implement, as CEN members. ESCS is developed on principles for soil classification following EN ISO 14688-2, and it generally complies with guidelines defined by the USCS soil classification system per the US standard ASTM D2487. ISO 14688-2: 2017 specifies the basic principles for the classification of those material characteristics most commonly used for soils for engineering purposes. It is intended to be read in conjunction with ISO 14688-1, which gives rules for the identification of soils. The relevant characteristics could vary and therefore, for particular projects or materials, more detailed subdivisions of the descriptive and classification terms could be appropriate.

In addition to the different soil classification systems used in different countries introduced above, we can also classify the soils based on their geological origin, structure, grain size, etc. The origin of soil may refer either to its constituents or to the agencies responsible for its present status. Based on constituents, a soil may be classified as inorganic soil and organic soil. Based on the agencies responsible for their present state, soils may be classified into different types including residual soils, transported soils, alluvial or sedimentary soils, aeolian soils, glacial soils, lacustrine soils and marine soils. Based on structure type, soils can be categorized as soils of single-grained structure, soils of honey-comb structure, soils of flocculent structure, etc. In the grain-size classification, soils are designated according to the grain-size. Terms such as gravel, sand, silt, and clay are used to indicate certain ranges of grain-sizes. Since natural soils are mixtures of all particle-sizes, it is preferable to call these fractions as sand size, silt size, etc.

4.7.2 Soil classification standards in China

In China, several standards about soil classification were separately issued by different departments such as Ministry of housing and urban rural development of China, Ministry of transport of China, Ministry of water resources of China, etc. Two main standards are introduced in this section: "Standard for engineering classification of soil" in GB/T 50145—2007 and "Code for design of building foundation" in GB 50007—2011.

4.7.2.1 Soil classification per GB/T 50145—2007

The Chinese soil classification system in GB/T 50145—2007 is in many respects similar to the USCS. Soils can be firstly divided into organic soils with the organic matter content bigger than 5% and otherwise inorganic soils. For inorganic soils, we can further classify them into three groups including extra-coarse-grained soils, coarse-grained soils, and fine-grained soils, as shown in Table 4-4. It should be noted that for clay particle, the grain size is less than 0.002mm was applied in the USCS.

Table 4-4 Soil classification based on grain size

Group name	Grain name		Range of grain size d/mm
Extra-coarse-grained soil	Boulder		$d>200$
	Cobble		$60<d\leqslant200$
Coarse-grained soil	Gravel	Coarse gravel	$20<d\leqslant60$
		Medium gravel	$5<d\leqslant20$
		Fine gravel	$2<d\leqslant5$

Continued

Group name	Grain name		Range of grain size d/mm
Coarse-grained soil	Sand	Coarse sand	$0.5 < d \leqslant 2$
		Medium sand	$0.25 < d \leqslant 0.5$
		Fine sand	$0.075 < d \leqslant 0.25$
Fine-grained soil	Silt		$0.005 < d \leqslant 0.075$
	Clay		$d \leqslant 0.005$

Ⅰ. Classification of extra-coarse-grained soil

Extra-coarse-grained soil can be categorized by the fraction of boulder and cobble with grain size greater than 60mm, as listed in Table 4-5.

Table 4-5 Classification of extra-coarse-grained soil

Group name	Grain size fraction		Symbol	Sub-group name
	Boulder and cobble	Fraction relation		
Extra-coarse-grained soil	>75%	Boulder Cobble	B	Boulder
		Boulder Cobble	Cb	Cobble
Extra-coarse-grained mixed soil	50%~75%	Boulder Cobble	BSl	Boulder with soil
		Boulder Cobble	CbSl	Cobble with soil
Mixed extra-coarse grained soil	15%~50%	Boulder Cobble	SlB	Soil with boulder
		Boulder Cobble	SlCb	Soil with cobble

Ⅱ. Classification of coarse grained soil

Soil with a grain size greater than 0.075mm and the coarse-grain content is more than 50% by weight is classified as coarse-grained soil. Coarse-grained soil can be categorized by two types: gravel and sand. If a soil fraction with the grain size greater than 2mm is more than 50% by weight, the soil is classified as gravel. A soil with the grain size greater than 2mm and the fraction is less than 50% by weight, the soil is categorized as sand.

Gravel or sand can be further categorized according to the gradation of soil and the fine-grain content. Fine-grain refers to the particle size smaller than 0.075mm in a soil mass. The gravel is classified to different subgroups according to the fine-grain content and the gradation of the coarse-grained soil, as shown in Table 4-6.

Table 4-6 Classification of gravel

Group name		Grain size fraction		Symbol	Sub-group name
		Fine-grain content	Gradation		
Gravel	Clean gravel	<5%	$C_u \geqslant 5$ $C_c = 1 \sim 3$	GW	Well-graded gravel

Continued

Group name		Grain size fraction		Symbol	Sub-group name
		Fine-grain content	Gradation		
Gravel	Clean gravel	<5%	Above requirements cannot be met at the same time	GP	Poorly-graded gravel
	Gravel with fines	5%~15%	N/A	GF	Gravel with fines
	Gravel with high fines	15%~50%	Predominantly clay	GC	Clayey gravel
			Predominantly silt	GM	Silty gravel

1) If the fine-grain content by weight of gravel is less than 5%, then the soil group name is clean gravel. If the coefficient of uniformity is greater than or equal to 5, and the coefficient of curvature is between 1 and 3, the gravel is classified as well-graded gravel (GW); if the coefficient of uniformity is less than 5, and/or the coefficient of curvature is less than 1 or greater than 3, the soil is a poorly-graded gravel (GP).

2) If the fine-grain content by weight of gravel is greater than or equal to 5% and less than 15%, the gravel can be classified as gravel with fines (GF).

3) If the content of fine-grained soil particles is more than or equal to 15% and less than 50%, the gravel belongs to gravel with high fines. According to the relations between silt content and clay content, the gravel can be clayey gravel (GC) or silty gravel (GM).

The subgroups of sand can be categorized by the same way as it is for gravel, where symbol G needs to be replaced by S. Sand is classified according to its fine-grain content and the gradation, as shown in Table 4-7.

Table 4-7 Classification of sand

Group name		Grain size fraction		Symbol	Sub-group name
		Fine-grain content	Gradation		
Sand	Clean sand	<5%	$C_u \geqslant 5$ $C_c = 1 \sim 3$	SW	Well-graded sand
			Above requirements cannot be met at the same time	SP	Poorly-graded sand
	Sand with fines	5%~15%	N/A	SF	Sand with fines
	Sand with high fines	15%~50%	Predominantly clay	SC	Clayey sand
			Predominantly silt	SM	Silty sand

Chapter 4 Engineering Properties of Soil

Ⅲ. Classification of fine-grained soil

The Chinese code GB/T 50145—2007 defines fine-grained soil as having a physical dominance of fines, which is similar to the Unified Soil Classification System (USCS) in the U. S. code ASTM D2487. Fine-grained soil is defined as the fine-grain (<0.075mm) content by weight is greater than or equal to 50%. Fine-grained soil can be categorized as below according to the coarse-grain content or organic content:

1) Fine-grained soil. When the coarse-grain content is less than or equal to 25% by weight, it is categorized as fine-grained soil;

2) Fine-grained soil with coarse particles. When the coarse-grain content is less than 50% and greater than 25% by weight, it is categorized as fine-grained soil with coarse particles;

3) Organic soil. When the organic matter content is larger than or equal to 5% and less than 10%, it is categorized as organic soil.

Figure 4-17 shows the plasticity chart used in the Chinese code GB/T 50145—2007, which is modified based on the Casagrande's plasticity chart. The plasticity chart can be used to differentiate the plasticity and organic characteristics of the fine-grained soils based on liquid limit (w_L) and plasticity index (I_p) of the soils. The plasticity index of the soil is the numerical difference between liquid limit and plastic limit of soil.

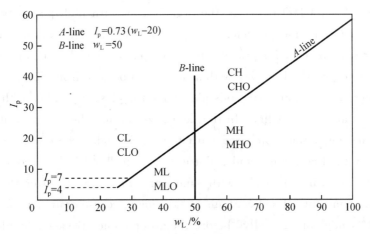

Figure 4-17 Plasticity chart

The axes of the plasticity chart are plasticity index and liquid limit; therefore, the plasticity characteristics of a particular soil can be represented by a point on the chart. Classification letters are allotted to the soil according to the zone within which the point lies. The chart is divided into two ranges of liquid limit by B-line, low (L) and high (H). Silt (M) plots below the A-line and Clay (C) above the A-line on the plasticity chart, i.e. silts exhibit plastic properties over a lower range of water content than clays having the same liquid limit. Clay is a fine-grained soil that can be made to exhibit plasticity (putty-like properties) within a range of water contents and that exhibits considerable strength when air dry. Silt is soil passing the 0.075mm sieve that is non-plastic or very slightly plastic and that exhibits little or no strength when air dry. In Figure 4-17, the A-line is a sloped line beginning at $I_p=4$ and $w_L=25.5$ with an equation of $I_p=0.73(w_L-20)$ and B-line with $w_L=50$. The A-line generally separates the more claylike materials from silty materials.

Table 4-8   Classification of fine-grained soil

Plasticity index I_p	Liquid limit w_L	Group symbol	Group name
$I_p \geqslant 0.73(w_L-20)$ and $I_p \geqslant 7$	$w_L \geqslant 50\%$	CH	Clay with high plasticity
	$w_L < 50\%$	CL	Clay with low plasticity
$I_p < 0.73(w_L-20)$ or $I_p < 4$	$w_L \geqslant 50\%$	MH	Silt with high plasticity
	$w_L < 50\%$	ML	Silt with low plasticity

Table 4-8 lists the classification of fine-grained soil based on the plasticity chart. The letter denoting the dominant size fraction is placed first in the group symbol. For example, CH refers to the clay with high plasticity or fat clay. CL is the clay with low plasticity or lean clay. It should be noted the group symbol may consist of two or more letters. For example, for organic silts and clays, we can add a symbol (O) as a suffix to any group. As shown in Figure 4-17, CHO refers to the organic clay with high plasticity. Besides, for fine-grained soil with coarse particles (with 25%~50% coarse-grain content), add a symbol (S) for sand or (G) for gravel after group name, whichever is predominant. For example, CLS refers to the lean clay with sand. It can be seen from Figure 4-17, there is a mixed zone where both CL and ML soils plot when the plasticity index in the range of 4 to 7 and plots on or above the A-line. In this zone, we can use the group symbol CL-ML.

It should be noted that compared with the U. S. code, there are small differences when we classify the fine-grained soils. For example, in USCS, if soil contains 15 to 29% coarse grains (>0.075mm), add "with sand" or "with gravel" after the group name, whichever is predominant. If soil contains 30% or more coarse grains, predominantly sand or gravel, add "sandy" or "gravelly" before the group name. Besides, highly organic soils (e.g. peat) are classified as PT.

【Example 4-5】 There are three soils, A, B and C, and their grain size distribution curves are shown in Figure 4-18. The liquid limit of soil C is 47% and its plastic limit is 24%. Try to classify the three soils per Chinese code GB/T 50145—2007.

【Solution】

(1) For soil A

① From grain size distribution curve of soil A, we can see that the soil A doesn't contain any particle that is greater than 60mm. The percentage of grain size greater than 0.075mm is around 99% (>50%), thus soil A is a coarse-grained soil.

② The fraction of particle size larger than 2mm for soil A is around 60% (>50%) by weight, so soil A can be classified as gravel.

③ The fine-grain content obtained from curve A is about 1%, which is less than 5%, thus soil A is a clean gravel.

④ From the curve A, d_{10}, d_{30} and d_{60} are 0.32mm, 1.65mm, and 3.55mm, respectively.

The coefficient of uniformity C_u is
$$C_u = \frac{d_{60}}{d_{10}} = \frac{3.55}{0.32} = 11.1$$

The coefficient of curvature C_c is
$$C_c = \frac{(d_{30})^2}{d_{10}d_{60}} = \frac{1.65^2}{0.32 \times 3.55} = 2.40$$

Since $C_u > 5$, $C_c = 1 \sim 3$, we can classify soil A a well-graded gravel (GW).

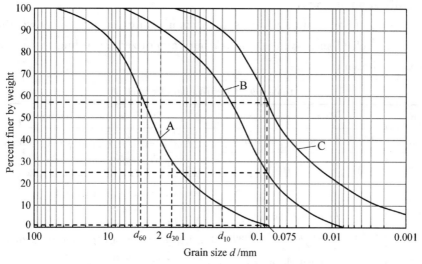

Figure 4-18 Grain size distribution curve for [Example 4-5]

(2) For soil B

① From curve B, we can get the fine-grain ($d<0.075$mm) content is about 25%, hence the coarse-grain ($d \geqslant 0.075$mm) content is 75% ($>50\%$). Soil B can be classified as a coarse-grained soil.

② From curve B, the fraction of particles larger than 2mm is about 10% ($<50\%$), hence soil B can be further classified as sand. Since the fine-grain content is about 25%, which is in the range between 15% and 50%, soil B can be further classified as sand with high fines.

③ Since the predominant fines in soil B is silt, we can classify soil B a silty sand (SM).

(3) For soil C

① From curve C, we can get the fine-grain ($d<0.075$mm) is about 57% ($>50\%$), hence the coarse-grain ($d \geqslant 0.075$mm) content is 43% ($>25\%$). Soil B is therefore can be classified as a fine-grained soil with coarse particles.

② Since the liquid limit of soil C is 47% and the plastic limit is 24%, we can get the plasticity index is $I_p = 47 - 24 = 23$. According to the plasticity chart, soil B is above A-line and on the left of B-line, hence soil C is classified as lean clay (CL).

③ From curve C, we can get the predominant coarse grain is sand since the fraction of particles larger than 2mm is 0. Therefore we need to add a symbol S after the group name, soil C is finally classified as lean clay with sand (CLS).

4.7.2.2 Soil classification per GB 50007—2011

The classification of soil is relatively simple per GB 50007—2011 "Code for Design of Building Foundation" in China. Soils can be categorized by four categories according to particle shape, grain size fraction or the plasticity index. They are gravel, sand, silt and clay.

I. Classification of gravel

If the fraction of soil grain size larger than 2mm is more than 50% by weight, the soil is classified as gravel. The gravel can be further categorized according to the grain shape and fraction, as shown in Table 4-9.

Table 4-9 Classification of gravel

Group name	Particle shape	Grain size fraction
Round boulder	Mainly round and sub-round	Fraction of grain size greater than 200mm is greater than 50% by weight
Angular boulder	Angular shope	
Round cobble	Mainly round and sub-round	Fraction of grain size greater than 20mm is greater than 50% by weight
Angular cobble	Angular shape	
Round gravel	Mainly round and sub-round	Fraction of grain size greater than 2mm is greater than 50% by weight
Angular gravel	Angular shape	

II. Classification of sand

If the fraction of soil grain size larger than 2mm is less than or equal to 50%, and the coarse-grain content ($d \geqslant 0.075$mm) is more than 50%, the soil is classified as sand. Sand can be classified as different group names based on the grain size fraction, as listed in Table 4-10.

Table 4-10 Classification of sand

Group name	Grain size fraction
Gravel sand	Fraction of grain size greater than 2mm is between 25% and 50% by weight
Coarse sand	Fraction of grain size greater than 0.5mm is more than 50% by weight
Medium sand	Fraction of grain size greater than 0.25mm is more than 50% by weight
Fine sand	Fraction of grain size greater than 0.075mm is more than 85% by weight
Silty sand	Fraction of grain size greater than 0.075mm is more than 50% by weight

III. Classification of silt

If the plasticity index is less than or equal to 10, and the fraction of grain size less than

0.075mm is greater than 50% by weight, the soil can be classified as silt.

Ⅳ. Classification of clay

If the plasticity index of soil is greater than 10 and the fraction of grain size less than 0.075mm is greater than 50% by weight, the soil can be classified as clay. The clay can be further categorized according to the plasticity index in Table 4-11.

In addition, there are many types of special soils in nature, such as soft soil, collapsible loess, expansive soil, red clay, saline soil, frozen soil, artificial fill. The special soils have special engineering properties, the classification of which can be found in the related codes.

Table 4-11 Classification of clay

Soil type	Range of plasticity index (I_p)
Clay	$I_p > 17$
Silty clay	$10 < I_p \leqslant 17$

Knowledge expansion

About Borrow Pits

A borrow pit is a land-use involving the excavation or digging of material for use as fill at another site and includes the pit area, stockpiles, haul roads, entrance roads, scales, crusher, and all related facilities. The hole left behind after the material has been harvested from a construction site is called a "borrow pit". Frequently, construction crews will dig borrow pits in order to gather gravel, soil, and sand for use in another location.

There are many things that can be done with a borrow pit once a construction crew has finished digging inside of the initial hole. Some pits are used as landfills, while other pits may form recreational areas. In fact, it is not uncommon for a municipality to fill a pit with water, resulting in a small manmade pond or a large lake. Other pits may be turned into wildlife habitats by adding certain elements, such as water, to the pit area. Rarely do municipalities decide to fill a pit in with extra sand or other material, though this is possible.

The digging of a borrow pit falls under the engineering discipline known as earthworks. Earthworks projects consist of engineering feats that include transporting large amounts of soil or rock from one area to another. Borrow pit construction may seem relatively easy to accomplish, though this type of digging actually requires an extensive amount of analysis prior to the first dig. Engineers must be sure that the amount of soil dug from a pit area will not disrupt the earth. This specific type of engineering, called geotechnical engineering, is a complex process. Prior to the

Engineering Geology and Soil Mechanics

invention of the computer, geotechnical engineers were forced to calculate the degree to which the earth would shift during digging. Today, computer programs make these types of calculations simpler.

Since massive quantities of earth must be moved in order to build roads, railways, canals, buildings, and other structures, the invention of various industrial tools has made this task easier. Bulldozers, loaders, production trucks, graders, and many other large pieces of equipment are often used to move soil from one place to another. Without these machines, digging a borrow pit would take years instead of months or weeks to accomplish.

A borrow pit's volume really depends upon the construction project at hand. While major roads and freeways may take multiple tons of gravel to build, small projects may not require much soil. How much volume needs to be excavated from the borrow pit to meet the construction project requirements can be solved based on the knowledge of phase relationships in soil and compaction of soil introduced in this chapter. We also need to understand that water can be added or removed from soil and the mass of the soil solids can not be changed during the compaction process.

Exercises

[4-1] What is the physical index of soil? How to define them? What indexes can be directly measured by laboratory testing?

[4-2] What is the relative density of non-cohesive soil? How to define the relative density of non-cohesive soil?

[4-3] What is the consistency of cohesive soil? How many states the cohesive soil can have based on water content?

[4-4] What is the plasticity index and the liquidity index? How to use them?

[4-5] How to classify a soil for engineering purposes?

[4-6] What is coarse-grained soil? What is fine-grained soil? What are the differences in the engineering properties between them?

[4-7] What are the principles for the engineering classification of the soil? Why there are different standards in China?

[4-8] A soil has a unit weight of $19kN/m^3$ and a water content of 11%. The value of G_s is 2.70. Calculate the void ratio and degree of saturation of the soil. What would be the values of density and water content if the soil were fully saturated at the same void ratio?

[4-9] A soil specimen is 39mm in diameter and 80mm in height and in its natural condition weighs 172.0g. When dried completely in an oven the specimen weighs 130.5g. The value of G_s is 2.71. What is the degree of saturation of the specimen?

[4-10] Calculate the dry unit weight, the saturated unit weight and the buoyant unit

weight of a soil having a void ratio of 0.70 and a value of G_s of 2.72. Calculate also the unit weight and water content at a degree of saturation of 75%.

[4-11] A soil sample from a borrow pit has a natural void ratio is 1.15. The soil will be used for a highway project where a total of 100000m³ of soil is needed in its compacted state with the compacted void ratio of 0.73. How much volume has to be excavated from the borrow pit?

[4-12] A soil sample ($G_s = 2.70$) having the maximum dry unit weight of 19kN/m³ and the optimum water content of 14% obtained from the standard proctor test. The soil was considered for grading work at a park. The soil is to be brought in from a borrow pit where its void ratio is 0.80 and degree of saturation is 0.40. At the park site, the soil will be compacted to a dry unit weight that exceeds 95% of the maximum dry unit weight, and the dimensions of the compacted fill region are approximately 250m by 250m by 4m deep. Please slove:

(1) What is the moisture content of the soil in the borrow pit?

(2) How much volume of soil must be taken from the borrow pit?

(3) How many 5-ton truckloads of the soil will need to be hauled to the business park site?

(4) If the optimum moisture content from the standard Proctor test is to be used in the field compaction, how much water must be added to each ton of moist soil from the borrow pit?

Chapter 5 Permeability and Seepage

5.1 Introductory case

Soil as a three-phase system that with many pores inside, and most of the pores are connected. When there is a pressure difference between two points in the soil, the water flows from the higher pressure side to the lower pressure side through the interconnected pores. The phenomenon that water flows through the soil under the action of pressure is commonly referred to seepage. Correspondingly, water penetrate through the soil is known as the permeability, which is one of the basic mechanical properties of soil.

Seepage is a common phenomenon found in geotechnical engineering, such as slopes, earth dams, and building foundations, as shown in Figure 5-1.

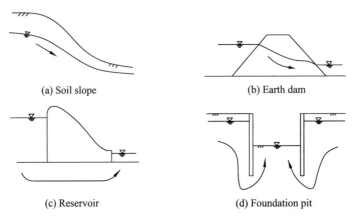

Figure 5-1 Schematics of seepage

Many engineering problems or even accidents are associated with seepage. For example, the Teton Dam in the U.S. failed abruptly on June 5, 1976, as shown in Figure 5-2. It released huge amount of water, then flooded farmland and towns downstream with the eventual loss of 14 lives and with a cost estimated to be nearly $1 billion. Another example is the Jiujiang dike failure on August 7, 1998, as shown in Figure 5-3. The dike is situated in an urban area of Jiujiang city in China with a flood-wall on top of the dike. The cause of dike failure was piping with subsequent sliding of the inner slope, resulting in a

Chapter 5　Permeability and Seepage

Figure 5-2　Teton dam failure

Figure 5-3　Jiujiang dike failure

final failure width of 64 m and a maximum scouring depth of 7 m. Figure 5-4 shows an example of a building foundation pit collapse at Haizhucheng Plaza in Guangzhou city of China on July 21, 2005. The failure of the retaining structure caused the hotel building failure on the south side of the pit with the loss of 3 lives. Seepage in the weak and strongly weathered soil layers around the foundation pit is one of the reasons for this accident.

Figure 5-4　Building foundation pit collapse

When water flows through the soil, it interacts with the soil leading to a variety of engineering problems, which can be divided into water problems and soil problems.

The problem of water is the engineering problem caused by the water. For example, groundwater infiltration problems of foundation, tunnel, mining and other excavation. The problems of water loss due to seepage of a dam, and the problems of groundwater pollution caused by coastal seawater intrusion and/or sewage infiltration. These are the problems caused by changes in the amount (water inflow, seepage), quality (water quality) and storage location of water (groundwater level).

The soil problem refers to changes in the internal stress state of the soil caused by water, resulting the change in soil structure and strength. Soil slope, earth dam, foundation will have a gradually change in soil structure and stress state in foundation due

133

to seepage, which will lead to damage. Soil moisture and soil strength will change because of water migration in unsaturated slope, which will lead to a landslide.

The permeability of soil has important influence on the deformation and strength of soil. Therefore, it is of great significance to study the seepage of water in soil and the stability of soil with seepage.

This chapter mainly discusses the permeability characteristics of the soil, the laboratory method in determination of permeability coefficient and field measurement, the seepage law of the water in the soil and the seepage field, the seepage force and the effective stress principle, and the seepage failure. It provides the necessary theoretical knowledge for engineering design, construction and other practical problems.

5.2 Permeability of soil and Darcy's Law

5.2.1 Permeability of soil

The permeability of the soil is related to the parameters such as soil porosity, water content, particle composition, etc. It is also related to the interaction between the particles. For non-cohesive soil, the arrangement of the soil particles is mainly controlled by the weight of the soil particles. The main factors affecting the permeability are the gradation of soil particles and the porosity of soil. For cohesive soil, the arrangement of the soil particles is mainly controlled by the interaction between the soil particles, where the permeability of the soil is not only affected by the composition of particles and porosity but also the mineral composition of soil, water film on particle surface and the chemical properties of aqueous solution. In addition, the permeability of soil is also related to the pore water fluid properties.

5.2.2 Darcy's Law

In fluid dynamics, the state of water flow is divided into laminar flow and turbulent flow. Laminar flow (or streamline flow) occurs when the fluid flows in the tube, where the particles move smoothly along the direction that is parallel to the tube axis. Turbulent flow is relative to the laminar flow, the inertia between the particles of the fluid accounted for the main position, where the fluid particles are not linear but an irregular flow. The head loss in the laminar flow state is proportional to the flow rate, yet the head loss in the turbulent flow state is almost proportional to the square of the flow rate. The results show that, in most cases, the flow velocity of water in soil pores is slow that belongs to laminar flow, which is one of the basic assumptions in soil mechanics.

When the groundwater flows via the pores of the soil, the loss of hydraulic energy will occur due to the seepage force. In order to study the permeability of water in the soil, the French engineer Henry Darcy developed the test device in 1856 shown in Figure 5-5 carrying out a large number of experimental studies with the permeability of homogeneous sand. The relationship between the seepage flow and the hydraulic gradient is obtained, where the seepage velocity is proportional to the hydraulic gradient.

$$q = kiA \quad (5\text{-}1)$$
or
$$v = ki \quad (5\text{-}2)$$

where, q—flow rate of water through the porous media, cm^3/s;

A—cross-sectional area, cm^2;

v—average flow velocity, or Darcy velocity, $v = \dfrac{q}{A}$, cm/s;

k—permeability coefficient or hydraulic conductivity, cm/s;

i—hydraulic gradient, $i = \dfrac{h_1 - h_2}{L} = \dfrac{\Delta h}{L}$;

Δh—head loss over a flow length L, cm or m;

L—length of flow, cm or m.

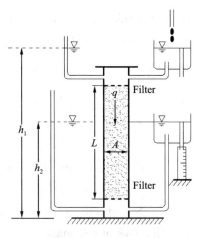

Figure 5-5 Schematic of Darcy's penetration test device

Eq. (5-2) is the expression of Darcy's Law, which is a linear relationship equation. The equation describes the basic law of water flow in the soil. It should be noted that the average flow velocity is not the actual velocity of the water in the pores, but the apparent velocity through the bulk of the porous medium. Actual velocity, also known as seepage velocity (v_s), is obtained using the average velocity divided by the soil porosity (n).

5.2.3 Influencing factors of soil permeability

I. Soil size and gradation

Soil size and gradation have significant effects on soil permeability. For gravel, Darcy's Law is valid only when the hydraulic gradient is low. When the fine-grain (clay and silt) content increased in sand, the permeability of sand will reduce greatly. According to the experimental data, the effective grain size of uniform sand is between 0.1mm and 3.0mm, where the permeability coefficient is proportional to the square of the effective particle size. The empirical expression is written as

$$k = C_1 d_{10}^2 \quad (5\text{-}3)$$

where, k—permeability coefficient of sand, cm/s;

d_{10}—effective grain size, cm;

C_1—a constant, value as 100~150.

II. Void ratio and dynamic viscosity

The void ratio (e) of soil and the dynamic viscosity of the water determine the permeability coefficient. As the density of soil increases, the void ratio becomes smaller, and the permeability of the soil decreases. The dynamic viscosity of water varies with temperature. Thus soil with the same density may have different permeability coefficient at different temperatures. The permeability coefficients used in engineering projects are often measured with the water at the temperature of 20℃ as a standard. The permeability coefficient measured at other temperatures should be corrected to the permeability coefficient at 20℃. According to some existing research results, we can get the following empirical relationship in Eq. (5-4) that permeability coefficient is a function of soil porosity and dynamic viscosity.

$$k = \frac{C_2}{S_s^2} \times \frac{n^3}{(1-n)^2} \times \frac{\gamma_w}{\eta} \quad (5\text{-}4)$$

where, n—soil porosity, $n = e/(1+e)$;

γ_w—unit weight of water, about 9.8kN/m^3;

η—dynamic viscosity of water, $\text{Pa} \cdot \text{s}$, value as $1.01 \times 10^{-3} \text{Pa} \cdot \text{s}$;

C_2—a constant that reflet the particle shape in the actual flow direction of water, about 0.145;

S_s—specific surface area of soil particles, cm^{-1}.

III. Gas in soil mass

When the soil is not saturated, which may have some air bubbles not connected with the atmosphere. This will block the seepage path (gas seal phenomenon). The more bubbles in the soil, the lower is the permeability of soil. The permeability of the soil is related to the amount of enclosed gas in the soil mass.

5.2.4 Permeability of layered soil

Natural deposited clay is often composed of different layers with different permeability. The permeability of the layered soil is related to the permeability of soil in each layer and the direction of seepage. When the permeability of the layered soil is studied, the permeability coefficient of each soil layer and the thickness of each layer are used to calculate the average permeability coefficient in horizontal or vertical direction.

As shown in Figure 5-6, assuming that the soil can be divided into several layers, the permeability coefficient of each soil layer is $k_1, k_2, k_3, \cdots, k_n$. The total thickness of soil layer is H, and the thickness of each soil layer is $H_1, H_2, H_3, \cdots, H_n$. First, when the seepage flow parallel to the layer (in the x direction), the length of L is taken as the flow path, the hydraulic gradient is i; the flow rate of each layer is $q_{1x}, q_{2x}, \cdots, q_{nx}$, so the total flow rate should be the sum of the flow rate of each layer.

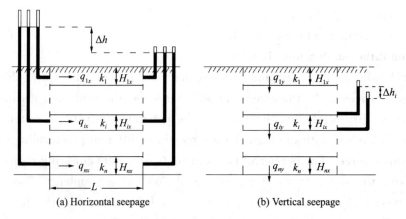

(a) Horizontal seepage (b) Vertical seepage

Figure 5-6 Seepage in the layered soil

$$q_x = q_{1x} + q_{2x} + q_{3x} + \cdots + q_{nx} = \sum_{i=1}^{n} q_{ix} \tag{5-5}$$

According to Darcy's Law, total flow rate can be expressed as

$$q_x = k_x j H \tag{5-6}$$

where, k_x—the average permeability coefficient of soil in horizontal seepage;

j—the average hydraulic gradient of soil in horizontal seepage.

For the seepage flow under this condition, the head loss through the same distance of each soil layer is equal, so the hydraulic gradient of each soil layer and the average hydraulic gradient of soil should be equal. Thus the flow rate of the i_{th} soil layer is

$$q_{ix} = k_i j H_i \tag{5-7}$$

Substituting Eq. (5-6) and Eq. (5-7) into Eq. (5-5)

$$k_x j H = k_1 j H_1 + k_2 j H_2 + k_3 j H_3 + \cdots + k_n j H_n = \sum_{i=1}^{n} k_i j H_i$$

After removing j, the average permeability coefficient k_x in the x direction is

$$k_x = \frac{\sum_{i=1}^{n} k_i H_i}{H} \tag{5-8}$$

The average permeability coefficient of soil in horizontal seepage is equivalent to the weighted average of the thickness. The average permeability coefficient is mainly affected by the permeability coefficient of the soil layer with the largest permeability.

For the case of seepage flow perpendicular to the layer (in the y direction), a similar method can be used to get the permeability coefficient. The total hydraulic gradient of soil is j, and the hydraulic gradient flowing through each soil layer is $j_1, j_2, j_3 \cdots$. The total flow rate q_y should be equal to the flow rate of each soil layer, $q_1, q_2, q_3 \cdots$, i. e., $q_y = q_1 = q_2 = q_3 = \cdots = q_n$.

From Darcy's Law,

$$k_y j A = k_1 j_1 A = k_2 j_2 A = k_3 j_3 A = \cdots = k_n j_n A \tag{5-9}$$

Both sides divided by A,

$$k_y j = k_i j_i$$

Hence

$$j_i = \frac{k_y j}{k_i} = \frac{h k_y}{H k_i} \tag{5-10}$$

where, j—average hydraulic gradient of soil with total thickness H, $j = \frac{h}{H}$;

A—cross-sectional area of seepage.

The total head loss is equal to the sum of the loss at each layer, so

$$h = H_1 j_1 + H_2 j_2 + H_3 j_3 + \cdots + H_n j_n = \sum_{i=1}^{n} j_i H_i \tag{5-11}$$

Substituting Eq. (5-10) into Eq. (5-11)

$$h = \frac{h k_y}{H} \sum_{i=1}^{n} \left(\frac{H_i}{k_i} \right) \tag{5-12}$$

Rearrage the equation,

$$k_y = \frac{H}{\sum_{i=1}^{n}\left(\frac{H_i}{k_i}\right)} \quad (5-13)$$

The average permeability coefficient k_y in the vertical direction is mainly affected by the permeability coefficient of the soil with the lowest permeability. The permeability coefficient of horizontal seepage is always larger than that of vertical seepage. Because of the possibility that the particles are arranged in certain direction, the permeability coefficient of each layer in horizontal seepage and vertical seepage cannot be necessarily the same.

5.2.5 Method for determination of soil permeability coefficient

The permeability coefficient of soil is a mechanical index which is commonly used in engineering. It is the basis for judging the magnitude of soil permeability and choosing filling materials for dams. According to permeability, the dam foundation can be divided into: ① strong permeable layer, with permeability coefficient greater than 10^{-3} cm/s; ② medium permeable layer with a permeability coefficient between 10^{-5} cm/s and 10^{-3} cm/s; ③ weak permeable layer, with the permeability coefficient less than 10^{-5} cm/s. When choosing the dam fills, the soil with low permeability coefficient ($k < 10^{-6}$ cm/s) is commonly used to fill the impervious body (core wall) of dam, and the soil with high permeability coefficient ($k > 10^{-3}$ cm/s) is used to fill other parts of the dam.

Due to the wide distribution of soil particle size, the variation in the soil permeability coefficient is very large. From the coarse gravel to the clay, with the decrease of particle size and porosity, the permeability coefficient increases from 10^{-9} cm/s to 1.0 cm/s. According to a large number of experimental studies, the reference values for the permeability coefficients of common soils are shown in Table 5-1.

Table 5-1 Permeability coefficient of soil

Soil type	Permeability coefficient $k/(\text{cm/s})$
Gravel	$6.0 \times 10^{-2} \sim 1.8 \times 10^{-1}$
Goarse sand	$2.4 \times 10^{-2} \sim 6.0 \times 10^{-2}$
Medium sand	$6.0 \times 10^{-3} \sim 2.4 \times 10^{-2}$
Fine sand	$1.2 \times 10^{-3} \sim 6.0 \times 10^{-3}$
Silty sand	$6.0 \times 10^{-4} \sim 1.2 \times 10^{-3}$
Silt	$6.0 \times 10^{-5} \sim 6.0 \times 10^{-4}$
Silty clay	$1.2 \times 10^{-6} \sim 6.0 \times 10^{-5}$
Clay	$< 1.2 \times 10^{-6}$

The accuracy of the permeability coefficient directly affects the seepage calculation and the rationality of the seepage control scheme. In general, the methods of determining the permeability coefficient mainly include empirical estimation, laboratory test, field test, and inversion method. The empirical method and the inversion method mainly depend on the existing data fit or numerical analysis, while the test results are relatively straightforward and more accurate. This section mainly introduces the laboratory test and field test methods to determine the soil permeability coefficient.

5.2.5.1 Laboratory test method

After taking soil samples from boreholes, one can carry out permeability tests in the labs. There are numbers of instruments and methods for measuring the permeability coefficient of soil. According to the principle, it can be divided into constant head method and falling head method, as shown in Figures 5-7a and 5-7b, respectively.

I. Constant head method

In the process of test, the head remains unchanged. It is suitable for non-cohesive soil, where the device is shown in Figure 5-7a.

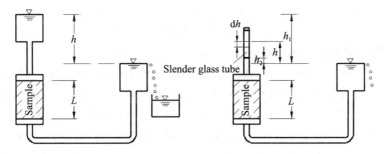

(a) Test device of constant head method (b) Test device of falling head method

Figure 5-7 Schematics of test methods to determine permeability coefficient

The thickness, cross-sectional area and head loss of the specimen were measured before the test. Using a measuring cylinder and stopwatch to measure the flow volume through the sample within a fixed time. According to Darcy's Law,

$$V = kiAt = k\frac{h}{L}At \tag{5-14}$$

The permeability coefficient k is

$$k = \frac{VL}{Aht} \tag{5-15}$$

where, L—sample thickness;
$\quad\quad A$—cross-sectional area;
$\quad\quad h$—head loss;
$\quad\quad V$—the volume of flow within a given time t.

II. Falling head method

Because the permeability coefficient of cohesive soil is very low, the flow rate through the sample is very slow. Constant head method is difficult to directly measure it accurately, so we can consider use the falling head method, as shown in Figure 5-7b.

The volume increment (dV) of water flow through the slender glass tube (Figure 5-

7b) during as small time increment (dt) is

$$dV = -a\,dh \tag{5-16}$$

where, a—crosss-sectional area of slender glass tube;

dh—head loss increment.

The minus sign indicates that the amount of water increases with decreasing dh.

According to Darcy's Law, the volume increment of water flow through the soil sample during the dt period is

$$dV = k\frac{h}{L}A\,dt \tag{5-17}$$

By combining Eq. (5-16) and Eq. (5-17),

$$a\,dh = -k\frac{h}{L}A\,dt \tag{5-18}$$

By integrating both sides, and defining the time t_1 and t_2 corresponding to the head loss h_1 and h_2, the permeability coefficient of soil is

$$k = \frac{aL}{A(t_2 - t_1)}\ln\frac{h_1}{h_2} \tag{5-19}$$

5.2.5.2 Field test method

Laboratory test method has the advantages of simple equipment and low cost. However, because soil permeability and soil structure have a relationship, it is difficult for soil sample in the lab to fully represent the field circumstance of soil. In addition, the disturbance of the sampling makes it difficult to obtain a undisturbed soil sample. Thus it is necessary to carry out field test for some important projects. The field test in determining the permeability coefficient is mostly carried out in the borehole. There are various field test methods including pumping water test, water injection test, lugeon method, tracer dilution method, etc. In this part, we mainly introduce the pumping water test and tracer dilution method to determine the permeability of soil in the field.

I. Pumping water test

Figure 5-8 shows the schematic diagram of pumping test in a well of field. Before the test, we need to drill a well and two observation boreholes at different distances from the center of the well. Then water was pumped from the well with a constant pumping rate. The original groundwater table (the horizontal dotted line) will gradually go downward to form a funnel-like groundwater surface (the inclined dash line). By determination of stable water level of the well and two observation boreholes, one can draw the piezometric water surface, which is in line with the present groundwater table. Assuming that the flow is horizontal, the cross section of the seepage flow to the well should be a series of cylindrical surfaces. The amount of water and the water level in the well will become stable after a period of time. If the measured flow rate is q, the distance between the observation borehole and the well axis is r_1 and r_2 respectively, and the height of water level in the borehole is h_1 and h_2. Then the average permeability coefficient k of the permeable soil layer can be obtained by Darcy's Law.

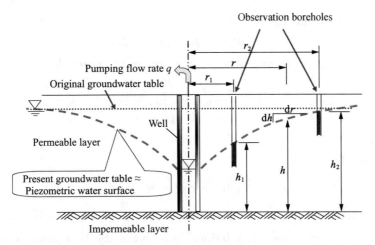

Figure 5-8 Schematic of pumping test in the field

We can choose one cylindrical surface that radius is r and the height of the water level is h, then the cross-sectional area A can be expressed as: $A = 2\pi rh$. Assuming that the hydraulic gradient across the cross section is a constant and is equal to the slope of the present groundwater table, so $i = \dfrac{dh}{dr}$ (Figure 5-8).

Based on Darcy's Law we can get the pumping water volume per unit time, known as the flow rate (q),

$$q = Aki = 2\pi rh \cdot k \dfrac{dh}{dr} \tag{5-20}$$

$$q \dfrac{dr}{r} = 2\pi kh \cdot dh \tag{5-21}$$

Integrating Eq. (5-21)

$$q\int_{r_1}^{r_2} \dfrac{dr}{r} = 2\pi k \int_{h_1}^{h_2} h\, dh \tag{5-22}$$

The permeability coefficient of soil can be obtained as

$$k = \dfrac{q}{\pi(h_2^2 - h_1^2)} \ln \dfrac{r_2}{r_1} \tag{5-23}$$

II. Tracer dilution method

Tracer dilution method is a measurement of flow velocity based on the dilution theory. The rate of water flow is determined using a single well dilution method by placing appropriate tracers in the hydrographic wells. If the hydraulic gradient near the observation hole is measured at the same time, the permeability coefficient k can be calculated using the Darcy' Law in Eq. (5-2).

The process of the calculating the water flow velocity by single hole dilution method is based on some assumptions: ① the groundwater flow uniformly; ② the flow direction is perpendicular to the borehole axis; ③ the concentration of tracer can only be reduced by the water infiltration dilution. In the condition that the tracer and the water in the borehole are evenly mixed, the increment volume of water (dQ) containing tracer flowing out from the borehole in a small time increment (dt) is

$$-dQ - CQdt = 2r_1 V_c C dt \tag{5-24}$$

In the dt time, the change in concentration of tracer in the borehole is

$$dC = \frac{dQ}{\pi r_1^2} = -\frac{2V_c}{\pi r_1} C dt \tag{5-25}$$

$$\frac{dC}{C} = -\frac{2V_c}{\pi r_1} dt \tag{5-26}$$

By integrating the both sides,

$$\ln \frac{C}{C_0} = -\frac{2V_c}{\pi r_1} t \tag{5-27}$$

The effect of drilling on the natural soil seepage state is not considered in the derivation process. If we know the relationship between the flow velocity in the borehole and the velocity in natural soil is $V_c = \alpha V_f$.

$$V_f = \frac{\pi r_1}{2\alpha t} \ln \frac{C_0}{C} \tag{5-28}$$

where, V_f—flow velocity in natural soil;

V_c—flow velocity in the borehole;

α—a coefficient that reflect the effect of drilling;

r_1—inner radius of the borehole;

C—concentration of the tracer at a given time (t), and $C = C_0$ at $t = t_0$.

In actual measurement, it is commonly to use radioactive isotope (such as iodine-131) as a tracer. The concentration of the tracer C at a given time (t) is expressed in terms of pulse count rate N, so Eq. (5-28) can be written as

$$V_f = \frac{\pi r_1}{2\alpha t} \ln \frac{N_0}{N} \tag{5-29}$$

5.3 Basic equations of seepage and net characteristics

The seepage problems involved in the previous are one-dimensional seepage under simple boundary conditions, i.e., soil seepage on the macroscopic scale is one directional. However, the seepage problems encountered in the engineering are more complex because the boundary conditions are complex, where the flow is often two-dimensional or three-dimensional. For example, the seepages in earth dams are three-dimensional. The flow characteristics for those complex conditions can be expressed in the form of differential equation, namely, the continuity equation of flow. The continuity equation is the most basic equation to describe the seepage of soil, which can be established according to the law of mass conservation. To establish the basic equation of seepage field and solve the specific seepage problem, it is necessary to combine the specific fixed-solution conditions. The three-dimensional seepage is briefly discussed by introducing the basic equation of soil seepage.

5.3.1 Continuity equation of flow

A small soil element of dimensions dx, dy and dz element in three-dimensional is shown in Figure 5-9. The difference between the mass of water entered in and expelled out of the element per unit time is equal to the change of water mass in the element per unit time according to the law of mass conservation. Therefore, the following equation can be established.

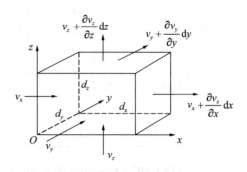

Figure 5-9 Schematic of a differential element

$$\rho_w [v_x dydz + v_y dxdz + v_z dxdy] - \rho_w \left[\left(v_x + \frac{\partial v_x}{\partial x} dx \right) dydz + \left(v_y + \frac{\partial v_y}{\partial y} dy \right) dxdz + \left(v_z + \frac{\partial v_z}{\partial z} dz \right) dxdy \right] = \frac{\partial (n\rho_w dxdydz)}{\partial t} dt \quad (5\text{-}30)$$

where, n—soil porosity;

ρ_w—water density.

Simplify the Eq. (5-30),

$$-\rho_w \left(\frac{\partial v_x}{\partial x} + \frac{\partial v_y}{\partial y} + \frac{\partial v_z}{\partial z} \right) dxdydz = \frac{\partial (n\rho_w dxdydz)}{\partial t} dt \quad (5\text{-}31)$$

Regardless of the compression of the soil particles, the change of the water mass in the soil consists of two parts: one is the reduction of the compressed pore in soil and the other is due to the expansion of the water. The right-hand side of Eq. (5-31) is

$$\frac{\partial (n\rho_w dxdydz)}{\partial t} dt = \rho_w \left(\frac{1}{\Gamma} + n \frac{1}{\Gamma_w} \right) \frac{\partial u}{\partial t} dxdydzdt \quad (5\text{-}32)$$

where, Γ—volumetric compression modulus of soil;

Γ_w—volumetric compression modulus of water;

u—pore water pressure.

$$u = \rho_w g(h-z) \quad (5\text{-}33)$$

so
$$\frac{\partial u}{\partial t} = \rho_w g \frac{\partial h}{\partial t} \quad (5\text{-}34)$$

Eq. (5-31) can be further simplified as

$$-\left(\frac{\partial v_x}{\partial x} + \frac{\partial v_y}{\partial y} + \frac{\partial v_z}{\partial z} \right) = \rho_w g \left(\frac{1}{\Gamma} + n \frac{1}{\Gamma_w} \right) \frac{\partial h}{\partial t} \quad (5\text{-}35)$$

If $S_s = \rho_w g \left(\frac{1}{\Gamma} + n \frac{1}{\Gamma_w} \right)$ is defined as a unit water storage, Eq. (5-35) becomes

$$-\left(\frac{\partial v_x}{\partial x} + \frac{\partial v_y}{\partial y} + \frac{\partial v_z}{\partial z} \right) = S_s \frac{\partial h}{\partial t} \quad (5\text{-}36)$$

Table 5-2 shows the empirical values of S_s recommended in engineering applications.

Table 5-2 Empirical values of unit water storage S_s

Type of soil	$S_s/(1/m)$	Type of soil	$S_s/(1/m)$
Plastic soft clay	$2.6\times10^{-3}\sim2\times10^{-2}$	Dense sand	$1.3\times10^{-4}\sim2\times10^{-4}$
Hard clay	$1.3\times10^{-3}\sim2.6\times10^{-3}$	Sandy gravel	$4.9\times10^{-5}\sim1\times10^{-4}$
Medium hard clay	$6.9\times10^{-4}\sim1.3\times10^{-3}$	Rock with fissures	$3.3\times10^{-6}\sim6.9\times10^{-5}$
Loose sand	$4.9\times10^{-4}\sim1\times10^{-3}$	Intact rocks	$<3.3\times10^{-6}$

I. Continuity equation of compressible fluid

For compressible fluid, the first term on the right-hand side of Eq. (5-35) is zero. So the continuity equation is

$$-\left(\frac{\partial v_x}{\partial x}+\frac{\partial v_y}{\partial y}+\frac{\partial v_z}{\partial z}\right)=\rho_w g n \frac{1}{\Gamma_w}\frac{\partial h}{\partial t} \tag{5-37}$$

II. Continuity equation of incompressible fluid

For incompressible fluid, the whole right-hand side of the Eq. (5-35) is zero, so the continuity equation is

$$\frac{\partial v_x}{\partial x}+\frac{\partial v_y}{\partial y}+\frac{\partial v_z}{\partial z}=0 \tag{5-38}$$

5.3.2 Basic equation of seepage field

The steady-state seepage field and the unsteady seepage field are two conditions. If the basic elements of the seepage field, such as water head, velocity etc. are not time dependent, it is called steady-state seepage field. Otherwise it is unsteady seepage field. The following is a brief discussion on steady-state seepage field.

Eq. (5-38) is a continuity equation without considering the compressibility of soil and water nor considering the change of water mass in soil over time, which is a steady-state seepage problem. Substituting the Darcy's Law equation concerning the three-dimensional problem into Eq. (5-38), the basic equation of the steady-state seepage field for heterogeneous and anisotropic soil can be established.

$$\frac{\partial}{\partial x}\left(k_{xx}\frac{\partial h}{\partial x}+k_{xy}\frac{\partial h}{\partial y}+k_{xz}\frac{\partial h}{\partial z}\right)+\frac{\partial}{\partial y}\left(k_{yx}\frac{\partial h}{\partial x}+k_{yy}\frac{\partial h}{\partial y}+k_{yz}\frac{\partial h}{\partial z}\right)+\frac{\partial}{\partial z}\left(k_{zx}\frac{\partial h}{\partial x}+k_{zy}\frac{\partial h}{\partial y}+k_{zz}\frac{\partial h}{\partial z}\right)=0 \tag{5-39}$$

If the soil is heterogeneous and anisotropic, and the directions of x, y, and z are the main seepage directions, Eq. (5-39) will be simplified as

$$\frac{\partial}{\partial x}\left(k_x\frac{\partial h}{\partial x}\right)+\frac{\partial}{\partial y}\left(k_y\frac{\partial h}{\partial y}\right)+\frac{\partial}{\partial z}\left(k_z\frac{\partial h}{\partial z}\right)=0 \tag{5-40}$$

If the soil mass is homogeneous and isotropic, Eq. (5-40) can be further simplified into a Laplace's equation.

$$\frac{\partial^2 h}{\partial x^2}+\frac{\partial^2 h}{\partial y^2}+\frac{\partial^2 h}{\partial z^2}=0 \tag{5-41}$$

5.3.3 Description of seepage field by flow net

From the basic equation of seepage, combined with boundary conditions and initial conditions, the distribution of water head at any point in the seepage field can be obtained by mathematical means (such as analytical method or numerical analysis). In order to visually plot the overall contour of the seepage in soil, the calculation results are usually presented in flow net. The flow net is a very useful means for two-dimensional seepage analysis.

Ⅰ. Stream function and potential function

For two-dimensional seepage, if the pore water is incompressible, the continuity equation of water flow can be expressed as

$$\frac{\partial v_x}{\partial x}+\frac{\partial v_y}{\partial y}=0 \tag{5-42}$$

The necessary and sufficient condition for Eq. (5-42) is that there is a function ψ that satisfies $d\psi=-v_y dx+v_x dy$, then

$$v_x=\frac{\partial \psi}{\partial y}, \quad v_y=-\frac{\partial \psi}{\partial x} \tag{5-43}$$

where ψ is called stream function, which is constant along any streamline.

A function ϕ, called velocity potential is used to obtain

$$v_x=-\frac{\partial \phi}{\partial x}, \quad v_y=-\frac{\partial \phi}{\partial y} \tag{5-44}$$

Substituting Eq. (5-44) into Eq. (5-42), we can get

$$\frac{\partial^2 \phi}{\partial x^2}+\frac{\partial^2 \phi}{\partial y^2}=0 \tag{5-45}$$

Eq. (5-45) is the two-dimensional Laplace's equation, where ϕ is the potential function. The flow function and the potential function are related to each other. The relationship between the two functions can be obtained from the differential method.

$$v_x=-\frac{\partial \phi}{\partial x}=\frac{\partial \psi}{\partial y}, \quad v_y=-\frac{\partial \phi}{\partial y}=\frac{\partial \psi}{\partial x} \tag{5-46}$$

By multiplying the above two equations, we can get

$$\frac{\partial \phi}{\partial x}\frac{\partial \psi}{\partial x}+\frac{\partial \phi}{\partial y}\frac{\partial \psi}{\partial y}=0 \tag{5-47}$$

For determined equipotential line and streamline, the slopes are

$$\left(\frac{dz}{dx}\right)_\phi=-\frac{\partial \phi}{\partial x}/\frac{\partial \phi}{\partial y}, \quad \left(\frac{dz}{dx}\right)_\psi=-\frac{\partial \psi}{\partial x}/\frac{\partial \psi}{\partial y} \tag{5-48}$$

Substituting Eq. (5-48) into Eq. (5-47), on the intersection points of equipotential lines and streamlines, there is a relationship as below.

$$\left(\frac{dz}{dx}\right)_\phi=-\frac{1}{\left(\frac{dz}{dx}\right)_\psi} \tag{5-49}$$

The slopes of the two sets of curves are reciprocal, indicating that the equipotential lines and streamlines are orthogonal to each other. The seepage problem described by Laplace's equation should satisfy the following conditions: ① seepage is a steady-state flow; ② seepage follows Darcy's Law; ③ soil mass is homogeneous.

II. General characteristics of the flow net

The seepage field is often described by a flow net. It is a curvilinear net formed by the combinations of streamlines and equipotential lines.

For a homogeneous isotropic soil, the flow net has the following characteristics:

① The increment of the stream function between adjacent streamlines in a flow net is the same;

② The head loss between adjacent equipotential lines is the same;

③ The streamline is perpendicular to the equipotential line;

④ For each grid in a flow net, the ratio of length and width is the same;

⑤ The flow rate of each flow channel is the same.

Figure 5-10 shows a flow net of a foundation pit for a typical homogeneous isotropic soil. For heterogeneous soil or anisotropic soil, the nature of the flow net will change. According to the continuity of flow, for heterogeneous soil, the hydraulic gradient for soil of high permeability will be larger than that for soil of low permeability. For anisotropic soils, the equipotential lines and streamlines are not orthogonal to each other.

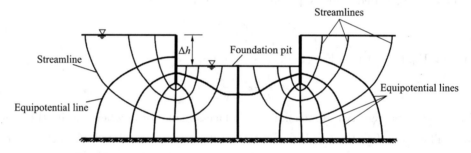

Figure 5-10 Schematic of flow net in a homogeneous isotropic soil

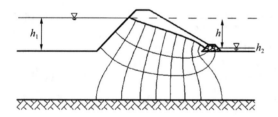

Figure 5-11 The flow net of a homogeneous dam

The flow net can be used to qualitatively determine the seepage profile of soil. For example, as shown in Figure 5-11, the flow distribution of the seepage for a earth-rock dam. It can be seen that streamlines are densest in the place close to the drainage prism, the hydraulic gradient is the largest, and the seepage velocity is also the largest; whereas the streamlines are more sparse away from the drainage prism, the hydraulic gradient is low, and the seepage velocity is also low. One can also quantitatively calculate the water head in the seepage field, hydraulic gradient, flow rate, pore water pressure and seepage force according to the flow wet.

5.4 Seepage force and seepage deformation

5.4.1 Seepage force

Seepage failure will occur in the course of soil seepage. The reason is that the drag force due to flow water on the soil particles is greater than its critical value. The drag force is known as the seepage force. Figure 5-12 shows the schematic of seepage failure test. It can be seen that a sand sample with a thickness of L was set up above a fine sieve that connected to a moveable water container (Figure 5-12a). If the top surface of the container is on the same elevation of water surface above the sample, then no water flow (seepage)

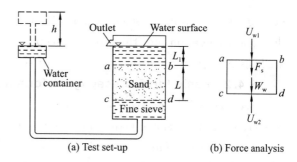

Figure 5-12 Schematic of seepage failure test

occurs. If the water container is lifted, the water level difference h will be produced, causing the water flow through the sand sample upward to the outlet until h reduces to zero. The higher the water container is lifted, the greater is h and the flow velocity. When the water container is lifted to a certain height, seepage failure will occur with sand flow upward to the water surface.

As shown in Figure 5-12b, we can set the sectional area of the sand sample is A. When water flows from the cd surface through the sample to the ab surface. The resistance that flow overcomes throughout the sand sample is F_s. The total water pressure decrement is $\Delta U_w = U_{w2} - U_{w1} = \gamma_w h A$. Because flow velocity is generally very small, the inertial force of the flowing water is negligible. According to the equilibrium condition, the total seepage force J acting on the sample should be equal to the resistance of sand F_s, which is in the opposite direction, i.e., $J = \gamma_w h A$.

The seepage force acting on unit volume of sand sample should be

$$j = \frac{J}{AL} = \frac{\gamma_w h A}{AL} = \gamma_w i \qquad (5\text{-}50)$$

Seepage force is a body force, with N/cm^3 or kN/m^3 as the unit. The direction of seepage force is consistent with the direction of seepage flow, the seepage force increases with the increase of hydraulic gradient.

Figure 5-13 shows the effect of seepage on a dam foundation. If the direction of seepage is from top to bottom, which is consistent with the direction of soil gravity (point 1 in Figure 5-13), the seepage force will compress the soil. On the other hand, if the direction of seepage is bottom-up, that is opposite to the gravity of the soil (point 4 in Figure 5-13), the seepage force will lift soil particles. The greater the seepage force is, the greater is the lift effect and the more unstable for the soil mass. Once the upward seepage force is greater than the soil's buoyant unit weight, the soil particles will be gushed up by seepage, which is the essence of

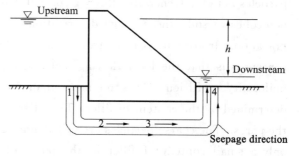

Figure 5-13 Sketch of seepage on a dam foundation

seepage deformation. Seepage is much likely to occur in the slope of dam foundation, earth dam and foundation pit, the effect of seepage force must also be taken into account in the analysis of stability and the mechanism of seepage deformation.

5.4.2 Basic form of seepage deformation

Seepage will bring two types of damage to a soil mass: one is the local stability failure of soil, i.e., the seepage water flows out and takes away the fine particles of soil resulting in deformation and failure of soil; another is the overall stability failure, i.e. the overall sliding or collapse of soil will occur under the action of seepage. Local stability failure, if not be carefully prevented, will lead to the overall stability failure. This section only deals with the problem of seepage induced local stability failure, often referred to seepage deformation. Seepage deformation is the phenomenon of deformation and failure of soil under seepage. It consists of two types: flowing soil and piping.

1) Flowing soil. Flowing soil or sand is a phenomenon that happened in the process of seepage. Particles in a certain range of soil are moved away at the same time. Flowing soil occurs mainly in the surface but not inside the soil body. The boiling sand phenomenon, which is often encountered when excavating a channel or a foundation pit, belongs to the this type of problem.

2) Piping. Piping is a phenomenon caused by fine particles in non-cohesive soil move out through the pores of coarse particles. It may occur at the surface or inside the soil, occurring mainly in sandy gravel soils that are lack of intermediate particle sizes.

In some soils, piping may happen when the hydraulic gradient is low. In other soils, even if the hydraulic gradient increase to the threshold of flowing soil, piping will not happen. The two modes of seepage deformation are different phenomena that are caused by action of the seepage force with a certain hydraulic gradient and the loss of soil particles through the soil. There is a fundamental difference between them where hydraulic gradient at failure is also different. It is necessary to explore the causes that affect them to distinguish the types in practice.

5.4.3 Determination of flowing soil and piping

The type of seepage deformation has a great relationship with the soil nature. Cohesive soil has cohesion between particles, where the structural connection between soil particles is tight. Generally, flowing soil will happen instead of piping in cohesive soil. For non-cohesive soil, the seepage deformation of sand or gravel are related to the particle gradation. In recent years, many scholars have studied this problem. The current consensus is that the local flowing soil and piping are more likely to happen when the uniformity coefficient C_u is no more than 10 in homogeneous sand, which is mainly determined by the content of filler. The filler refers to the small particles freely located in the pores. The coarse particles are called aggregates. The results show that when there is only a small content of filler in the pores of aggregates, a low hydraulic gradient is sufficient to yield piping due to the smaller resistance of filler. In contrast, if the content of filler in the aggregate pores increases to fullfil the pores of aggregates, such as well-graded soil, the flowing soil will happen instead of piping. The necessary condition for the piping is that the fine particles are free to move in the pores of the aggregate, otherwise flowing soil will occur. At present, there are some suggestions for determining the turn-point size for aggregate and filler: ① for the soil lack of intermediate grain size like sandy gravel, the particle size corresponding to the turn-point (or breakpoint) of the grain size distribution curve can be used as the turn-point size; ② for most well-graded soils, according to the measured particle size flowed away, 0.1mm can be used as the turn-point size to tell aggregate from filler roughly.

Seepage failure of soil is related to the hydraulic gradient. When the hydraulic gradient exceeds the allowable hydraulic gradient of soil, seepage failure will occur. The hydraulic gradient at which the soil begins to undergo seepage deformation is called the critical hydraulic gradient. It can be determined by seepage failure test (Figure 5-12) or by calculation. However, the calculation method is not accurate at present; therefore, the critical hydraulic gradient should be determined by experiment and /or practical observations for important engineering projects.

Ⅰ. Flowing soil

Taking a sand sample element with unit volume for analysis (Figure 5-12). The buoyant unit weight of soil is

$$\gamma' = (G_s - 1)\gamma_w /(1+e) = \gamma_{sat} - \gamma_w \tag{5-51}$$

When the vertical upward seepage volumetric force j is equal to the soil's buoyant unit weight, i.e.

$$j = i\gamma_w = (G_s - 1)\gamma_w /(1+e) = \gamma_{sat} - \gamma_w = \gamma' \tag{5-52}$$

The effective weight of the soil is zero, i.e., the soil is in the critical condition, resulting in flowing soil phenomenon. The hydraulic gradient at the critical condition obtained from Eq. (5-52) is

$$i_{cr} = (G_s - 1)/(1+e) = (G_s - 1)(1-n) = \gamma'/\gamma_w = (\gamma_{sat} - \gamma_w)/\gamma_w \tag{5-53}$$

where, i_{cr}—hydraulic gradient at critical state or critical hydraulic gradient;

G_s—specific gravity of soil solids;

e—void ratio;

n—porosity;

γ_{sat}—saturated unit weight;

γ'—buoyant unit weight;

γ_w—unit weight of water.

Equation (5-53) is proposed by Karl Terzaghi in 1948. It can be seen that when the void ratio or porosity and G_s are known, or γ' is known, the critical hydraulic gradient of soil is obtained, which is generally between 0.8 and 1.2. When designing a dam foundation or a earth-rock dam, the hydraulic gradient in the seepage area should be lower than the critical hydraulic gradient, and an appropriate factor of safety should be given. Because the duration of flowing soil is short and soil will fail quickly, the critical hydraulic gradient calculated by Eq. (5-53) should be divided by a factor of safety (usually 2~2.5) as the allowable hydraulic gradient. Some references show that the allowable hydraulic gradient is $[i]=0.27 \sim 0.44$ for homogeneous sand; for gravel material with filler content greater than 30%, $[i]=0.3\sim0.4$.

When using Eq. (5-53), it should be noted that it applies only to the non-cohesive soil where the flow is from the bottom to top and the soil is in unconfined condition. Many experiments show that the calculated critical hydraulic gradient is lower than the actual one, since it does not consider the constraint of the surrounding soil in the flowing soil region. If the constraint is taken into account, the calculated results are closer to the actual critical hydraulic gradient.

Due to the presence of cohesion in cohesive soil, the critical hydraulic gradient is large, especially with a protective layer as the boundary condition, where the critical hydraulic gradient is greatly increased. In addition, the mechanisms for flowing soil in non-cohesive and cohesive soil are different. For cohesive soil, it is not only the result of seepage force, but also related to the degree of hydration disintegration of soil surface (i. e. water stability) and the pore size at the outlet of the seepage flow. The water stability in soil is directly related to the content and composition of clay minerals in the soil. According to large number of testing data from China Institute of Water Resources and Hydropower Research, the critical hydraulic gradient of flowing soil due to upward seepage in cohesive soil is estimated by the following equation:

$$i_{cr} = \frac{24(1-n)}{[(1-n_L)-0.79(1-n)](1+CD_0^2)} \tag{5-54}$$

where, n—porosity;

n_L—porosity of soil at liquid limit;

D_0—average pore size at the seepage outlet in cohesive soil, it can be estimated by $D_0 = 0.63nd_{40}$, mm;

C—coefficient that reflects the hydration ability and hydration degree of soil, its value varies from 0.06 to 0.15.

The flowing soil generally occurs in the outlet (exit) of seepage. As long as the hydraulic gradient at exit (i_e) of the seepage is obtained, it is possible to judge the possibility of flowing soil: when $i_e < i_{cr}$, the soil is in a stable state; when $i_e = i_{cr}$, the soil is in the critical state; when $i_e > i_{cr}$, the soil is in the flowing state.

The way to determinate i_e is to get the average hydraulic gradient in the flow net grid at the seepage outlet. If the head loss of the grid at the exit of the seepage is Δh, the average length of the grid in the flow direction is ΔL, the hydraulic gradient at exit is:

$$i_e = \frac{\Delta h}{\Delta L} \tag{5-55}$$

To ensure the safety of the construction, the critical hydraulic gradient is divided by a factor of safety that is greater than 1 for the design as the allowable hydraulic gradient $[i]$.

Generally, the hydraulic gradient at exit i_e is lower than the allowable hydraulic gradient $[i]$, i. e.

$$i_e \leqslant [i] = \frac{i_{cr}}{F_s} \tag{5-56}$$

where, F_s—factor of safety, generally in the range of $2 \sim 3.5$.

II. Piping

Piping is the phenomenon that the fine particles in the soil are taken out along the pores of the skeleton particles. According to the equilibrium condition of soil particles, many scholars have given various calculation methods for the critical hydraulic gradient. China Institute of Water Resources and Hydropower Research gave an empirical equation of the critical hydraulic gradient for piping

$$i_{cr} = 2.2(G_s - 1)(1-n)^2 \frac{d_5}{d_{20}} \tag{5-57}$$

where d_5 and d_{20} are particle sizes related to 5% and 20% finer, respectively.

III. Measures to prevent seepage deformation

To prevent the seepage deformation, we can take measures from two aspects: first way is to reduce the hydraulic gradient, where it is possible to lengthen the seepage path or reduce the water head. There are many ways to reduce hydraulic gradients. For example, we can set up a relief well or ditch at the downstream of construction; we can extend the impervious length for tailing dam or hydraulic fill dam. Second way is to set the filter layer, thus water has a smooth way to flow out, or cover with permeable weights at the exit side, thus water can flow without carrying soil particles.

For embankment dams, the conventional types of seepage control and drainage features include impervious core, inclined/vertical filter with horizontal filter, network of inner longitudinal drain and cross drains, horizontal filter, transition zones/transition filters, intermediate filters, and rock toe. For example, the inclined or vertical filter abutting downstream face of either impervious core or downstream transition zone is provided to collect seepage emerging out of core/transition zone and thereby keeping the downstream shell relatively dry. In the eventuality of hydraulic fracturing of the impervious core, it prevents the failure of dam by piping. The horizontal filter collects the seepage from the inclined/vertical filter or from the body of the dam, in the absence of inclined/vertical filter, and carries it to toe drain. It also collects seepage from the foundation and minimizes the possibility of piping along the dam seat. When the filter material is not available in the required quantity at a reasonable cost, a network of inner longitudinal and inner cross drains is preferred to inclined/vertical filters and horizontal filters. This type of drainage feature is generally adopted for small dams, where the quantity of seepage to be drained away is comparatively small. When the transition zones/filters are placed on either side of the impervious core, they help to minimize failure by internal piping, cracking that may develop in the core or by migration of fines from the core material. The principal function of the rock toe is to provide drainage. It also protects the lower part of the downstream slope of an earth dam from tail water erosion.

5.4.4 Effective stress principle

I. Pore water pressure and effective stress in saturated soils

Due to the seepage effect, the water head in the soil will change, causing the change of pore water pressure. The pore water pressure (u) at a given point is constant in all directions with the value of

$$u = \gamma_w h \tag{5-58}$$

From Eq. (5-58), as long as the height (h) of the piezometric level at any given point is known, the pore water pressure can be obtained using the unit weight of water (γ_w) times h.

In practice, the sum of the contact forces at the interparticle surfaces of a study plane in the normal direction is marked as N_s, which is divided by the total area of the plane (including the interparticle contact area A_s and pore area) A. We can obtain the average stress. This stress is defined as the effective stress

$$\sigma' = \frac{N_s}{A} \tag{5-59}$$

As shown in Figure 5-14, we assume that the total area of a saturated soil element is A, which is under the action of normal stress (σ). The sum of the normal forces on the study plane is equal to the sum of the force acted on the pore water and the force acted on the interparticle contact surface, as shown in the following equation.

$$\sigma A = N_s + (A - A_s) u \tag{5-60}$$

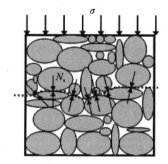

Figure 5-14 Schematic of stress transmission in soil mass

Substituting Eq. (5-59) into Eq. (5-60),

$$\sigma = \frac{N_s}{A} + \left(1 - \frac{A_s}{A}\right) u = \sigma' + (1 - \alpha) u \tag{5-61}$$

where, α is the ratio of the interparticle contact area and the total area.

Experimental studies have shown that the interparticle contact area A_s is very small, thus α is only a few percent. If we assume that $\alpha = 0$, Eq. (5-61) can be simplified as

$$\sigma = \sigma' + u \tag{5-62}$$

Eq. (5-62) is the effective stress principle of saturated soils, which was firstly proposed by Karl Terzaghi in the early 1920s. It shows the relationship between total stress, effective stress and pore water pressure. When the total stress remains unchanged, pore water pressure and effective stress can be transformed to each other. Generally, the total stress can be calculated and the pore water pressure can be measured or calculated. Therefore, the effective stress can be obtained by Eq. (5-62).

II. Pore water pressure and effective stress in hydrostatic condition

As shown in Figure 5-15, a clay layer with thickness of h_2 is above a sand layer. It is assumed that the clay layer is fully saturated and compressed under self-weight. The distance from the clay layer surface to the water surface is h_1. The total stress of soil on

a-a plane is

$$\sigma = \gamma_w h_1 + \gamma_{sat} h_2 \qquad (5\text{-}63)$$

Pore water pressure is

$$u = \gamma_w h_w = \gamma_w (h_1 + h_2) \qquad (5\text{-}64)$$

According to the principle of effective stress, the effective stress on a-a plane is

$$\sigma' = \sigma - u = (\gamma_w h_1 + \gamma_{sat} h_2) - \gamma_w (h_1 + h_2) = (\gamma_{sat} - \gamma_w) h_2 = \gamma' h_2 \qquad (5\text{-}65)$$

Thus, in hydrostatic conditions, the pore water pressure at a given depth is equal to the water unit weight γ_w times height of piezometric level at that depth, and the pore water pressure follows a triangle distribution, as shown in Figure 5-15b. The effective stress is proportional to the soil depth and dependent on buoyant unit weight γ'. It also follows a triangle distribution shown in Figure 5-15b.

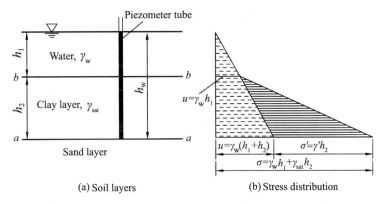

Figure 5-15　Pore water pressure and effective stress in hydrostatic condition

III. Pore water pressure and effective stress with seepage

If waterflow (seepage) occurs in the soil mass, the pore water pressure and the effective stress are different from that in hydrostatic condition. As shown in Figure 5-16a, the downward seepage will cause a water head loss (h) in piezometer. The pore water pressure on b-b plane is the same as that in hydrostatic condition, and the pore water pressure on a-a plane will decrease due to the loss of the water head.

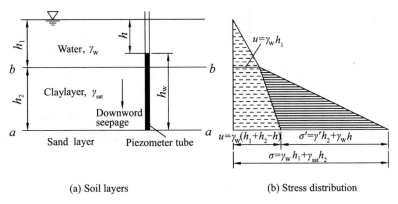

Figure 5-16　Pore water pressure and effective stress with downward seepage

$$u = \gamma_w h_w = \gamma_w (h_1 + h_2 - h) \tag{5-66}$$

where, h—water head loss between $b-b$ and $a-a$ plane.

The total stress and effective stress in the condition of downward seepage can be calculated according to effective stress principle. The total stress on $a-a$ plane remains the same as Eq. (5-63). Thus the effective stress on $a-a$ plane is obtained as

$$\sigma' = \sigma - u = \gamma' h_2 + \gamma_w h \tag{5-67}$$

The distribution of pore water pressure and effective stress are shown in Figure 5-16b.

Compared with the hydrostatic condition, the total stress on $a-a$ plane remains unchanged and the pore water pressure decreases when there is a downward seepage. The reduction of the pore water pressure transfers to the increment in effective stress since the total stress is unchanged.

When we analyze an upward water flow shown in Figure 5-17a, we can get an increment of water head due to upward seepage. The total stress on $a-a$ plane remains the same as Eq. (5-63). The while the pore water pressure is

$$u = \gamma_w h_w = \gamma_w (h_1 + h_2 + h) \tag{5-68}$$

Thus, according to the effective stress principle, the effective stress on $a-a$ plane is

$$\sigma' = \sigma - u = \gamma' h_2 - \gamma_w h \tag{5-69}$$

The distribution of pore water pressure and effective stress are shown in Figure 5-17b. Compared with the hydrostatic condition, when there is a upward seepage, the total stress on $a-a$ plane is unchanged, the pore water pressure increases with the decrease of effective stress. Thus, it indicates that the increase in pore water pressure is equal to the reduction in effective stress when the total stress is unchanged.

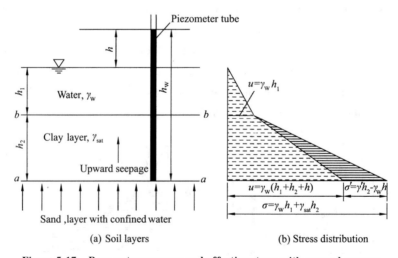

Figure 5-17 Pore water pressure and effective stress with upward seepage

If the water head increment h (Figure 5-17a) continues to increase until equal to $\gamma' h_2 / \gamma_w$, i.e., the effective stress reduced to zero. Eq. (5-69) can be rewritten as

$$\gamma' = \gamma_w \frac{h}{h_2} = \gamma_w i_{cr} \tag{5-70}$$

Eq. (5-70) is the same as the critical condition in the previous section about flowing soil. According to the principle of effective stress, it can be considered that the critical condition of the flowing soil is where the effective stress is equal to zero.

IV. Determination of pore water pressure and effective stress with flow net

According to the above definition of the pore water pressure, once the flow net is drawn, the pore water pressure at any point in the seepage field can be obtained by multiplying the water height (or pressure head) in the piezometer at that point and unit weight of water (γ_w).

Figure 5-18 shows the flow net in a homogeneous earth dam built on impermeable bedrock. Let's study the pore water pressures at three points (A, B, C) on an equipotential line. If the head loss on each grid is Δh, the equipotential line studied should have the piezometric water level lower than the upstream water level and the head loss is $2\Delta h$. Point A is the intersection between equipotential line studied and the seepage line. The seepage line is a zero pressure line when we assume that the atmospheric pressure is zero. Thus the height of the water head at point A should be zero, and the pore water pressure at point A is equal to zero. Points B and C are also located on the equipotential line studied, hence the piezometric water level of B or C should be the same with point A. Therefore, point B should have the water head of h_b as shown in Figure 5-18. The pore water pressure at point B is equal to $\gamma_w h_b$. Similarly, the height of water head at point C is h_c, and the pore water pressure is equal to $\gamma_w h_c$. It should be noted that when the calculated point is below the downstream water table, for example, point C. The pore water pressure calculated from the height of the water head is composed of two parts: one is the pore water pressure generated by the downstream water table and this part of the pore water pressure is usually referred to as static pore water pressure ($=\gamma_w h_2$); the second part is caused by the seepage, that is, the excess part of pore water pressure than static pore water pressure. This part of the pore water pressure is called ultra-static pore water pressure or excess pore water pressure [$\gamma_w(h_c - h_2)$]. It should be noted that for steady seepage, soil pore water pressure would not change overtime and the pore water pressure is hydrostatic. While for unsteady seepage, it causes excess pore pressure, which can be positive or negative, resulting in the corresponding changes in the effective stress of soil.

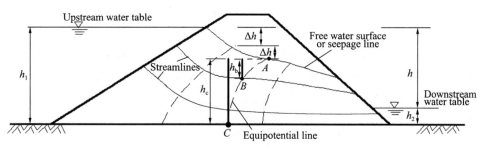

Figure 5-18　Flow net in a homogeneous earth dam

It should also be noted that the excess pore water pressure of the soil can not only be caused by unsteady seepage, but also caused by the external loading. The excess pore water pressure caused by loading will change over time, which also follows the effective stress principle.

【Example 5-1】 As shown in Figure 5-19, if the specific gravity of soil G_s is 2.68 and the porosity n is 38.0%, please solve: (1) the pore water pressure and effective stress at point a; (2) does the flowing soil occur at exit surface 1-2? (3) the seepage force on grid 9, 10, 11, 12.

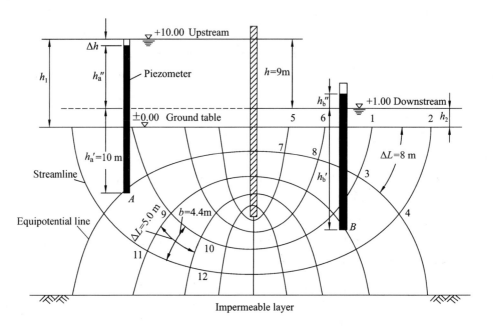

Figure 5-19 Seepage flow under a reservoir gate foundation

【Solution】

(1) According to Figure 5-19, the difference of water level between upstream and downstream is $h=9$m. The number of equipotential line interval $N=10$, then the head loss between adjacent two equipotential line $\Delta h = \dfrac{h}{10} = \dfrac{9}{10} = 0.9$m. The point A is on the second equipotential line, therefore, the water head loss with respect to the water level of upstream $\Delta h = 0.9$m. We can get $h_a'' = h_1 - \Delta h = 9 - 0.9 = 8.1$m. From Figure 5-19, we have $h_a' = 10$m. The height of piezometric level at point A is $h_w = h_a' + h_a'' = 10 + 8.1 = 18.1$m. Therefore, the pore water pressure at point A is: $u = \gamma_w h_w = 9.8 \times 17.2 = 168.56$kPa.

The static pore water pressure caused by the downstream water level is
$$u' = \gamma_w h_a' = 9.8 \times 10 = 98 \text{kPa}$$

The excess pore water pressure caused by seepage is
$$u'' = \gamma_w h_a'' = 9.8 \times 7.2 = 70.56 \text{kPa}$$

The total stress at point A is $\sigma = \gamma_w h_1 + \gamma_{sat}(h_a' - h_2)$.

The saturated unit weight is

Chapter 5 Permeability and Seepage

$$\gamma_{sat} = \rho_w[1+(G_s-1)(1-n)] \times 9.8 = [1+(2.68-1)(1-0.38)] \times 9.8 = 20\text{kN/m}^3$$

The total stress is

$$\sigma = 9.8 \times 10 + 20 \times (10-1) = 98 + 180 = 278\text{kPa}$$

Thus, according to the effective stress principle, the effective stress at point A is

$$\sigma' = \sigma - u = 278 - 168.56 = 109.44\text{kPa}$$

(2) As shown in Figure 5-19, the average seepage length for grid 1, 2, 3, 4 is $\Delta L = 8\text{m}$, and the head loss for any grid is $\Delta h = 0.9\text{m}$, then the average hydraulic gradient is

$$i = \frac{\Delta h}{\Delta L} = \frac{0.9}{8} = 0.1125$$

The gradient is approximately equal to the hydraulic gradient at exit surface 1-2, i_e. The critical hydraulic gradient of the soil is

$$i_{cr} = (G_s - 1)(1-n) = (2.68-1)(1-0.38) = 1.04 > i_e$$

Therefore, the flowing soil will not occur at exit surface 1-2.

(3) For grid 9, 10, 11, 12, the average seepage length $\Delta L = 5.0\text{m}$, and the average distance $b = 4.4\text{m}$ between the two streamlines. The head loss $\Delta h = 0.8\text{m}$, so the seepage force acting on the grid is

$$J = \gamma_w \frac{\Delta h}{\Delta L} b \Delta L = \gamma_w b \Delta h = 9.8 \times 4.4 \times 0.8 = 34.5\text{kN/m}$$

Knowledge expansion

Traffic Flow and Water Flow: Another Way to Consider Flow Net

Figure 5-20a shows the flow net of smooth traffic and uniform seepage field. Figure 5-20b shows the flow net of block traffic and seepage field in the turbulent environment. It can be seen that the equipotential lines narrowed at the middle section, indicating that the pressure gradient reached a maximum. But how is the pressure represented by two contactless vehicles? It is not the change in the potential difference in physics sense, but it depends on driver's psychological condition. Because of the reflection on potential difference in driver's mind, the speed of vehicle will change.

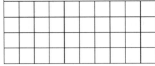

(a) Uniform flow condition   (b) Turbulent flow condition

Figure 5-20 Schematic of traffic flow and water flow net

Imagine that the two vehicles crash on a four-lane city road, two lanes will be occupied, leading to a rapid rise in degree of block traffic. The distance between two vehicles passing on this accident site would be 2.0m, but the distance behind the accident site would be 0.2m, indicating the speed of the vehicle would increase after passing through the accident site. As shown in Figure 5-20, the characteristics of the traffic flow net and water flow net with same boundary conditions are very similar.

Flow velocity varies directly with the potential difference and inversely with the resistance. The water flow, traffic flow, pedestrian flow, blood flow and the operation of society are all follow this rule. Moreover, the flow of stars in the universe and the forms that substance exists (such as particles, anti-matters, anti-particles, and dark matters) also follows this universal law.

Exercises

[5-1] What is Darcy's Law? What assumptions does Darcy's Law have?

[5-2] Is the flow velocity calculated from Darcy's Law the same with the actual flow velocity in soil? Why?

[5-3] What are the characteristics of the permeability coefficient of natural soils? Please introduce the reason.

[5-4] Why we do falling head test and constant head test? What are the advantages and disadvantages of field permeability testing and laboratory testing?

[5-5] What is a flowing soil? What is a piping? What are the necessary conditions for producing flowing soil and piping?

[5-6] What are the characteristics of seepage deformation?

[5-7] What is the pore water pressure? What is excess pore water pressure? What is the difference between pore water pressure under hydrostatic condition and steady seepage? How to determine the pore water pressure under steady seepage using flow net.

[5-8] How to define the critical hydraulic gradient? How to determine the critical hydraulic gradient of soil?

[5-9] A soil sample is with a length of 40cm, the cross-sectional area of 124cm^2, the soil samples at both ends are fixed and the head difference is 75cm, there is water flowing out of soil at 92cm^3/min. What is the permeability coefficient k of the soil sample?

[5-10] A soil sample 10 cm in diameter is placed in a tube 1 m long. A constant supply of water is allowed to flow into one end of the soil at A and the outflow at B, as shown in Figure 5-21. The average amount of water collected is 1 cm^3 for every 10 seconds. The tube is inclined as shown in the figure. Determine the (a) hydraulic gradient, (b) flow rate, (c) average flow velocity, (d) seepage velocity, if $e = 0.6$, and (e) permeability coefficient.

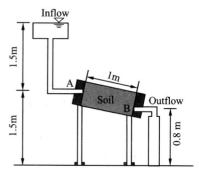

Figure 5-21 Schematic diagram of Exercise [5-10]

Chapter 6 Stresses in Soil

6.1 Introductory case

Stress is the force per unit area on a body. Stress in soil is a measure of the internal forces in a soil mass between soil particles. These internal forces are a reaction to the external loads. Soil is a particulate system with individual particles bonded together by weak or negligible cementing or cohesive forces. Hence, when external loads are applied on a soil mass from the structure, the soil particles tend to undergo displacement relative to one another. Most of the loads from earth or earth-supported structures on soils cause compression of the soil. Soils deform under the load of foundation structures. The total vertical displacement that occur at foundation level is termed as settlement. The cause of foundation settlement is the reduction of volume air void ratio in the soil. Structures will suffer damages due to settlement of its foundation. Settlement that occurs at different rates between different portions of a building is termed differential settlement. Differential settlement occurs if there is difference in soils, loads, or structural systems between parts of a building. Consequently, the frame of the building may become distorted, floors may slope, walls and glass may crack, and doors and windows may not work properly. Figure 6-1 is an example.

Figure 6-1 Tilt of a building due to the differential settlement

Engineering Geology and Soil Mechanics

In engineering, a foundation is the element of a structure which connects it to the ground, and transfers loads from the structure to the ground. The superstructure, foundation and ground soil form a whole working system. Usually, the soil layer with a certain thickness under the foundation of a building is called bearing stratum, and the soil layer under the bearing stratum is called underlying soil layer. To calculate foundation settlement, it is required to estimable the vertical stress increase in the soil mass due to the net load applied on the foundation.

According to the effective stress principle, the total stress in a soil mass can be divided into effective stress and pore pressure; according to the causes, the stress in a soil mass can be divided into self-weight stress and induced stress.

① Effective stress. It is defined as the total force in the soil grains divided by the gross cross-sectional area over which the force acts. Effective stress carried by the grain skeleton of the soil. When a load is applied to soil, it is carried by the water in the pores as well as the solid grains. The increase in pressure within the pore water causes drainage (flow out of the soil), and the load is transferred to the solid grains. The rate of drainage depends on the permeability of the soil. The strength and compressibility of the soil depend on the stresses within the solid granular fabric. These are called effective stresses.

② Pore pressure. It is the pressure carried by pore fluid (air and water) in soil voids. Fore saturated soils, pore pressure mainly refers to the pore water pressure. Sometime it is otherwise termed as neutral stress. The pressure developed in pore are depends upon the depth of ground water and seepage flow condition. Pore water pressure may consist of hydrostatic pressure, capillary pressure, seepage or pressure resulting from applied loads to soils which drain slowly.

③ Self-weight stress, also known as geostatic stress. It is defined as the vertical stress produced by the self-weight of overlying soil. In most cases, the soil has been compressed and becomes stable under its own weight, and the self-weight stress of the soil is the effective stress carried by the soil skeleton. For newly deposited soil or recent artificial fill, part of the weight of the soil is carried by the soil skeleton, and the other part is carried by the pore fluid. The soil has not been compressed and stabilized under its own weight. When the pore pressure carried by the pore fluid is converted into effective stress, the soil will be further compressed until it becomes stable.

④ Induced stress. It is defined as the vertical stress increase or additional stress in the soil mass due to the external loads applied on the ground. The sources of the external loads include the construction of new buildings, surface surcharge load by large-scale filling or excavation, earthquake or vibration, groundwater table change, seepage, etc. In a broad sense, when the soil is in a relatively stable state, any stress increased in the soil mass due to external loads can result in an induced or additional stress. When the soil has been subjected to the act of external load in the past and the deformation of the soil has been finished, the stress in the soil can be called the in-situ stress, which is an effective stress.

Chapter 6 Stresses in Soil

When the soil has not been subjected to the action of external load, the in-situ stress in the soil mass is the self-weight stress. Any stress increase in the soil mass (induced stress) produced by external load change can be divided into effective stress and pore pressure. It should be noted that the induced or additional stress is the main external cause of change of deformation and strength of the soil mass, no matter whether the external cause is loading or unloading. The magnitude of the induced stress is related to the size and distribution of the external load, such as a point load, circularly or triangularly loaded area, vertical line load, strip load, etc. This chapter mainly introduces the calculation method of self-weight stress and induced stress.

6.2 Self-weight stress in the ground

6.2.1 Natural state of ground

Natural ground with a horizontal surface can be assumed as a semi-infinite body, as shown in Figure 6-2. It can be found that any vertical plane is a symmetric plane and any vertical line is also a symmetric axis. The stresses acting on any plane at a point in a soil mass can be resolved into normal and tangential (shear) components. Figure 6-3 shows the stresses in the element of a soil mass. It can be found that there are three normal stresses and six shear stresses. The plane on which the shear stress is zero is called the principal plane and the normal stress on the principal plane is called the principal stress. There are three principal planes at any point in a soil mass, which are mutually perpendicular, and hence three principal stresses on these principal planes. In the semi-infinite ground, the horizontal ground surface can be regarded as one principle plane since the shear stress on the ground surface is zero. Therefore, the stresses applied on two symmetric vertical planes are also principle planes. Thus, every point in a soil mass is subjected to three principal stresses due to external loads. The maximum principal stress is called the major principal stress and the minimum principal stress is called the minor principal stress. For the ground soil, the vertical stress at any depth below the ground surface produced by the overlying soil self-weight is the major principal stress, designated by the symbol σ_{sz}. Two horizontal stresses, designated by the symbols σ_{sx} and σ_{sy}, are minor principal stresses.

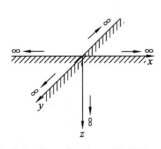
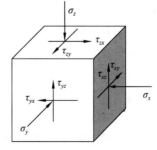

Figure 6-2 Concept of semi-infinite ground   Figure 6-3 Stresses in an element of soil mass

Since the natural ground is assumed to be semi-infinite, the horizontal stresses in any direction at the same depth are equal ($\sigma_{sx} = \sigma_{sy}$). We can get that the deformation of soil mass in the initial state at any depth is $\varepsilon_x = \varepsilon_y = 0$. In the case of no external load, the natural soil is under the geostatic state, also known as no lateral deformation state. The ratio of the lateral stress σ_{sx} or σ_{sy} and the vertical stress σ_{sz} of the soil mass at a given depth is

$$K_0 = \frac{\sigma_{sx}}{\sigma_{sz}} \tag{6-1}$$

K_0 is the coefficient of earth pressure at rest. If the soil layer conditions are determined, then the value of K_0 is a constant. We can obtain K_0 values through K_0 tests. Table 6-1 lists the typical ranges of K_0 for different type of soils.

Table 6-1 Values of K_0 for different soils

Soil category	Loose sand	Dense sand	Compacted clay	Normally consolidated clay	Overconsolidated clay
Range of K_0	0.40~0.45	0.45~0.50	0.80~1.50	0.50~0.60	1.00~4.00

6.2.2 Calculation of self-weight stress

The vertical stress of soil at a point below certain depth of ground surface is caused by the effective weight of the overlying soil above this point. It is the vertical effective self-weight stress, designated by the symbol σ_{sz}.

For homogeneous soil layer above the groundwater table, as shown in Figure 6-4, the pore water pressure at depth z is equal to 0. At this time, the weight of the overlying soil layer is assumed taken by the soil skeleton, i.e., the self-weight stress is equal to the total stress.

$$\sigma_{sz} = \gamma \cdot z \tag{6-2}$$

where, σ_{sz}—self-weight stress, kPa;
γ—unit weight of soil, kN/m³;
z—the depth of the interest point, m.

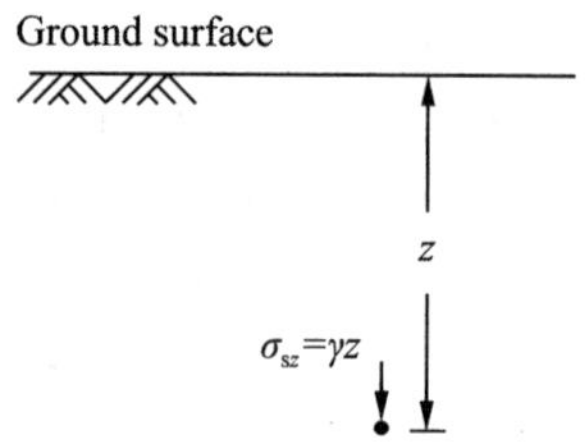

Figure 6-4 Self-weight stress in homogeneous soil layer

According to Eq. (6-2), the self-weight stress of soil increases linearly with depth.

In homogeneous soil layer submerged in water, the self-weight stress σ_{sz} is calculated by the effective weight of overlying soil. As shown in Figure 6-5, assuming the water table is above the soil layer surface at depth of h_0, the self-weight stress at depth z is

$$\sigma_{sz} = \gamma' \cdot z \tag{6-3}$$

According to the effective stress principle, the pore water pressure and the total stress are calculated respectively. Then the pore water pressure is subtracted from the total stress, thus the effective self-weight stress is obtained. The hydrostatic pressure

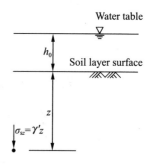

Figure 6-5 Self-weight stress in homogeneous soil layer submerged in water

of water is called static pore water pressure, noted as u_0. According to the hydraulic principle, the pore water pressure at a given point is equal in all directions. As shown in Figure 6-5, assuming that the water table is horizontal and above the soil layer surface with height of h_0, thus the pore water pressure u_0 at depth z is

$$u_0 = \gamma_w(z+h_0) \tag{6-4}$$

The total stress σ_{szt} at depth z is

$$\sigma_{szt} = \gamma_{sat} z + \gamma_w h_0 \tag{6-5}$$

The self-weight stress σ_{sz} at depth z is obtained using the effective stress principle

$$\sigma_{sz} = \sigma_{szt} - u_0 = \gamma_{sat} z + \gamma_w h_0 - \gamma_w(z+h_0) = \gamma' z \tag{6-6}$$

where, γ_{sat} — saturated unit weight of soil, kN/m^3;

γ_w — unit weight of water, approximately $9.8 kN/m^3$;

γ' — buoyant unit weight of soil, kN/m^3.

For a soil strata composed of multilayers, if the total depth is $h_1 + h_2 + \cdots + h_n$ and all above the groundwater table, the self-weight stress of the interest point can be expressed as

$$\sigma_{sz} = \gamma_1 h_1 + \gamma_2 h_2 + \cdots + \gamma_n h_n = \sum_{i=1}^{n} \gamma_i h_i \tag{6-7}$$

where, γ_i — unit weight of soil for i^{th} layer;

h_i — the thickness of ith layer.

For a condition with some soil layers above the groundwater table and some layers below the water table, the self-weight stress at point M in Figure 6-6 is

$$\sigma_{sz} = \gamma_1 h_1 + \gamma_2 h_2 + \gamma'_3 h_3 \tag{6-8}$$

Considering the effects groundwater and soil layers, the self-weight stress σ_{sz} at any point below the ground surface can be expressed as

$$\sigma_{sz} = \sum_{i=1}^{n} \gamma_i h_i \tag{6-9}$$

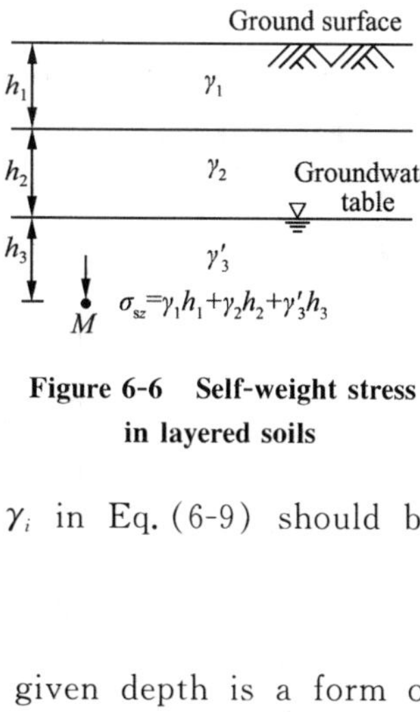

Figure 6-6 Self-weight stress in layered soils

Note that for soil mass below the groundwater table, γ_i in Eq. (6-9) should be replaced by γ'.

6.2.3 Discussion on self-weight stress

According to the definition, the self-weight stress at a given depth is a form of effective stress caused by the effective weight of the overlying soil. In application, the concept should be clarified, and the following situations need to be paid attention:

(1) Division of weight of overlying soil and water into self-weight stress (effective stress) and pore water pressure depends on the position of the groundwater table or the flow field induced by seepage. For static water condition, effective stresses at any point below the groundwater level may be computed using the total unit weight of soil above the water level and buoyant unit weight below the water level. Pore water pressure is equal to

the static head times the unit weight of water. If there is steady seepage, pore pressure is equal to the piezometric head times the unit weight of water, and the effective stress is obtained by subtracting the pore water pressure from the total stress.

(2) Eq. (6-9) can be used to directly calculate the self-weight stress at any point below or above the groundwater level when there is an unconfined aquifer in the ground. We can also calculate the total stress and pore water pressure and indirectly get the self-weight stress by subtracting the pore water pressure from the total stress according to the principle of effective stress.

(3) If there is perched aquifer above the calculation point, the saturated unit weight was used to obtain the self-weight stress for the soil layer with perched aquifer and the confining soil layer below the perched aquifer. If the calculation point is in the perched aquifer, Eq. (6-9) can be used to calculate the self-weight stress.

(4) When the calculation point is below the confined aquifer, we can first calculate the pore water pressure distribution and then calculate the total stress distribution along depth. Note that saturated unit weight was used to obtain the total stress below the water table. The self-weight stress is obtained by subtracting the pore water pressure from the total stress.

(5) For the newly deposited soil or new fills, we need to consider the effect of stress history on the self-weight stress. Since the soil mass is unstable and the soil layer is still under consolidation, the self-weight stress may be less than the buoyant unit weight times the depth of the calculation point.

【Example 6-1】 The unit weights of layered soils with groundwater table are shown in Figure 6-7, and the thickness of each soil layer is also marked in the figure (assuming $\gamma_w = 10 \text{kN/m}^3$).

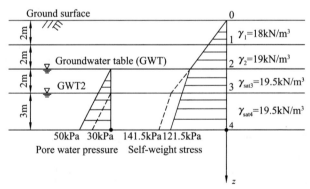

Figure 6-7 Schematic diagram of [Example 6-1]

(1) Calculate the self-weight stress and pore water pressure on the interface of soil layer, and plot the stress distribution along the depth;

(2) If the groundwater table drops by 2m, plot the distribution of self-weight stress and pore water pressure along the depth.

[Solution]

(1) Before the drop of the groundwater table:

The self-weight stress on the interface of soil layers is calculated respectively as follows

$$\sigma_{s0} = 0$$
$$\sigma_{s1} = \gamma_1 h_1 = 18 \times 2 = 36 \text{kPa}$$
$$\sigma_{s2} = \gamma_1 h_1 + \gamma_2 h_2 = 36 + 19 \times 2 = 74 \text{kPa}$$
$$\sigma_{s3} = \gamma_1 h_1 + \gamma_2 h_2 + \gamma_3' h_3 = 74 + 9.5 \times 2 = 93 \text{kPa}$$
$$\sigma_{s4} = \gamma_1 h_1 + \gamma_2 h_2 + \gamma_3' h_3 + \gamma_4' h_4 = 93 + 9.5 \times 3 = 121.5 \text{kPa}$$

The pore water pressure can be obtained by

$$u_{03} = \gamma_w h_w = 10 \times 2 = 20 \text{kPa}$$
$$u_{04} = \gamma_w h_w = 10 \times (2+3) = 50 \text{kPa}$$

(2) After the drop of groundwater table:

$\sigma_{s0}, \sigma_{s1}, \sigma_{s2}$ is unchanged and we can get

$$\sigma_{s3} = \gamma_1 h_1 + \gamma_2 h_2 + \gamma_3 h_3 = 74 + 19.5 \times 2 = 113 \text{kPa}$$
$$\sigma_{s4} = \gamma_1 h_1 + \gamma_2 h_2 + \gamma_3 h_3 + \gamma_4' h_4 = 113 + 9.5 \times 3 = 141.5 \text{kPa}$$
$$u_{00} = u_{01} = u_{02} = u_{03} = 0$$
$$u_{04} = 10 \times 3 = 30 \text{kPa}$$

Self-weight stress and pore water pressure distribution before (solid line) and after (dash line) the drop of groundwater table are plotted in Figure 6-7. From the distribution of stress along depth, we know that: ① the self-weight stress of each layer of soil varies linearly with increasing depth; ② different stress distribution is obtained in each layer where a turning point is presented at the interface due to the different unit weight for each layer; ③ the distribution of pore water pressure is linear with depth, regardless of soil properties; ④ the increase in self-weight stress caused by the drop of the groundwater level is equal to the decrease of pore water pressure. Therefore, the rise or drop of water table will cause decrease or increase in the soil self-weight stress. The decrement or increment of soil self-weight stress due to the change of water table can be regarded as induced stress. For example, long-term pumping of groundwater in some areas will result in the lowering of groundwater table, thus increasing the self-weight stress. This is equivalent to apply an induced stress in the soil mass, resulting in the land subsidence.

6.3 Contact pressure under foundations

A foundation usually transmit load to soil underneath, a response of which soil will exert a reaction pressure to the foundation at the contact surface between the soil and foundation. The stability of structure is majority depends upon soil-foundation interaction. Even though they are of different physical nature, they both must be act together to get required stability. It is important to know about the contact pressure developed between soil and foundation and its distribution in different conditions. Generally loads from the structure are transferred to the soil through foundation or footing. A reaction to this load, soil exerts an upward pressure on the bottom surface of the footing which is termed as contact pressure.

6.3.1 Distribution of contact pressure

The distribution of contact pressure is related to a variety of factors. Obviously, contact pressure distribution is affected the load properties (magnitude, distribution, etc.) acting on the foundation. The experimental and theoretical studies have shown that the distribution of contact pressure is also related to the stiffness of foundation, shape and depth of embedment. In addition, the nature of the soil also influences the distribution of the contact pressure.

The two extreme conditions for foundations are flexible foundation (stiffness $EI=0$) and rigid foundation (stiffness $EI=\infty$). For a foundation with small stiffness, it can be regarded as a flexible foundation, where the contact pressure distribution and magnitude are identical to the load distribution and magnitude acting on foundation surface. For example, the foundations of earth dams, embankments and large diameter oil or water tanks can be regarded as flexible foundations. The contact pressure distribution is exactly the same as that of the load distribution, as shown in Figure 6-8.

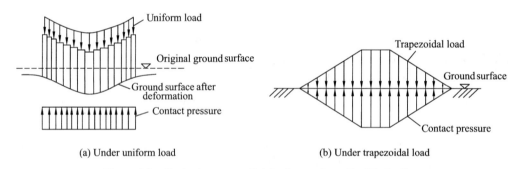

Figure 6-8 Contact pressure distribution under a flexible footing

For a foundation with large stiffness, it can be regarded as a rigid foundation, such as pier foundation, box foundation, concrete dam, concrete retaining wall, etc. The contact pressure distribution of rigid foundation and magnitude are related to the soil properties. For example, in the construction of rigid strip foundation resting on sandy soil, if the foundation is under a concentrated load, the contact pressure is maximum at center and gradually reduces to zero at edges because sandy soil has no cohesion, which is similar to a parabolic distribution. When the load continues to increase until the ultimate load, the distribution of contact pressure is still like a parabola while the peak value increases significantly, as shown in Figure 6-9a. For a rigid strip foundation built on cohesive soil under a concentrated load, the settlement is uniform but contact pressure varies, the contact pressure at edges is maximum and at center is minimum when the load is small. The distribution of contact pressure is similar to an inverted bowl shape. When the load increases, the values of stresses at edges gradually increase and become finite when plastic flow occurs in real soils. The shape of distribution changes from the inverted bowl shape to parabolic shape as shown in Figure 6-9b. This means the actual contact pressure is generally not a linear distribution because the foundation is not absolutely rigid.

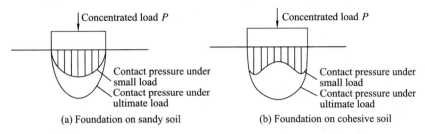

Figure 6-9 Contact pressure distribution under a rigid footing

Note that the foundation size is small compared to the semi-infinite ground. According to Saint-Venant principle, the error of contact pressure in the calculation is acceptable when we assume a linear distribution of contact pressure. This is a simplified assumption that is often used in conventional design of foundation.

In engineering, the real rigid foundation does not exist, and the stiffness of foundation is generally between the two extreme conditions, which is called an elastic foundation. The strip foundation, raft foundation, dam foundation, box foundation, etc. can be slightly bent after loading, where the stress redistribution will occur. Thus they can be regraded as elastic foundation. More precise methods can be used such as Winkler elastic foundation approach. In this section, we only introduce the calculation method of contact pressure under rigid foundation based on the assumption of linear distribution.

6.3.2 Contact pressure distribution under rigid foundation

I. Contact pressure under vertical concentrated load

Figure 6-10 shows the foundation under the action of vertical concentrated load P that transferred from the superstructure to the foundation surface. The contact pressure p at the contact surface between the soil and foundation is assumed as uniformly distributed, which can be written as

$$p = \frac{P+G}{A} = \frac{F_v}{A} \qquad (6\text{-}10)$$

where, F_v—sum of P and G;

P—load transferred from the superstructure to the foundation surface;

A—area of the foundation base for a rectangular shape, $A = l \cdot b$, l is length and b is the breadth;

G—self-weight of foundation (including the weight of backfill soil on the foundation).

As we known, $G = \gamma_G A d$, where γ_G is the average unit weight of foundation and backfill soil. $\gamma_G = 20 \text{kN/m}^3$ is generally used but the buoyancy shall be subtracted for the part below the groundwater level; d is the buried depth calculated from the designed ground surface; when the ground elevation on both sides of the foundation is different, the average value of buried depth shall be taken.

When the ratio of foundation length and breadth $l/b \geqslant 10$, it is called a strip foundation. If the load is uniformly distributed along the length, it can be considered as a plane condition. The calculation of contact pressure only needs considering a unit length, i.e., $l = 1\text{m}$ where the contact pressure along the length is the same. Eq. (6-10) can aslo be used to compute the contact pressure of strip foundation, where $l=1\text{m}$, $A=b$; P or G with the unit of kN/m, which is the load or weight per unit length.

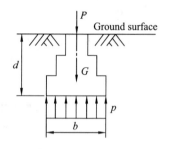

Figure 6-10 Distribution of contact pressure under vertical concentrated load

II. Contact pressure under vertical eccentric load

As shown in Figure 6-11, a rectangular rigid foundation with a vertical eccentric load F_v, the base pressure is not uniform with maximum and the minimum pressure at edges are p_{max} and p_{min} respectively. Based on the assumption of linear distribution of base pressure, we can calculate the contact pressure at two edges according to eccentric loading equation in material mechanics.

$$p_{\substack{max \\ min}} = \frac{F_v}{lb} \pm \frac{M}{W} \qquad (6\text{-}11)$$

where, F_v—vertical load on the foundation base;

M—moment acting on a rectangular base;

W—resistance moment of the cross section of rectangular base plane, $W = \dfrac{lb^2}{6}$;

b—breadth of foundation;

l—length of foundation.

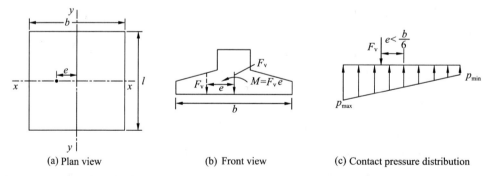

(a) Plan view (b) Front view (c) Contact pressure distribution

Figure 6-11 Distribution of contact pressure for a rectangular foundation under vertical eccentric load

Define the load eccentricity $e = \dfrac{M}{F_v}$, Eq. (6-11) can be converted to

$$p_{\substack{max \\ min}} = \frac{F_v}{lb}\left(1 \pm \frac{6e}{b}\right) \qquad (6\text{-}12)$$

Based on Eq.(6-12), if $e < \dfrac{b}{6}$, the distribution of the contact pressure is in a trapezoidal shape, as shown in Figure 6-11c. If $e = \dfrac{b}{6}$, then $p_{min} = 0$. If $e > \dfrac{b}{6}$, then the minimum contact pressure will be smaller than 0, i. e., tensile pressure occurs. However, the contact pressure between the foundation base and subsoil will not be tensile pressure. Thus the contact pressure will show redistribution. We should avoid this condition occurring in the design. Sometimes this condition may happen for very high building foundation. In this case, according to the condition that the total vertical load on the foundation base is equal to the total contact pressure, the maximum contact pressure at the the base edge is

$$p_{max} = \frac{2F_v}{3al} \qquad (6\text{-}13)$$

Where $a = b/2 - e$ and $e > \frac{b}{6}$. The minimum contact pressure is 0 at a distance of $3a$ from base edge. Besides, for a strip foundation, we can take one meter along the length direction to calculate the contact pressure, i. e., $l = 1\text{m}$ in Eq. (6-12).

For a rectangular foundation base under a two-way eccentric load F_v (Figure 6-12), the contact pressure at any point (x, y) of the foundation base can be obtained following biaxial eccentric compression equation in material mechanics as below

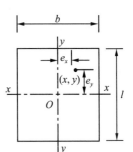

Figure 6-12 A rectangular foundation due to a two-way eccentric load

$$p_{(x,y)} = \left(\frac{F_v}{lb} \pm \frac{M_x}{W_x} \pm \frac{M_y}{W_y}\right) = \frac{F_v}{lb}\left(1 \pm \frac{6e_y}{l} \pm \frac{6e_x}{b}\right) \qquad (6\text{-}14)$$

where, M_x—moment to the x axis caused by the eccentric load, $M_x = F_v \cdot e_y$;

M_y—moment to the y axis caused by the eccentric load, $M_y = F_v \cdot e_x$;

e_x—eccentricity along the x axis;

e_y—eccentricity along the y axis.

III. Contact pressure under inclined eccentric load

As shown in Figure 6-13, when the foundation is subjected to an inclined eccentric load R, which can be decomposed into the vertical component F_v and the horizontal component F_h. $F_v = R\cos\beta$ and $F_h = R\sin\beta$. β is the angle between the directions of the vertical and inclined loads. The horizontal contact pressure can be obtained.

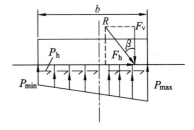

Figure 6-13 Distribution of contact pressure under inclined eccentric load

For rectangular foundation $\qquad p_h = \frac{F_h}{lb} \qquad (6\text{-}15)$

For strip foundation $\qquad p_h = \frac{F_h}{b} \qquad (6\text{-}16)$

The contact pressure in horizontal direction is always assumed as a uniform distribution.

Ⅳ. Induced stress at base

Before constructing a building, the self-weight stress of natural ground already exists. In general, the deformation caused by the self-weight stress is completed. Therefore, only the induced stress(vertical stress increase) caused by the building load can cause the new deformation of the ground soil. In engineering practice, the foundation is always embedded below the ground surface for a certain depth (at least 0.5m), since we need to consider the effects of atmosphere, temperature, animals, plants and other factors on the foundation stability and durability. Accordingly, foundation excavation is needed to remove the soil.

For a shallow foundation, the self-weight stress is calculated from the original ground, and the induced stress is obtained from the bottom of the foundation. Therefore, the removed soil due to excavation can be regarded as a vertical pressure unloading that is equivalent to apply a uniform vertical pressure $\sigma_c = \gamma_0 d$. Before calculating the induced stress of the foundation, the negative load should be superimposed with the contact pressure, which is called the net induced stress of the foundation. It is a simplified approximate calculation method, because the partial excavated soil mass does not belong to the semi-infinite space unloading.

When the contact pressure is uniformly distributed, as shown in Figure 6-14a, the net induced stress at base is

$$p_n = p - \sigma_c = p - \gamma_0 d \tag{6-17}$$

When the contact pressure follows a trapezoidal distribution, as shown in Figure 6-14b, the maximum and minimum net induced stresses are

$$(p_n)_{max} = p_{max} - \gamma_0 d \tag{6-18}$$

$$(p_n)_{min} = p_{min} - \gamma_0 d \tag{6-19}$$

where, γ_0 — average unit weight of natural soil;

d — depth of excavation;

p_n — net induced stress at base (uniformly distributed);

p — contact pressure at base (uniformly distributed);

$(p_n)_{max}$ — maximum net induced stress at base edge;

$(p_n)_{min}$ — minimum net induced stress at base edge.

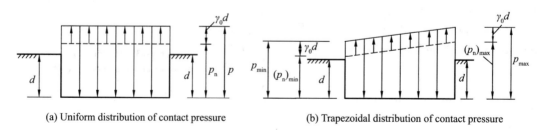

(a) Uniform distribution of contact pressure   (b) Trapezoidal distribution of contact pressure

Figure 6-14   Distribution of induced stress at base

Under the same building load, the larger the embedment depth of foundation is, the smaller is the induced stress at foundation base. Thus increasing the depth will reduce the settlement of foundation. When the load transferred from superstructure plus foundation weight is equal to the weight of the excavated soil, the induced stress at base is zero, thus the ground deformation is zero if there is no rebound of ground soil. When the ground soil rebounds significantly, it can use the approximate method for calculating the induced stress at base, where $\alpha \gamma_0 d$ can be used instead of $\gamma_0 d$. The coefficient α represents the degree of rebound, which is in range of 0 to 1 according to the experience.

Note that when the excavated section area is large and the time after the excavation is long, the ground soil will rebound significantly. In this condition, the self-weight stress of soil and induced stress should be calculated from the bottom of the foundation. The influence of excavation depth is not considered at this time.

6.4 Induced stress calculation for a space problem

The calculation of induced stress of ground soil can be divided into two problem types: space problem and plane problem. If the induced stress in the ground soil is a function of three-dimensional coordinates (x, y, z), it is called a space problem. For example, the calculation of induced stress under a rectangular foundation ($l/b<10$) or a circular foundation belongs to a space problem. If the induced stress in the ground soil is only a function of two coordinates (x, z), it is called a plane problem. For example, the calculation of induced stress under strip foundation ($l/b \geqslant 10$) belongs to a plane problem. Flood control dyke, retaining wall and highway embankment can be regarded as plane problems.

The ground is assumed to be a semi-infinite, homogeneous and isotropic linear elastic body. When calculating the induced stress in ground soil, the theoretical solution of half space elastic problem in elastic mechanics can be directly applied.

6.4.1 Induced stress due to a point load

When the vertical concentrated force F acts on the surface of the semi-infinite elastic ground, any point $M(x, y, z)$ inside the ground soil will have a stress increase, i. e., the induced stress caused by the action of F. As shown in Figure 6-15, the origin of the coordinate point O is taken as the action point of concentrated force, J. Boussinesq determined the solutions of stress increase, which is called the basic solutions of induced stresses. There are six independent stress components at point M.

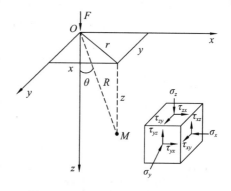

Figure 6-15 Induced stress at any point M due to a concentrated force

$$\sigma_x = \frac{3F}{2\pi} \left\{ \frac{x^2 z}{R^5} + \frac{1-2\mu}{3} \left[\frac{1}{R(R+z)} - \frac{(2R+z)x^2}{(R+z)R^5} - \frac{z}{R^3} \right] \right\} \qquad (6\text{-}20)$$

Engineering Geology and Soil Mechanics

$$\sigma_y = \frac{3F}{2\pi}\left\{\frac{y^2 z}{R^5} + \frac{1-2\mu}{3}\left[\frac{1}{R(R+z)} - \frac{(2R+z)y^2}{(R+z)R^5} - \frac{z}{R^3}\right]\right\} \quad (6-21)$$

$$\sigma_z = \frac{3F}{2\pi R^2}\cos^3\theta = \frac{3F}{2\pi}\frac{z^3}{R^5} \quad (6-22)$$

$$\tau_{xy} = \tau_{yx} = \frac{3F}{2\pi}\left[\frac{xyz}{R^5} - \frac{1-2\mu}{3}\frac{(2R+z)xy}{(R+z)^2 R^3}\right] \quad (6-23)$$

$$\tau_{xz} = \tau_{zx} = \frac{3F}{2\pi}\frac{xz^2}{R^5} \quad (6-24)$$

$$\tau_{yz} = \tau_{zy} = \frac{3F}{2\pi}\frac{yz^2}{R^5} \quad (6-25)$$

Under the concentrated force F, the radial and vertical displacement equations are

$$u(r,z) = \frac{F(1+\mu)}{2\pi E}\left[\frac{rz}{R^3} - (1-2\mu)\frac{r}{R(R+z)}\right] \quad (6-26)$$

$$w(r,z) = \frac{F(1+\mu)}{2\pi E}\left[\frac{z^2}{R^3} + \frac{2(1-\mu)}{R}\right] \quad (6-27)$$

The vertical displacement at any point ($z=0$) on the ground surface is

$$w(r,0) = \frac{F(1-\mu^2)}{\pi E r} \quad (6-28)$$

where, F—concentrated load at the coordinate origin point O;

z—depth of point M;

r—distance between point M and the action line of concentrated load, $r = \sqrt{x^2 + y^2}$;

R—distance between the point M and the point O, $R = \sqrt{x^2 + y^2 + z^2}$;

θ—angle between the MO line and the z axis;

E, μ—modulus of elasticity and Poisson's ratio of ground soil, respectively.

In the above six independent stress components, the vertical induced stress component σ_z due to the point load is the main reason for the vertical compression of the soil. This will lead to the foundation settlement and uneven settlement, so people pay special attention to the variation law of vertical induced stress component in the ground soil.

Therefore, according to the geometric relations, σ_z can be rewritten as

$$\sigma_z = \frac{3F}{2\pi}\frac{1}{R^2}\cos^3\theta = \frac{3F}{2\pi}\frac{z^3}{R^5} \quad (6-29)$$

From Figure 6-15, $R = \sqrt{r^2 + z^2}$, Eq. (6-29) can be further written as

$$\sigma_z = \frac{3F}{2\pi}\frac{z^3}{R^5} = \frac{3F}{2\pi z^2}\frac{1}{\left[1+\left(\frac{r}{z}\right)^2\right]^{5/2}} = K_F\frac{F}{z^2} \quad (6-30)$$

where $K_F = \dfrac{3}{2\pi\left[1+\left(\dfrac{r}{z}\right)^2\right]^{5/2}}$, which is called the vertical stress increase coefficient; it is a function of r and z.

According to the Eq. (6-29) or Eq. (6-30), the distribution law of induced stress due to a point load is shown in Figure 6-16. Figure 6-16 is usually called stress bulbs. It can be seen that the distribution of the vertical induced stress σ_z under the vertical concentrated force F in ground soil spreads away from the load point. ① when $r=0$, i. e., along the z axis, the value of σ_z decreases gradually with the increase of depth z; ② at the same depth, i. e., on the same horizontal plane, the induced stress shows the maximum value along the z axis line of the concentrated force F and decreases gradually with the increase of r; ③ when r is bigger than zero, σ_z increases with z near the ground surface and then decreases gradually; ④ when $r \to 0$ and $z \to \infty$, we get $\sigma_z \to \infty$. Thus the point where the concentrated force F acting on is a mathematical singularity point.

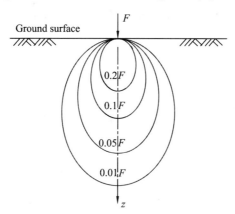

Figure 6-16 Distribution of induced stress in soil under vertical concentrated force

When a vertical distributed load is acting on a certain ground area, as shown in Figure 6-17. The load distribution area can be divided into several small areas, where the distributed load on each distribution area is taken by a concentrated force instead. So there are several concentrated forces F_i ($i = 1, 2, 3, \cdots, n$). By Eq. (6-30), the induced stress at point M with depth z can be obtained for each concentrated force F_i. Because the induced stress is proportional to the load, the superposition principle can be used to calculate the total induced stress produced by all concentrated forces at point M:

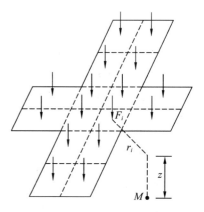

Figure 6-17 Equivalent load method to calculate σ_z

$$\sigma_{zM} = \frac{1}{z^2} \sum_{i=1}^{n} K_{Fi} F_i \tag{6-31}$$

where, K_{Fi} — induced stress coefficient for i^{th} concentrated force.

Eq. (6-31) is the expression of equivalent load method. When the base area is irregular, it is more convenient to calculate the induced stress in ground soil by this method. When $n \to \infty$, the right-hand side of the equation is in an integral form.

6.4.2 Induced stress below a rectangular base under uniform pressure

I. Induced stress below the corner of a rectangular base

Establish the coordinate system as shown in Figure 6-18, where the resultant force

$dF = p_n dx dy$ on the area $dxdy$ of ground surface can be regarded as a concentrated force. It will produce a vertical induced stress at depth z below the corner point O. By Eq. (6-29), we get

$$d\sigma_z = \frac{3 p_n dx dy \times z^3}{2\pi R^5} \tag{6-32}$$

By geometric relation, $R = \sqrt{x^2 + y^2 + z^2}$ is substituted into above equation and integrated for the entire base area. Thus we can get the vertical induced stress at point M with depth z below the corner point O under the vertical uniformly distributed pressure p_n is

$$\begin{aligned} \sigma_z &= \int_0^b \int_0^l \frac{3 z^3 p_n dx dy}{2\pi (\sqrt{x^2 + y^2 + z^2})^5} \\ &= \frac{p_n}{2\pi} \left[\frac{mn}{\sqrt{1+m^2+n^2}} \left(\frac{1}{m^2+n^2} + \frac{1}{1+n^2} \right) + \tan^{-1} \left(\frac{m}{n\sqrt{1+m^2+n^2}} \right) \right] \\ &= K_s p_n \end{aligned} \tag{6-33}$$

where $K_s = \frac{1}{2\pi} \left[\frac{mn}{\sqrt{1+m^2+n^2}} \left(\frac{1}{m^2+n^2} + \frac{1}{1+n^2} \right) + \tan^{-1} \left(\frac{m}{n\sqrt{1+m^2+n^2}} \right) \right]$. K_s is the vertical induced stress coefficient below the corner of a rectangular area under a vertical distributed load. It is a function of m and n. l is the length of the rectangle, and b is the breadth of the rectangle, where $l \geq b$, $m = l/b$, $n = z/b$. z is the depth of point M. The value of K_s can be obtained by Eq. (6-33), or from Table 6-2. p_n is the net contact pressure at base. We can draw the distribution of induced stress (stress contours) shown in Figure 6-19. It can be seen that the induced stress becomes smaller as the depth increases. Since σ_z is proportional to p_n, the principle of superposition can be applied.

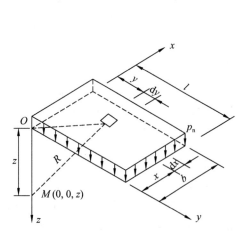

Figure 6-18 Schematic of induced stress below corner under an uniformly distributed pressure

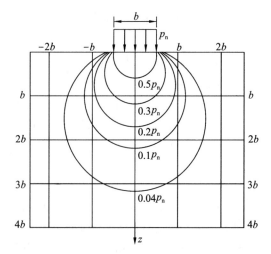

Figure 6-19 Distribution of induced stress below a square base

Table 6-2 Induced stress coefficient K_s at a given depth z below the corner of a rectangular base under uniform pressure

$n=z/b$	$m=l/b$										
	1.0	1.2	1.4	1.6	1.8	2.0	3.0	4.0	5.0	6.0	10.0
0.2	0.2486	0.2489	0.2490	0.2491	0.2491	0.2491	0.2492	0.2492	0.2492	0.2492	0.2492
0.4	0.2401	0.2420	0.2429	0.2434	0.2437	0.2439	0.2442	0.2443	0.2443	0.2443	0.2443
0.6	0.2229	0.2275	0.2300	0.2315	0.2324	0.2329	0.2339	0.2341	0.2342	0.2342	0.2342
0.8	0.1999	0.2075	0.2120	0.2147	0.2165	0.2176	0.2196	0.2200	0.2202	0.2202	0.2202
1.0	0.1752	0.1851	0.1911	0.1955	0.1981	0.1999	0.2034	0.2042	0.2044	0.2045	0.2046
1.2	0.1516	0.1626	0.1705	0.1758	0.1793	0.1818	0.1870	0.1882	0.1885	0.1887	0.1888
1.4	0.1308	0.1423	0.1518	0.1569	0.1613	0.1644	0.1712	0.1730	0.1735	0.1738	0.1740
1.6	0.1123	0.1241	0.1329	0.1436	0.1445	0.1482	0.1567	0.1590	0.1598	0.1601	0.1604
1.8	0.0969	0.1083	0.1172	0.1241	0.1294	0.1334	0.1434	0.1463	0.1474	0.1478	0.1482
2.0	0.0840	0.0947	0.1034	0.1103	0.1158	0.1202	0.1314	0.1350	0.1363	0.1368	0.1374
2.2	0.0732	0.0832	0.0917	0.0984	0.1039	0.1084	0.1205	0.1248	0.1264	0.1271	0.1277
2.4	0.0642	0.0734	0.0812	0.0879	0.0934	0.0979	0.1108	0.1156	0.1175	0.1184	0.1192
2.6	0.0566	0.0651	0.0725	0.0788	0.0842	0.0887	0.1020	0.1073	0.1095	0.1106	0.1116
2.8	0.0502	0.0580	0.0649	0.0709	0.0761	0.0805	0.0942	0.0999	0.1024	0.1036	0.1048
3.0	0.0447	0.0519	0.0583	0.0640	0.0690	0.0732	0.0870	0.0931	0.0959	0.0973	0.0987
3.2	0.0401	0.0467	0.0526	0.0580	0.0627	0.0668	0.0806	0.0870	0.0900	0.0916	0.0933
3.4	0.0361	0.0421	0.0477	0.0527	0.0571	0.0611	0.0747	0.0814	0.0847	0.0864	0.0882
3.6	0.0326	0.0382	0.0433	0.0480	0.0523	0.0561	0.0694	0.0763	0.0799	0.0816	0.0837
3.8	0.0296	0.0348	0.0395	0.0439	0.0479	0.0516	0.0645	0.0717	0.0753	0.0773	0.0796
4.0	0.0270	0.0318	0.0362	0.0403	0.0441	0.0474	0.0603	0.0674	0.0712	0.0733	0.0758
4.2	0.0247	0.0291	0.0333	0.0371	0.0407	0.0439	0.0563	0.0634	0.0674	0.0696	0.0724
4.4	0.0227	0.0268	0.0306	0.0343	0.0376	0.0407	0.0527	0.0597	0.0639	0.0662	0.0692
4.6	0.0209	0.0247	0.0283	0.0317	0.0348	0.0378	0.0493	0.0564	0.0606	0.0630	0.0663
4.8	0.0193	0.0229	0.0262	0.0294	0.0324	0.0352	0.0463	0.0533	0.0576	0.0601	0.0635
5.0	0.0179	0.0212	0.0243	0.0274	0.0302	0.0328	0.0435	0.0504	0.0547	0.0573	0.0610
6.0	0.0127	0.0151	0.0174	0.0196	0.0218	0.0238	0.0325	0.0388	0.0431	0.0460	0.0506
7.0	0.0094	0.0112	0.0130	0.0147	0.0164	0.0180	0.0251	0.0306	0.0346	0.0376	0.0428
8.0	0.0073	0.0087	0.0101	0.0114	0.0127	0.0140	0.0198	0.0246	0.0283	0.0311	0.0367
9.0	0.0058	0.0069	0.0080	0.0091	0.0102	0.0112	0.0161	0.0202	0.0235	0.0262	0.0319
10.0	0.0047	0.0056	0.0065	0.0074	0.0083	0.0092	0.0132	0.0167	0.0198	0.0222	0.0280

Note: the vertical induced stress at $z=0$ can directly use the value of net contact pressure at base.

Since the Boussinesq's solution at $z=0$ is a mathematical singularity, for the space problem or plane problem, the induced stress at the base surface ($z=0$) due to the uniform or triangular load can directly use the value of net contact pressure p_n.

II. Calculation of induced stress using the corner point method

Eq. (6-33) can be used to calculate the vertical induced stress at arbitrary depth z below the corner of a rectangular base. When to calculate the induced stress at a given depth below a point O inside or outside the rectangular base area, we can divide the area into four rectangles, making O become a shared corner for each rectangular. Using Eq. (6-33) to calculate the vertical induced stress of each rectangle at a given depth for point O, then by superposition we can get the induced stress of the whole rectangular area. This method of solving induced stress is called the corner-point method.

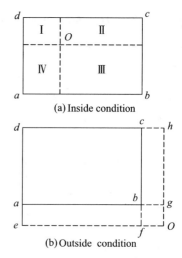

(a) Inside condition

(b) Outside condition

Figure 6-20 Schematic of Corner-point method

As shown in Figure 6-20a, to calculate the induced stress at a given depth below the point O inside the rectangular area. We can divide the base area is into four rectangles. Then the point O becomes the shared corner of four rectangles I, II, III and IV. Eq. (6-33) can be used to calculate the induced stress below the shared point O of four rectangles σ_{zI}, σ_{zII}, σ_{zIII} and σ_{zIV}. Then the total induced stress can be obtained:

$$\sigma_z = \sigma_{zI} + \sigma_{zII} + \sigma_{zIII} + \sigma_{zIV} \quad (6\text{-}34)$$

As shown in Figure 6-20b, to calculate the induced stress at a given depth below the point O outside the rectangular area, the original base area $abcd$ can be expanded to a new rectangle $Oedh$. The point O can be regarded as the shared corner of four rectangles $Oedh$, $Ofch$, $Oeag$, and $Ofbg$. The induced stress of each rectangle is obtained using Eq. (6-33). The superposition principle is applied to obtain the total induced stress:

$$\sigma_z = \sigma_{zOedh} - \sigma_{zOfch} - \sigma_{zOeag} + \sigma_{zOfbg} \quad (6\text{-}35)$$

It should be pointed out that K_s can be obtained from Table 6-2, where l is the length and b is the breadth for each rectangle.

【Example 6-2】 As shown in Figure 6-21, the length of a rectangular base $l = 4\text{m}$, the breadth $b = 2\text{m}$, the depth $d = 0.5\text{m}$, and the unit weight of the ground soil $\gamma_0 = 18\text{kN/m}^3$. A known uniform contact pressure $p = 140\text{kPa}$. Try to calculate the induced stress at point O,

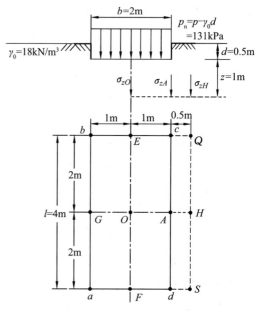

Figure 6-21 Schematic diagram of 【Example 6-2】

point A and point H below the base with depth $z=1\text{m}$.

【Solution】

(1) Calculate the net contact pressure p_n: $p_n = p - \gamma_0 d = 140 - 18 \times 0.5 = 131\text{kPa}$.

(2) Calculate the induced stress σ_{zO} at point O: the point O is the shared corner point of four rectangles OGbE, OGaF, OAdF, and OAcE. Since the of length and breadth are the same for four rectangles, we have

$$l/b = 2/1 = 2, \ z/b = 1/1 = 1$$

Check Table 6-2, $K_s = 0.1999$. According to the principle of superposition, we have

$$\sigma_{zO} = 4K_s p_n = 4 \times 0.1999 \times 131 = 104.75\text{kPa}$$

(3) Calculate the induced stress σ_{zA} at point A: the point A is the shared corner point of two rectangles AcbG and AdaG. The two rectangles are with the same length and breadth, so the coefficients of induced stress are the same.

$$l/b = 2/2 = 1, \ z/b = 1/2 = 0.5$$

Check Table 6-2 and use interpolation method, $K_s = 0.2315$. According to the principle of superposition, we have

$$\sigma_{zA} = 2K_s p_n = 2 \times 0.2315 \times 131 = 60.65\text{kPa}$$

(4) Calculate the induced stress σ_{zH} at point H: the H point is the shared corner point of four rectangles HGbQ, HSaG, HAcQ, and HAdS. For two rectangles HGbQ and HsaG, the length and breadth are the same, so

$$l/b = 2.5/2 = 1.25, \ z/b = 1/2 = 0.5$$

Check Table 6-2 and use interpolation method, $K_{s1} = 0.2350$.

For the other two rectangles HAcQ and HAds, the length and breadth are the same, so

$$l/b = 2/0.5 = 4$$
$$z/b = 1/0.5 = 2$$

Check Table 6-2, $K_{s2} = 0.1350$.

According to the principle of superposition, we have: $\sigma_{zH} = (2K_{s1} - 2K_{s2})p_n = (2 \times 0.2350 - 2 \times 0.1350) \times 131 = 26.2\text{kPa}$.

From this example, we can see that below the same depth of the rectangular base, the vertical induced stress at the center reaches its maximum and decreases gradually to the edge.

6.4.3 Induced stress below a rectangular base under triangular pressure

As shown in Figure 6-22, the rectangular foundation base is under the triangular distributed pressure. The induced stress at point M in the the soil mass below the ground can be calculated by Eq. (6-31).

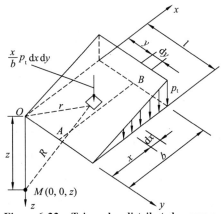

Figure 6-22 Triangular distributed pressure on a rectangular base

Taking the coordinate system shown in Figure 6-22, if the maximum contact pressure at base is p_t, an unit force acting on a small area $dxdy$ is $\frac{x}{b}p_t dxdy$. According to Eq. (6-29), the induced stress at point M right below the corner point O caused by the unit force is

$$d\sigma_z = \frac{3z^3 p_t x dx dy}{2\pi R^5 b} \qquad (6-36)$$

By geometric relation, $R = \sqrt{x^2 + y^2 + z^2}$ is substituted into Eq. (6-36) and integrated,

$$\sigma_z = \int_0^l \int_0^b \frac{3p_t}{2\pi b} \frac{xz^3 dx dy}{(\sqrt{x^2+y^2+z^2})^5} = K_t p_t \qquad (6-37)$$

where, $K_t = \frac{mn}{2\pi}\left[\frac{1}{\sqrt{m^2+n^2}} - \frac{n^2}{(1+n^2)\sqrt{1+m^2+n^2}}\right]$. K_t is the induced stress coefficient below the corner of the rectangular base under the triangular distributed pressure. K_t is a function of $m = l/b$, $n = z/b$, which can be obtained directly by Eq. (6-37) or Table 6-3. It should be noted that when the contact pressure under the foundation base follows the vertical trapezoidal distribution, the induced stress at any given depth below the corner point of foundation base can be divided into two parts, i.e., the sum of two induced stresses due to a uniformly distributed pressure and a triangular distributed pressure.

Table 6-3 Induced stress coefficient K_t at a given depth z below the corner of a rectangular base under triangular distributed pressure

$n=z/b$	$m=l/b$							
	0.2	0.4	0.6	0.8	1.0	1.2	1.4	1.6
0.2	0.0223	0.0280	0.0296	0.0301	0.0304	0.0305	0.0305	0.0306
0.4	0.0269	0.0420	0.0487	0.0517	0.0531	0.0539	0.0543	0.0545
0.6	0.0259	0.0448	0.0560	0.0621	0.0654	0.0673	0.0684	0.0690
0.8	0.0232	0.0421	0.0553	0.0637	0.0688	0.0720	0.0739	0.0751
1.0	0.0201	0.0375	0.0508	0.0602	0.0666	0.0708	0.0735	0.0753
1.2	0.0171	0.0324	0.0450	0.0546	0.0615	0.0664	0.0698	0.0721
1.4	0.0145	0.0278	0.0392	0.0483	0.0554	0.0606	0.0644	0.0672
1.6	0.0123	0.0238	0.0339	0.0424	0.0492	0.0545	0.0586	0.0616
1.8	0.0105	0.0204	0.0294	0.0371	0.0435	0.0487	0.0528	0.0560
2.0	0.0090	0.0176	0.0255	0.0324	0.0384	0.0434	0.0474	0.0507
2.5	0.0063	0.0125	0.0183	0.0236	0.0284	0.0326	0.0362	0.0393
3.0	0.0046	0.0092	0.0135	0.0176	0.0214	0.0249	0.0280	0.0307
5.0	0.0018	0.0036	0.0054	0.0071	0.0088	0.0104	0.0120	0.0135
7.0	0.0009	0.0019	0.0028	0.0038	0.0047	0.0056	0.0064	0.0073
10.0	0.0005	0.0009	0.0014	0.0019	0.0023	0.0028	0.0033	0.0037

Continued

$n=z/b$	$m=l/b$							
	1.8	2.0	3.0	4.0	6.0	8.0	10.0	
0.2	0.0306	0.0306	0.0306	0.0306	0.0306	0.0306	0.0306	
0.4	0.0546	0.0547	0.0548	0.0549	0.0549	0.0549	0.0549	
0.6	0.0694	0.0696	0.0701	0.0702	0.0702	0.0702	0.0702	
0.8	0.0759	0.0764	0.0773	0.0776	0.0776	0.0776	0.0776	
1.0	0.0766	0.0774	0.079	0.0794	0.0795	0.0796	0.0796	
1.2	0.0738	0.0749	0.0774	0.0779	0.0782	0.0783	0.0783	
1.4	0.0692	0.0707	0.0739	0.0748	0.0752	0.0752	0.0753	
1.6	0.0639	0.0656	0.0697	0.0708	0.0714	0.0715	0.0715	
1.8	0.0585	0.0604	0.0652	0.0666	0.0673	0.0675	0.0675	
2.0	0.0533	0.0553	0.0607	0.0624	0.0634	0.0636	0.0636	
2.5	0.0419	0.044	0.0504	0.0529	0.0543	0.0547	0.0548	
3.0	0.0331	0.0352	0.0419	0.0449	0.0469	0.0474	0.0476	
5.0	0.0148	0.0161	0.0214	0.0248	0.0283	0.0296	0.0301	
7.0	0.0081	0.0089	0.0124	0.0152	0.0186	0.0204	0.0212	
10.0	0.0041	0.0046	0.0066	0.0084	0.0111	0.0128	0.0139	

Note: the vertical induced stress at $z=0$ can directly use the value of net contact pressure at base

【Example 6-3】 As shown in Figure 6-23, the length and breadth of a rectangular foundation base is 4m and 2m, respectively. The base depth is 1.5m. The unit weight of ground soil is 19kN/m³ and the unit weight of foundation and overburden soil on average is 20kN/m³. An eccentric force transferred from the superstructure is 800kN with an eccentricity of 0.2m. Try to calculate the induced stresses at depth of 2m respectively below the point A and point B at base.

【Solution】

(1) Calculate the resultant force and eccentricity:

$$F_v = 800 + 20 \times 4 \times 2 \times 1.5 = 1040 \text{kN}$$

$$e = \frac{800 \times 0.2}{1040} = 0.154$$

(2) Calculate the contact pressure:

$$p_{max} = \frac{F_v}{lb}\left(1 + \frac{6e}{l}\right) = \frac{1040}{4 \times 2}\left(1 + \frac{6 \times 0.154}{4}\right) = 160.03 \text{kPa}$$

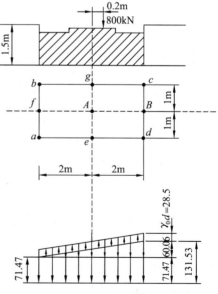

Figure 6-23 Schematic diagram of 【Example 6-3】

$$p_{\min}=\frac{F_v}{lb}\left(1-\frac{6e}{l}\right)=\frac{1040}{4\times2}\left(1-\frac{6\times0.154}{4}\right)=99.97\text{kPa}$$

(3) Calculate the net contact pressure:

$$(p_n)_{\max}=p_{\max}-\gamma_0 d=160.03-19\times1.5=131.53\text{kPa}$$
$$(p_n)_{\min}=p_{\min}-\gamma_0 d=99.97-19\times1.5=71.47\text{kPa}$$

The net contact pressure with a trapezoidal form can be divided into two parts: a uniformly distributed pressure and a triangular distributed pressure.

The uniformly rectangular distributed pressure: $p_{n1}=71.47\text{kPa}$ (downward).

The triangular distributed pressure: $p_{t1}=131.53-71.47=60.06\text{kPa}$ (downward).

Besides, we can also regard the trapezoidal distributed net contact pressure as the combination of a uniformly distributed pressure (downward) and a triangular distributed pressure (upward), i.e., using a uniformly distributed pressure minus a triangular distributed pressure with p_{t2} on the left side.

$$p_{n2}=131.53\text{kPa (downward)}$$
$$p_{t2}=131.53-71.47=60.06\text{kPa (upward)}$$

(4) Calculate the induced stress σ_{zA} of foundation at 2m depth below the point A: the point A is the common corner of four rectangulars $Aeaf, Afbg, AgcB$, and $ABde$. In order to simplify the calculation, the avarage net contact pressure of the trapezoidal distribution can be obtained by using symmetry and superposition principle.

$$\overline{p_n}=\frac{1}{2}[(p_n)_{\max}+(p_n)_{\min}]=\frac{1}{2}(131.53+71.47)=101.5\text{kPa}$$

Since the length and breadth of the four rectangles are the same, the induced stress coefficient K_s for each rectangle is the same. From the value of l, b, z for each rectangle: $l/b=2/1=2; z/b=2/1=2$. From Table 6-2, we get $K_s=0.1202$. Thus $\sigma_{zA}=4K_s p_n=4\times0.1202\times101.5=48.8\text{kPa}$.

(5) Calculate the induced stress σ_{zB} at 2m depth below the point B: the point B is the shared corner point of two rectangles $Bfbc$ and $Bfad$ with the same length and breadth, so the induced stress coefficients are the same.

① For uniformly distributed pressure, $p_{n2}=131.53\text{kPa}$. For each rectangle, $l=4\text{m}, b=1\text{m}, m=l/b=4/1=4, n=z/b=2/1=2$. Check Table 6-2, $K_s=0.1350$.

$$\sigma_{zBn}=2K_s p_{n2}=2\times0.1350\times131.53=35.51\text{kPa}$$

② For triangular distributed pressure, $p_t=60.06\text{kPa}$. According to Figure 6-22, for each rectangle, $l=1\text{m}, b=4\text{m}, m=l/b=1/4=0.25, n=z/b=2/4=0.5$. Check Table 6-3 and use interpolation method

$$K_t=\frac{0.0269+0.0259}{2}\times0.5\times\frac{0.4-0.25}{0.4-0.2}+\frac{0.0420+0.0448}{2}\times\frac{0.25-0.2}{0.4-0.2}=0.03065$$

$$\sigma_{zBt}=2K_t p_{t2}=2\times0.03065\times60.06=3.68\text{kPa}$$

③ Combined σ_{zBn} and σ_{zBt}, we get

$$\sigma_{zB}=35.51-3.68=31.83\text{kPa}$$

6.4.4 Induced stress below a rectangular base under horizontal pressure

As shown in Figure 6-24, when the rectangular base is due to horizontally uniformly distributed pressure p_h, the vertical induced stress at depth z below corner point 1 and corner point 2 is

$$\sigma_{z1} = -K_h p_h, \quad \sigma_{z2} = +K_h p_h \quad (6-38)$$

where, $K_h = \dfrac{m}{2\pi}\left[\dfrac{1}{\sqrt{m^2+n^2}} - \dfrac{n^2}{(1+n^2)\sqrt{1+m^2+n^2}}\right]$.

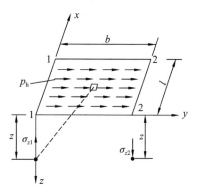

Figure 6-24 Vertical induced stress below the corner of a rectangular base under horizontal pressure

The negative symbol "−" means the vertical induced stress is upward and the positive symbol "+" means the vertical induced stress is downward. K_h is the induced stress coefficient at a given depth below the corner of rectangular base under the horizontal pressure. It is a function of $m = l/b$ and $n = z/b$. b is the breadth of rectangle along the load direction, and l is the length of the rectangle perpendicular to load direction. Table 6-4 lists the values of K_h.

From the knowledge of material mechanics, we can find that on the vertical line in the middle of breadth, the vertical induced stress caused by p_h is zero. The calculation of induced stress at any point below the base can use the corner-point method.

Table 6-4 Induced stress coefficient K_h at a given depth z below a rectangular base under horizontally uniformly distributed pressure

$n=z/b$	$m=l/b$										
	1.0	1.2	1.4	1.6	1.8	2.0	3.0	4.0	5.0	6.0	10.0
0.0	0.1592	0.1592	0.1592	0.1592	0.1592	0.1592	0.1592	0.1592	0.1592	0.1592	0.1592
0.2	0.1518	0.1523	0.1526	0.1528	0.1529	0.1529	0.1530	0.1530	0.1530	0.1530	0.1530
0.4	0.1328	0.1347	0.1356	0.1362	0.1365	0.1367	0.1371	0.1372	0.1372	0.1372	0.1372
0.6	0.1091	0.1121	0.1139	0.1150	0.1156	0.1160	0.1168	0.1169	0.1170	0.1170	0.1170
0.8	0.0861	0.0900	0.0924	0.0939	0.0948	0.0955	0.0967	0.0969	0.0970	0.0970	0.0970
1.0	0.0666	0.0708	0.0735	0.0753	0.0766	0.0774	0.0790	0.0794	0.0795	0.0796	0.0796
1.2	0.0512	0.0553	0.0582	0.0601	0.0615	0.0624	0.0645	0.0650	0.0652	0.0652	0.0652
1.4	0.0395	0.0433	0.0460	0.0480	0.0494	0.0505	0.0528	0.0534	0.0537	0.0537	0.0538
1.6	0.0308	0.0341	0.0366	0.0385	0.0400	0.0410	0.0436	0.0443	0.0446	0.0447	0.0447
1.8	0.0242	0.0270	0.0293	0.0311	0.0325	0.0336	0.0362	0.0370	0.0374	0.0375	0.0375
2.0	0.0192	0.0217	0.0237	0.0253	0.0266	0.0277	0.0303	0.0312	0.0317	0.0318	0.0318
2.5	0.0113	0.0130	0.0145	0.0157	0.0167	0.0176	0.0202	0.0211	0.0217	0.0219	0.0219
3.0	0.0070	0.0083	0.0093	0.0102	0.0110	0.0117	0.0140	0.0150	0.0156	0.0158	0.0159
5.0	0.0018	0.0021	0.0024	0.0027	0.0030	0.0032	0.0043	0.0050	0.0057	0.0059	0.0060
7.0	0.0007	0.0008	0.0009	0.0010	0.0012	0.0013	0.0018	0.0022	0.0027	0.0029	0.0030
10.0	0.0002	0.0003	0.0003	0.0004	0.0004	0.0005	0.0007	0.0008	0.0011	0.0013	0.0014

6.4.5 Induced stress below a circular base under uniform pressure

The radius of a circular base is r_0 and the uniformly distributed pressure is p_n, as shown in Figure 6-25. The unit force acting on a small unit area $rdrd\theta$ is $dF_v = p_n rdrd\theta$. l is the distance between the vertical projection of point M and center point O. The distance between unit force and point M is $R = (z^2 + r^2 + l^2 - 2lr\cos\theta)^{1/2}$, which can be substituted into Eq. (6-31) and integrated to obtain the induced stress at any point M below the circular base.

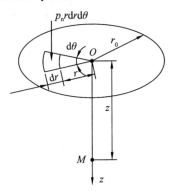

$$\sigma_z = \int_0^{2\pi}\int_0^{r_0} \frac{3p_n z^3}{2\pi} \frac{rdrd\theta}{(z^2 + r^2 + l^2 - 2lr\cos\theta)^{5/2}} = K_r p_n$$

(6-39)

Figure 6-25 Schematic of induced stress below a circular base

where K_r is the induced stress coefficient below the circular base under uniform pressure, as listed in Table 6-5; K_r is the function of z/r_0 and l/r_0, where r_0 is the radius of the circular area. In the conventional design, the vertical induced stress at $z=0$ can directly use the value of contact pressure $p_n(K_r=1.0)$.

Table 6-5 Induced stress coefficient K_r at a given depth z below a circular base under circular uniform pressure

z/r_0	l/r_0					
	0.0	0.4	0.8	1.2	1.6	2.0
0.0	1.000	1.000	1.000	0.000	0.000	0.000
0.2	0.993	0.987	0.890	0.077	0.005	0.001
0.4	0.949	0.922	0.712	0.181	0.026	0.006
0.6	0.864	0.813	0.591	0.224	0.056	0.016
0.8	0.756	0.699	0.504	0.237	0.083	0.029
1.0	0.646	0.621	0.465	0.236	0.091	0.036
1.2	0.547	0.593	0.434	0.235	0.102	0.042
1.4	0.461	0.425	0.329	0.212	0.118	0.062
1.8	0.332	0.311	0.254	0.182	0.118	0.072
2.2	0.246	0.233	0.198	0.153	0.109	0.074
2.6	0.187	0.179	0.158	0.129	0.098	0.071
3.0	0.146	0.141	0.127	0.108	0.087	0.067
3.8	0.096	0.093	0.087	0.078	0.067	0.055
4.6	0.067	0.066	0.063	0.058	0.052	0.045
5.0	0.057	0.056	0.054	0.050	0.046	0.041
6.0	0.040	0.040	0.039	0.037	0.034	0.031

6.5 Induced stress calculation for a plane problem

When the length of the foundation is much larger than the breadth, and the distribution of the load along the length does not change, we can regard this as a plane problem. The plane problem calculation is simpler than the space problem. In fact, for a rectangular foundation base with $l/b \geqslant 10$, it can be regarded as a plane problem, and the calculation error is very small.

6.5.1 Induced stress due to an uniformly distributed line load

When a linear load $\bar{p}$ (kN/m) acts on the surface of the semi-infinite space, as shown in Figure 6-26, the concentrated force acting on the small unit section dy can be written as $dF_v = \bar{p}dy$. Therefore, the induced stress at any point M can be derived from the Boussinesq solution.

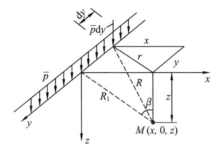

Figure 6-26 Calculation of induced stress under uniformly distributed line load

$$\sigma_z = \int_{-\infty}^{+\infty} \frac{3z^3 \bar{p} dy}{2\pi R^5} = \frac{2\bar{p}z^3}{\pi R_1^4} = \frac{2\bar{p}}{\pi R_1}\cos^3\beta = \frac{2\bar{p}}{\pi z}\cos^4\beta \tag{6-40}$$

$$\sigma_x = \frac{2\bar{p}x^2 z}{\pi R_1^4} = \frac{2\bar{p}}{\pi R_1}\cos\beta \sin^2\beta = \frac{2\bar{p}}{\pi z}\cos^2\beta \sin^2\beta \tag{6-41}$$

$$\tau_{xz} = \tau_{zx} = \frac{2\bar{p}xz^2}{\pi R_1^4} = \frac{2\bar{p}}{\pi R_1}\cos^2\beta \sin\beta = \frac{2\bar{p}}{\pi z}\cos^3\beta \sin\beta \tag{6-42}$$

where, $\cos\beta = \frac{z}{R_1}$, $\sin\beta = \frac{x}{R_1}$, Eq. (6-40)~Eq. (6-42) are the Flamant solution.

Since the linear load extends infinitely along the y axis with the same value, the stress states on any plane perpendicular to the y axis are the same. There are three independent components of the stress ($\sigma_x, \sigma_z, \tau_{xz}$), which is a plane problem. It can be extended to a certain breadth b as strip foundation. For plane problems, $\tau_{xz} = \tau_{zx}, \tau_{yz} = \tau_{zy} = 0$. If the soil is a elastic body, there is $\sigma_y = \mu(\sigma_x + \sigma_z)$, where μ is the Poisson ratio of soil.

6.5.2 Induced stress due to uniformly distributed pressure on a strip foundation

The breadth of the strip foundation is b, as shown in Figure 6-27a. The uniformly distributed net contact pressure p_n is acting on the surface of foundation base. The linear load on a small unit width dζ is $p_n d\zeta$. Applying Eq. (6-32) to integrate along the b, the vertical induced stress of any point M in the ground soil is expressed as

$$\sigma_z = \int_0^b \frac{2p_n}{\pi} \frac{z^3 d\zeta}{(x-\zeta)^2 + z^2} = K_s^z p_n \tag{6-43}$$

Similarly, the expressions for σ_x and τ_{xz} are as follows.

$$\sigma_x = K_s^x p_n \tag{6-44}$$

$$\tau_{xz} = K_s^\tau p_n \tag{6-45}$$

K_s^z, K_s^x and K_s^τ are respectively the induced stress coefficients of vertical normal stress, horizontal normal stress, and shear stress in ground soil subjected to vertical uniformly distributed pressure on a strip foundation. They are functions of $m = x/b$, and $n = z/b$, which can be directly checked by Table 6-6 or calculated according to the following equations

$$K_s^z = \frac{1}{\pi}\left[\arctan\left(\frac{m}{n}\right) - \arctan\left(\frac{m-1}{n}\right) + \frac{mn}{n^2+m^2} - \frac{n(m-1)}{n^2+(m-1)^2}\right] \tag{6-46}$$

$$K_s^x = \frac{1}{\pi}\left[\arctan\left(\frac{m}{n}\right) - \arctan\left(\frac{m-1}{n}\right) - \frac{mn}{n^2+m^2} + \frac{n(m-1)}{n^2+(m-1)^2}\right] \tag{6-47}$$

$$K_s^\tau = \frac{1}{\pi}\left[\frac{n^2}{n^2+(m-1)^2} - \frac{n^2}{n^2+m^2}\right] \tag{6-48}$$

In conventional design, the vertical induced stress at the base surface is directly adopted the value of p_n. The contours of vertical induced stress σ_z is shown in Figure 6-27b. The analysis of induced stress distribution characteristics is of great significance for the study of foundation stability.

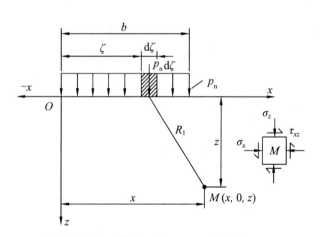
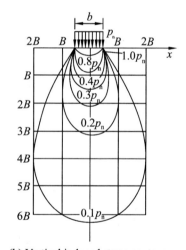

(a) Calculation of induced stress at point M (b) Vertical induced stress contours

Figure 6-27 Induced stress under uniformly distributed pressure acting on a strip foundation

Table 6-6 Induced stress coefficients K_s^z, K_s^x and K_s^τ at a given depth z below the strip foundation under vertical uniformly distributed pressure

$m=x/b$		\multicolumn{11}{c}{$n=z/b$}										
		0.0	0.1	0.2	0.4	0.5	0.6	0.8	1.0	1.2	1.4	2.0
−1.00	K_s^z	0.000	0.000	0.001	0.01	0.018	0.026	0.048	0.07	0.091	0.108	0.134
	K_s^x	0.004	0.032	0.06	0.017	0.012	0.132	0.132	0.134	0.124	0.109	0.07
	K_s^τ	0.000	−0.003	−0.009	−0.032	−0.045	−0.058	−0.08	−0.095	−0.103	−0.106	−0.095
−0.50	K_s^z	0.000	0.002	0.011	0.056	0.084	0.111	0.155	0.186	0.202	0.21	0.205
	K_s^x	0.008	0.082	0.147	0.208	0.211	0.204	0.177	0.146	0.117	0.094	0.049
	K_s^τ	0.000	−0.011	−0.038	−0.103	−0.127	−0.144	−0.158	−0.157	−0.147	−0.133	−0.096
−0.25	K_s^z	0.000	0.011	0.058	0.174	0.213	0.243	0.276	0.288	0.287	0.279	0.243
	K_s^x	0.021	0.18	0.027	0.274	0.248	0.221	0.169	0.127	0.096	0.073	0.035
	K_s^τ	−0.001	−0.042	−0.116	−0.199	−0.211	−0.212	−0.197	−0.175	−0.153	−0.132	−0.085
0	K_s^z	0.500	0.499	0.498	0.489	0.479	0.468	0.440	0.409	0.375	0.348	0.275
	K_s^x	0.494	0.437	0.376	0.269	0.224	0.188	0.13	0.091	0.067	0.011	0.02
	K_s^τ	−0.318	−0.315	−0.306	−0.274	−0.255	−0.234	−0.194	−0.590	−0.131	−0.108	−0.064
0.25	K_s^z	1.0	0.988	0.936	0.797	0.734	0.679	0.586	0.511	0.450	0.401	0.298
	K_s^x	0.935	0.685	0.466	0.215	0.185	0.143	0.087	0.055	0.037	0.026	0.010
	K_s^τ	−0.001	−0.039	−0.103	−0.159	−0.157	−0.147	−0.121	−0.096	−0.078	−0.061	−0.034
0.50	K_s^z	1.0	0.997	0.978	0.881	0.818	0.756	0.642	0.549	0.478	0.420	0.306
	K_s^x	0.848	0.752	0.538	0.260	0.182	0.129	0.070	0.040	0.026	0.017	0.006
	K_s^τ	0.000	0.000	0.000	0.000	0.000	0.000	0.000	0.000	0.000	0.000	0.000
0.75	K_s^z	1.0	0.988	0.936	0.797	0.734	0.679	0.586	0.511	0.450	0.401	0.298
	K_s^x	0.935	0.685	0.466	0.215	0.185	0.143	0.087	0.055	0.037	0.026	0.010
	K_s^τ	0.001	0.039	0.103	0.159	0.157	0.147	0.121	0.096	0.078	0.061	0.034
1.00	K_s^z	0.500	0.499	0.498	0.489	0.479	0.468	0.440	0.409	0.375	0.348	0.275
	K_s^x	0.494	0.437	0.376	0.269	0.224	0.188	0.13	0.091	0.067	0.011	0.02
	K_s^τ	0.318	0.315	0.306	0.274	0.255	0.234	0.194	0.590	0.131	0.108	0.064
1.25	K_s^z	0.000	0.011	0.058	0.174	0.213	0.243	0.276	0.288	0.287	0.279	0.243
	K_s^x	0.021	0.180	0.027	0.274	0.248	0.221	0.169	0.127	0.096	0.073	0.035
	K_s^τ	0.001	0.042	0.116	0.199	0.211	0.212	0.197	0.175	0.153	0.132	0.085
1.50	K_s^z	0.000	0.002	0.011	0.056	0.084	0.111	0.155	0.186	0.202	0.210	0.205
	K_s^x	0.008	0.082	0.147	0.208	0.211	0.204	0.177	0.146	0.117	0.094	0.049
	K_s^τ	0.000	0.011	0.038	0.103	0.127	0.144	0.158	0.157	0.147	0.133	0.096

Note: the induced stress at $z=0$ can directly use the value of net contact pressure at base.

6.5.3 Induced stress due to a vertical triangular distributed pressure

Figure 6-28 shows a triangular distributed pressure acting on a strip foundation base with the maximum pressure p_t. Linear load on a small unit breadth $d\zeta$ is $\dfrac{\zeta \cdot d\zeta}{b} p_t$. Using the Flamant solution and integrating along the breadth b of strip foundation base, the vertical induced stress at point M below the strip foundation is

$$\sigma_z = K_t^z p_t \qquad (6\text{-}49)$$

Similarly, the expressions for σ_x and τ_{xz} are as follows

$$\sigma_x = K_t^x p_t \qquad (6\text{-}50)$$
$$\tau_{xz} = K_t^\tau p_t \qquad (6\text{-}51)$$

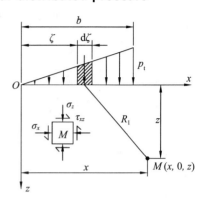

Figure 6-28 Calculation of induced stress at point M below a strip foundation under a triangular distributed pressure

K_t^z, K_t^x, and K_t^τ are respectively the induced stress coefficients of the vertical normal stress, horizontal normal stress, and shear stress in ground soil below strip foundation under the vertical triangular pressure. They are functions of $m = x/b$ and $n = z/b$, which can be calculated according to the following equations. Table 6-7 lists the values of K_t^z calculated by Eq. (6-52).

$$K_t^z = \frac{1}{\pi} \left\{ m \left[\arctan\left(\frac{m}{n}\right) - \arctan\left(\frac{m-1}{n}\right) \right] - \frac{(m-1)n}{(m-1)^2 + n^2} \right\} \qquad (6\text{-}52)$$

$$K_t^x = \frac{1}{\pi} \left\{ m \left[\arctan\left(\frac{m}{n}\right) - \arctan\left(\frac{m-1}{n}\right) \right] + n \ln\left[\frac{(m-1)^2 + n^2}{m^2 + n^2}\right] + \frac{(m-1)n}{(m-1)^2 + n^2} \right\} \qquad (6\text{-}53)$$

$$K_t^\tau = \frac{1}{\pi} \left\{ n \left[\arctan\left(\frac{m-1}{n}\right) - \arctan\left(\frac{m}{n}\right) \right] - \frac{n^2}{(m-1)^2 + n^2} \right\} \qquad (6\text{-}54)$$

Table 6-7 Vertical induced stress coefficient K_t^z at a given depth z below the strip foundation under vertical triangular pressure

$m = x/b$	$n = z/b$										
	0.0	0.1	0.2	0.4	0.5	0.6	0.8	1.0	1.2	1.4	2.0
-1.00	0.000	0.000	0.001	0.003	0.005	0.008	0.017	0.025	0.033	0.041	0.057
-0.50	0.000	0.000	0.002	0.014	0.022	0.031	0.049	0.065	0.076	0.084	0.089
-0.25	0.000	0.002	0.009	0.036	0.025	0.066	0.089	0.104	0.111	0.114	0.108
0	0.003	0.032	0.061	0.010	0.127	0.140	0.155	0.159	0.154	0.151	0.127
0.25	0.249	0.251	0.255	0.263	0.262	0.258	0.243	0.244	0.204	0.186	0.143
0.50	0.500	0.498	0.489	0.441	0.409	0.378	0.321	0.275	0.239	0.210	0.153
0.75	0.750	0.737	0.682	0.534	0.472	0.421	0.343	0.286	0.246	0.215	0.155
1.00	0.497	0.468	0.437	0.379	0.353	0.328	0.285	0.250	0.221	0.198	0.147
1.25	0.000	0.010	0.050	0.137	0.161	0.177	0.188	0.184	0.176	0.165	0.134
1.50	0.000	0.002	0.009	0.043	0.061	0.080	0.106	0.121	0.126	0.127	0.115

Note: the induced stress at $z = 0$ can use the value of contact pressure at base.

6.5.4 Induced stress due to a horizontally uniform pressure

As shown in Figure 6-29, a horizontally uniform pressure p_h is acting along the breadth b of a strip foundation base. At any point M below the foundation, the vertical induced stress can be calculated by the basic equation in elastic mechanics via integrating the basic equation along the breadth b

$$\sigma_z = K_h^z p_h \qquad (6\text{-}55)$$

Similarly, the expressions for σ_x and τ_{xz} are as follows

$$\sigma_x = K_h^x p_h \qquad (6\text{-}56)$$
$$\tau_{xz} = K_h^\tau p_h \qquad (6\text{-}57)$$

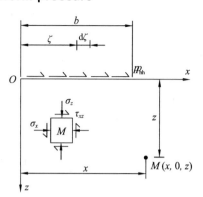

Figure 6-29 Calculation of vertical induced stress at point M under a horizontally uniform pressure

K_h^z, K_h^x, and K_h^τ are respectively the induced stress coefficients of the vertical normal stress, horizontal normal stress and shear stress in ground soil below strip foundation under the action of horizontally uniform pressure, which are a function of $m=x/b, n=z/b$, which can be calculated according to the following equations. Table 6-8 lists the values of K_h^z calculated by Eq. (6-58).

$$K_h^z = \frac{1}{\pi}\left[\frac{n^2}{n^2+(m-1)^2}-\frac{n^2}{n^2+m^2}\right] \qquad (6\text{-}58)$$

$$K_h^x = \frac{1}{\pi}\left[\frac{n^2}{n^2+m^2}-\frac{n^2}{(m-1)^2+n^2}+\ln\left[\frac{m^2+n^2}{(m-1)^2+n^2}\right]\right] \qquad (6\text{-}59)$$

$$K_h^\tau = \frac{1}{\pi}\left[\frac{n(m-1)}{(m-1)^2+n^2}-\frac{mn}{n^2+m^2}+\arctan\left(\frac{m}{n}\right)-\arctan\left(\frac{m-1}{n}\right)\right] \qquad (6\text{-}60)$$

Table 6-8 Vertical induced stress coefficient K_h^z at a given depth z below the strip foundation under horizontally uniformly distributed pressure

$m=x/b$	$n=z/b$										
	0.01	0.10	0.20	0.40	0.50	0.60	0.80	1.00	1.20	1.40	2.00
−0.50	0.000	−0.011	−0.038	−0.103	−0.127	−0.144	−0.158	−0.157	−0.147	−0.133	−0.096
−0.25	−0.001	−0.042	−0.116	−0.199	−0.211	−0.212	−0.197	−0.175	−0.153	−0.132	−0.085
0	−0.318	−0.315	−0.306	−0.274	−0.255	−0.234	−0.194	−0.159	−0.131	−0.108	−0.064
0.25	−0.001	−0.039	−0.103	−0.159	−0.157	−0.147	−0.121	−0.096	−0.078	−0.061	−0.034
0.50	0.000	0.000	0.000	0.000	0.000	0.000	0.000	0.000	0.000	0.000	0.000
0.75	0.001	0.039	0.103	0.109	0.157	0.147	0.121	0.096	0.078	0.061	0.034
1.00	0.318	0.315	0.306	0.274	0.255	0.234	0.194	0.159	0.131	0.108	0.064
1.25	0.001	0.042	0.116	0.199	0.211	0.212	0.197	0.175	0.153	0.132	0.085
1.50	0.000	0.011	0.038	0.103	0.127	0.144	0.158	0.157	0.147	0.133	0.096

For a plane problem, the induced stress calculation at any point below the foundation does not need to use the corner-point method. The calculation of the plane problem is simpler than that of the space problem. In practical engineering, especially in hydraulic structures, the resultant force F acting on the foundation base is often eccentric and inclined (Figure 6-30). The contact pressures the base include the vertical trapezoid distributed pressure and the horizontal distributed pressure p_h. To calculate the induced stress at a given depth, the trapezoid distributed pressure at the base

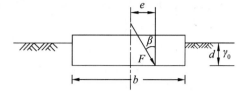

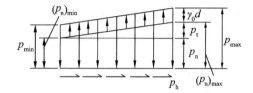

Figure 6-30 Decomposition of contact pressure at the foundation base for a plane problem

of foundation can be divided into a uniformly distributed pressure p_n and triangular distributed pressure with peak value of p_t. Then the induced stress under the actions of p_n, p_t, and p_h can be separately calculated. The total induced stress can be finally obtained using the superposition principle to combine each stress component.

【Example 6-4】 As shown in Figure 6-31, a retaining wall with the base width is 6m. The embedment depth of the wall is 1.5m. The weight of wall and other vertical load per meter is $F_v = 2400$kN/m, where the loading point is 3.2m away from the edge point A. The horizontal earth pressure force $F_h = 400$kN/m, where the acting point is 2.4m high from the base plane. The unit weight of ground soil is 19kN/m³. If it does not consider the influence of the backfill, please calculate the vertical induced stress at point M and point N with depth $z = 7.2$m below the base.

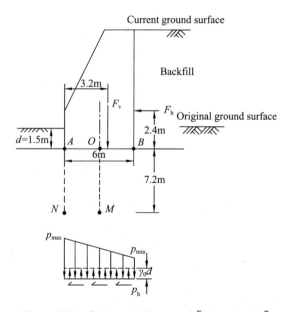

Figure 6-31 Schematic diagram of [Example 6-4]

[Solution]

(1) Calculate the eccentricity of resultant force acting on the base line: set the distance between the resultant force acting point and point A is x, according to the theorem of moment of resultant force,

$$F_v \cdot x = F_v \times 3.2 - F_h \times 2.4$$

$$x = \frac{3.2F_v - 2.4F_h}{F_v} = 3.2 - 2.4 \times \frac{400}{2400} = 2.8 \text{m}$$

Thus, the eccentricity of resultant force is $e = \frac{b}{2} - 2.8 = 0.2\text{m}$, which means the resultant force action point is located on the point to the left of midline with 0.2m.

(2) Calculate the contact pressure: according to Eq. (6-12), we have

$$p_{max} = \frac{F_h}{b}\left(1 + \frac{6e}{b}\right) = \frac{2400}{6}\left(1 + \frac{6 \times 0.2}{6}\right) = 480 \text{kPa}$$

$$p_{min} = \frac{F_h}{b}\left(1 - \frac{6e}{b}\right) = \frac{2400}{6}\left(1 - \frac{6 \times 0.2}{6}\right) = 320 \text{kPa}$$

By Eq. (6-15),

$$p_h = \frac{F_h}{b} = \frac{400}{6} = 66.7 \text{kPa}$$

(3) Calculate the net contact pressure:

$$p_n = p_{min} - \gamma_0 d = 320 - 19 \times 1.5 = 291.5 \text{kPa}$$

$$p_t = p_{max} - p_{min} = 480 - 320 = 160 \text{kPa}$$

(4) Calculate the induced stresses at point M and point N by calculating the sum of induced stress under the action of p_n, p_t, and p_h, which are listed in Table 6-9.

$$\sigma_{zM} = p_n \cdot K_s^z + p_t \cdot K_t^z + p_h K_h^z = 291.5 \times 0.420 + 160 \times 0.239 + 0$$
$$= 122.43 + 38.24 + 0 = 160.67 \text{kPa}$$

$$\sigma_{zN} = p_n K_s^z + p_t K_t^z + p_h K_h^z = 291.5 \times 0.375 + 160 \times 0.221 + 66.7 \times 0.131$$
$$= 109.31 + 35.36 + 8.74 = 153.41 \text{kPa}$$

Table 6-9 Calculation of induced stress under different types of pressure for [Example 6-4]

Pressure type	x		z		x/b		z/b		induced stress coefficient		Vertical induced stress at point	
	M	N	M	N	M	N	M	N	M	N	M	N
Vertical uniform distribution, p_n	3.0	6	7.2	7.2	0.5	1	1.2	1.2	$K_s^z =$ 0.420	$K_s^z =$ 0.375	122.43	109.31
Vertical triangular distribution, p_t	3.0	6	7.2	7.2	0.5	1	1.2	1.2	$K_t^z =$ 0.239	$K_t^z =$ 0.221	38.24	35.36
Horizontal uniform distribution, p_h	3.0	6	7.2	7.2	0.5	1	1.2	1.2	$K_h^z = 0$	$K_h^z =$ 0.131	0	8.74
Total induced stress	—	—	—	—	—	—	—	—	—	—	160.67	153.41

Note: the coordinate origin of x is set at point B, and the unit of induced stress is kPa.

Engineering Geology and Soil Mechanics

Knowledge expansion

Winkler Foundation Model

Soil has very complex mechanical behavior, because of its nonlinear, stress-dependent, anisotropic and heterogeneous nature. Hence, instead of modeling the subsoil in all its complexity, the subgrade is often replaced by a much simpler system called a subgrade reaction model. Winkler (1867) proposed a model that assumes the ratio between contact pressure at any given point and the associated vertical settlement, is linear and given by the coefficient of subgrade reaction K. In this model, the subsoil is replaced by fictitious springs whose stiffness is equal to K. The stiffness of the springs is named as the modulus of subgrade reaction. This modulus depends on some parameters like soil type, dimension, shape, embedment depth and type of foundation (flexible or rigid).

Winkler (1867) assumed the soil medium as a system of identical but mutually independent, closely spaced, discrete and linearly elastic springs. But, the simplified assumptions which this approach is based on caused approximations. One of the basic limitations is that, this model cannot transmit shear stresses which are derived from lack of spring coupling. Also, stress-strain behavior is assumed to be linear. Geometry, dimensions of the foundation and soil layering are assigned to be the most important effective parameters on K. Generally, the value of subgrade modulus can be obtained by some tests including plate load test, consolidation test, triaxial test, and CBR test.

Exercises

[6-1] What are the types of the soil stress according to the causes? How to define?

[6-2] What is the self-weight stress and what should be considered when calculating the self weight stress?

[6-3] What is a induced stress? What is the difference between a space problem and a plane problem?

[6-4] What is the flexible foundation? What is the rigid foundation? Tell the difference of contact pressure distribution for the two types of foundation.

[6-5] What is the net contact pressure? How to calculate the net contact pressure?

[6-6] How is the vertical induced stress distribution below foundation base? Why do two adjacent foundations will affect each other on the induced stress?

[6-7] What is the purpose for calculation of the induced stress below the foundation base?

[6-8] What is the corner-point method? How to apply the principle of superposition?

[6-9] How can the induced stress be consistent with the actual induced stress below the foundation?

[6-10] Figure 6-32 is a ground soil profile. The unit weight of each soil layer and the groundwater level are shown in the figure. Try to calculate the self-weight stress of the soil and draw the stress distribution diagram.

[6-11] Figure 6-33 shows a rectangular foundation with a buried depth of 1m. The load transmitted from the superstructure to the designed ground elevation is $P=2100$ kN, and the load eccentricity is $e=0.5$m. Try to calculate the vertical induced stress at the depth of 4m below the O, point A and point B, respectively.

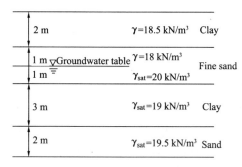

Figure 6-32 Schematic diagram of Exercise [6-10]

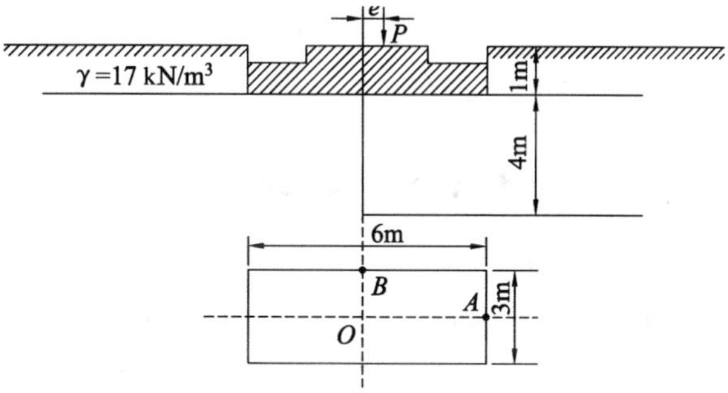

Figure 6-33 Schematic diagram of Exercise [6-11]

[6-12] Two squares A and B are two foundations and their dimensions as well as the net contact pressures at each foundation base are shown in Figure 6-34. Try to calculate the vertical induced stress at depth of 2m below the point O of foundation A.

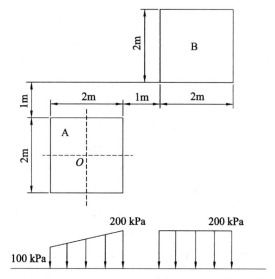

Figure 6-34 Schematic diagram of Exercise [6-12]

Chapter 7 Soil Consolidation and Settlement

7.1 Introductory case

To evaluate the suitability of foundation or earth structure, it is necessary to have design against bearing capacity failure and excessive settlement. For foundations on cohesive soils, the principal design criterion is typically in the control of settlements within the limits that is considered tolerable for the structure. As a result, once allowable settlements of foundation have been established, the total settlement over the service life of the structure is a major factor in the choice of foundation design.

When a soil layer is subjected to vertical stress, volume change can take place through rearrangement of soil grains, where certain amount of grain fracture may also take place. The volume of soil grains remains constant, so the change in total volume is due to change in volume of water. In saturated soils, it can happen only if water is squeezed out of the voids. The movement of water takes time, which is controlled by the permeability of the soil and the locations of free draining boundary surfaces. It is necessary to determine both the magnitude of volume change (or the settlement) and the time required for the volume change to occur. The settlement is depend on the applied stress, thickness of the soil layer, and the compressibility of the soil. When soil is loaded undrained, the pore pressure increases. As the excess pore pressure dissipates where water leaves the soil, settlement takes place. This process takes time, and the rate of settlement decreases over time. In coarse-grained soils (sand and gravel), volume change occurs immediately as pore pressures are dissipated rapidly due to high permeability. In fine-grained soils (silt and clay), slow seepage occurs due to low permeability.

Consolidation is a process by which soils decrease in volume. As Karl Terzaghi said: "consolidation is any process which involves a decrease in water content of saturated soil without replacement of water by air." In general, it is the process in which reduction in volume takes place by expulsion of water under long term static loads. It occurs when stress is applied to a soil that causes the soil particles to pack more tightly, therefore reducing the bulk volume. When it occurs in a soil that is saturated, consolidation is the gradual reduction in volume due to drainage of the pore water, the process continues until the excess pore water pressure set up by an increase in total stress and is completely dissipated; the simplest case is that in one-dimensional consolidation, in which a condition of zero lateral strain is implicit. The process of swelling is the reverse of consolidation, which is the gradual increase in volume of soil under negative excess pore water pressure.

The purpose of this chapter is to present the fundamental concepts regarding consolidation and settlement analysis for saturated, inorganic, cohesive soils. In addition, the recommended procedure for estimation of foundation settlements is described.

It is an important issue to estimate the foundation settlement in the practice of civil engineering. Because of the uneven load distribution on the foundation of a building or structure, and the natural inhomogeneity of foundation soils, differential settlement may occur, i. e., the vertical deformation at each part of the foundation soil may be different. The famous Leaning Tower of Pisa in Italy is a typical example (Figure 7-1). Two questions should be addressed for the foundation settlement: how much and how long, i. e., the magnitude and rate of settlement. The answers of the two questions can be found in this chapter.

Figure 7-1 Photograph of Leaning Tower of Pisa

The Leaning Tower of Pisa was built in 1173, which is the campanile or freestanding bell tower of the Cathedral of the Italian city of Pisa, known worldwide for its unintended tilt. It is located behind the Cathedral and is the third oldest structure in Pisa's Cathedral Square after the Cathedral and the Baptistery. The height of the tower is 55.86m from the ground on the low side and 56.67m on the high side. The width of the walls at the base is 4.09m and the top is 2.44m. Its weight is estimated at 14500 metric tons. The tower has a cylinder body and there are 15 giant marble columns to support the upper loads at the first floor. 30 pillars were set from the second to the seventh floor and 12 pillars of small size were at the top floor. The big clock was placed on the top floor. Tourists can climb spirally through the 294 steps to the top floor. Based on the calculation, the center of gravity of the tower is located 22.6m above the foundation. The area of the circular foundation is 285m^2, and the average pressure on the ground is 497kPa.

The tower has been leaning shortly after its initial construction in 1173. By 1178, the third floor was complete. It was observed at this stage that the tower was continuing leaning on the north side. Then the construction was on hold on for the next 94 years because of wars happened in neighboring countries. The construction was restarted in 1273 and completed in 1350. The tower leaned on the north side about 2.1m from the vertical in 1350. According to the testing results from 1829 to 1910, there are around 3.8 millimeters settlement was found at the top floor each year. The tower was shut down to tourists in 1990, but reopened up in 2001 because of extensive restoration efforts. Prior to restoration work that was performed between 1990 and 2001, the tower leaned at an angle of 7.5 degrees, whereas it leans at about 3.99 degrees presently. It means that the top of the tower is displaced horizontally 3.9m from the center.

7.2 Components of total settlement

The total settlement of a loaded soil has three components: elastic settlement, primary consolidation, and secondary compression.

During construction, surface loads from foundations or earth structures are transmitted to the underlying soil profile. As a result, stresses increase within the soil mass where the structure undergoes a time-dependent vertical settlement. In general, this time-settlement curve can be represented conceptually as shown in Figure 7-2. The total settlement, S_t, is calculated as the sum of the following three components

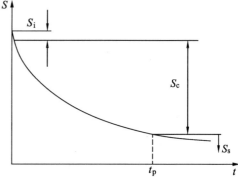

Figure 7-2 Time-settlement curve showing total settlement components

$$S_t = S_i + S_c + S_s \qquad (7-1)$$

where, S_i—immediate settlement (or elastic settlement);

S_c—primary consolidation (or simply consolidation) settlement;

S_s—secondary compression settlement.

Immediate (or elastic) settlement is to account of the change in shape at constant volume, i.e. due to vertical compression and lateral expansion. Primary consolidation (or simply consolidation) is to account of water from the voids, which is a function of the permeability and compressibility of soil. Secondary compression is to account of creep behaviour. Primary consolidation is the major component of settlement and it can be estimated.

Immediate settlement is time-independent and results from shear strains that occur at constant volume as the load is applied to the soil. Although this type of settlement component is not elastic, it is generally calculated using elastic theory for cohesive soils such as clays. Both consolidation and secondary compression settlement components are time-dependent and result from a reduction of void ratio and expulsion of water from the voids of the soil skeleton. For consolidation settlement, the rate of void ratio reduction is controlled by the rate at which water can flow out of the soil. Therefore, during consolidation, pore water pressure exceeds the steady state condition throughout the depth of the layer. Over time, the rate of consolidation settlement continuously decreases as effective stresses increase to reach the equilibrium values. Once the consolidation process is completed at time t_p (Figure 7-2), settlement continues in the form of secondary compression. During secondary compression, the rate of void ratio reduction is controlled by the rate of compression of soil skeleton. Thus, it is essentially a creep phenomenon that occurs at constant vertical effective stress and without excess pressure in the pore water.

The time-settlement relationship shown in Figure 7-2 is conceptually valid for all soil types. However, differences are in the magnitude of the components and the time at which they occur for different soils. For granular soils, such as sands, the permeability coefficient is sufficiently large where consolidation occurs nearly instantaneously with the applied load. In addition, although granular soils do exhibit creep effects, the secondary compression of which is generally insignificant. For cohesive soils, such as clays, permeability coefficient is very low and the consolidation of a thick deposit may require years or even decades to complete. Secondary compression can be substantial for cohesive soils. Different from sands and clays, peats and organic soils generally undergo rapid consolidation and extensive, long-term secondary compression. The first step in a settlement analysis is a comprehensive study of the changes in applied loads and the selection of live load pertinent to each of the three total settlement components. Often, insufficient attention is given to this aspect of the problem. In general, immediate settlements should be computed using live and dead loads of the structure. Consolidation and secondary compression settlements should be calculated using the dead load and permanent live load, with a reasonable fraction of the transient live load. The proper estimate of this fraction should be made after consulting with the structural engineer of the project.

7.2.1 Immediate (elastic) settlement

For saturated or nearly saturated cohesive soils, a linear elastic model is generally used for the calculation of immediate settlement. Although clays do not behave as linear elastic materials, the reason for the use of elastic theory is the availability of solutions for a wide variety of boundary conditions in foundation engineering problems. In general, the elastic approximation performs reasonably well in the case of saturated clays under monotonic loading conditions which is not approaching failure. In addition, in the same conditions, the elastic parameters can generally be assumed as approximately constant throughout homogeneous soil mass.

For non-cohesive soils, in which the equivalent elastic modulus depends on confinement, where the use of linear elastic theory coupled with the assumption of material homogeneity is inappropriate. Immediate settlement on granular soils is most often estimated using the procedure of Schmertmann method.

For those cases in which a linear elastic model is acceptable, solutions for stress distribution and surface deflection under a variety of flexible and rigid surface loading configurations can be found in literatures. One particularly useful relationship is provided herein for the immediate settlement of a circular or rectangular footing at the surface of a deep isotropic stratum. In this case, the immediate vertical displacement is given by

$$S_i = C_s pB \left(\frac{1-v^2}{E_u} \right) \tag{7-2}$$

where, S_i—immediate settlement of a point on the surface;

C_s—shape and rigidity factor;

p—equivalent uniform stress on the footing (total load/footing area);

B—characteristic dimension of the footing;

v—Poisson's ratio;

E_u—undrained elastic modulus (Young's modulus).

The factor C_s is a function of the shape and rigidity of the loaded area and the point on the footing for which the immediate settlement estimate is desired. Thus, Eq. (7-2) can be used for both rigid and flexible footings with the appropriate values of C_s given in Table 7-1. The characteristic footing dimension, B, is taken by convention as the diameter of a circular footing or the width a rectangular footing. For saturated cohesive soils, constant volume strain is usually assumed and Poisson's ratio, v, equals to 0.7. For soils that are nearly saturated, v will be less than 0.7. However, using $v=0.5$ is generally acceptable since the computed immediate settlement is not sensitive to small changes in v. Reliable evaluation of the remaining soil parameter, the undrained elastic modulus, E_u, is critical for an appropriate estimation of immediate settlement. In general, E_u is the slope of the undrained stress-strain curve on a stress path in field condition. Figure 7-3 illustrates the measurement of E_u from a plot of principal stress difference, $\Delta\sigma$, versus axial strain, ε_a, as typically obtained from an undrained triaxial test. Principal stress difference is defined as $\sigma_1 - \sigma_3$, where σ_1 and σ_3 are the major and minor principal stresses, respectively. The initial tangent modulus, E_{ui}, is determined from the initial slope of the curve. The secant modulus, E_{us}, is used instead of E_{ui} when there is obvious nonlinearity in the stress-strain relationship over the stress range. Generally, the secant modulus would be taken at some predetermined stress level, such as 50% of the principal stress difference at failure, $\Delta\sigma_f$, shown in Figure 7-3.

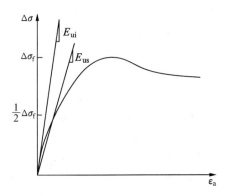

Figure 7-3　Definitions of the initial tangent modulus and secant modulus

Chapter 7 Soil Consolidation and Settlement

Table 7-1 Values of C_s

Shape and rigidity		Center	Corner	Edge/middle of long side	Average
Circle (flexible)		1.00		0.64	0.85
Circle (rigid)		0.79		0.79	0.79
Square (flexible)		1.12	0.56	0.76	0.95
Square (rigid)		0.82	0.82	0.82	0.82
Rectangle (flexible): length/width	2	1.53	0.76	1.12	1.30
	5	2.10	1.05	1.68	1.82
	10	2.56	1.28	2.10	2.24
Rectangle (rigid): length/width	2	1.12	1.12	1.12	1.12
	5	1.6	1.6	1.6	1.6
	10	2.0	2.0	2.0	2.0

As a first approximation, the undrained elastic modulus can be estimated from the undrained shear strength using

$$E_u = (500 \sim 1500)c_u \quad (7\text{-}3)$$

where, c_u—undrained shear strength determined from a field vane shear test.

In general, E_u depends strongly on the level of shear stress. The lower value in Eq. (7-3) corresponds to highly plastic clays where the applied stress is relatively large as compared to the soil strength. The higher value is for low plasticity clays under small shear stress. In addition, the E_u/c_u ratio decreases with increasing overconsolidation ratio at a given stress level. Thus, Eq. (7-3) can provide a rough estimation of E_u suitable only for preliminary design.

In situations where a field loading test is not warranted, the undrained modulus should be estimated from a consolidated undrained (CU) triaxial test in the laboratory. The following procedure is recommended:

1) Obtain the high quality of soil samples. If possible, use a large-diameter piston sampler, or excavate blocks by hand from a test pit. Optimally, the laboratory test should be performed on the same day as the field sampling operations.

2) Reconsolidate the specimen in the triaxial cell to the estimated initial in situ state. If possible, anisotropic K_0 consolidation is preferred to isotropic consolidation. Undrained modulus values determined from unconfined compression tests will significantly underestimate the actual value of E_u, thereby overestimate the immediate settlement.

3) Load and unload the specimen in undrained axial compression to the expected in situ stress level for a minimum of 5 cycles. For field loading conditions other than a structural foundation, a different laboratory stress path may be needed to better match the actual in situ stress path.

4) Obtain E_u from the fifth (or greater) cycle in similar fashion as shown in Figure 7-3.

For sensitive clays with low plasticity, CU triaxial tests will likely yield low values of E_u, even if the specimens are allowed to undergo appreciable aging and E_u is determined at a low stress level. For highly plastic clays and organic clays, CU tests may yield stress-strain curves that are indicative of in situ behavior. However, it may be difficult to represent the nonlinear behavior with a single modulus value.

The undrained elastic modulus is best measured directly from field tests. For near surface clay deposits having a consistency that does not vary greatly with depth, E_u may be obtained from a plate load test per relevant code at footing elevation and passed through several loading-unloading cycles. In this case, all the parameters in Eq. (7-2) are obtained except the factor of $(1-v^2)/E_u$, which can then be calculated. Because of the relatively little influence of the test, it may be advisable to use a selection of different size plates and then scale $(1-v^2)/E_u$ to the size of the prototype foundation. In situations where the loaded stratum is deep or displays substantial heterogeneity, plate load tests may not provide a representative value for E_u. Large-scale loading tests, for example, an embankment or a large tank of water may be better. In this case, the immediate settlement of the foundation is measured directly without requiring Eq. (7-2). Measurement of stress-strain behavior using field tests is preferred to laboratory tests because of the difficulties in determining the appropriate modulus in the laboratory. The most important of issue is the invariable disturbance of soil that occurs during sampling and testing. Of the many soil properties defined in geotechnical engineering, E_u is one of the most sensitive to sample disturbance.

For many foundations on cohesive soils, the immediate settlement is a relatively small part of the total settlment. Thus, a detailed study is seldom justified unless the structure is very sensitive to distortion, footing sizes and/or loads vary considerably, or the shear stresses imposed by the foundation are approaching a failure condition.

7.2.2 Consolidation settlement

Different from immediate settlement, consolidation settlement occurs as the result of volume compression within the soil. For granular soils, the consolidation process is sufficiently fast where consolidation settlement is generally included with immediate settlement. Cohesive soils have a much lower permeability coefficient, as a result, consolidation requires a much longer time to complete. In this case, consolidation settlement is calculated separately from immediate settlement, as suggested by Eq. (7-1).

When a load is applied to the ground surface, there is a tendency for volumetric compression of the underlying soils. For saturated materials, an increase in pore water pressure occurs immediately upon load application. Consolidation is then the process of reduction in volume due to the expulsion of water from the pores of the soil. The dissipation of excess pore water pressure is accompanied by an increase in effective stress and volumetric strain. Analysis of the resulting settlement is greatly simplified if it is assumed that such strain is one-dimensional, occurring only in the vertical direction. This assumption of one-dimensional compression is considered to be reasonable when ① the width of the loaded area exceeds four times of the thickness of the clay stratum, ② the depth to the top of the clay stratum is greater than twice the width of the loaded area, or ③ the compressible material lies between two stiffer soil strata whose presence tends to reduce the magnitude of horizontal strains.

Employing the assumption of one-dimensional compression, the consolidation settlement of a cohesive soil stratum is generally calculated in two steps:

1) Calculate the total (or "ultimate") consolidation settlement, S_c, upon the completion of the consolidation process.

2) Using the theory of one-dimensional consolidation, calculate the fraction of S_c that would have occurred by the end of the service life of the structure. This fraction is the component of consolidation settlement shown in Figure 7-2.

In reality, the total amount of consolidation settlement and the rate at which the settlement occurs is a coupled problem in which neither quantity can be calculated independently. However, in geotechnical engineering practice, total consolidation settlement and rate of consolidation are almost always computed independently since there is no widely accepted procedures to solve the coupled problem. In this section, the calculation of total consolidation settlement will be presented first, followed by procedures to calculate the rate at which this settlement occurs.

The total one-dimensional consolidation settlement, S_c, results from a change in void ratio, Δe, over the depth of the consolidating layer. The basic equation for calculating the total consolidation settlement for a single compressible layer is

$$S_c = \frac{\Delta e H_0}{1 + e_0} \tag{7-4}$$

where, e_0 —initial void ratio;

H_0 —initial height of the compressible layer.

Consolidation settlement is sometimes calculated using H_0 for the entire consolidating stratum where the stress is acting at the mid of the height. This procedure will underestimate the actual settlement, and the error will increase with the increase in thickness of the clay. As Δe generally varies with depth, settlement calculations can be improved by dividing the consolidating stratum into n sublayers for analysis purposes. Eq. (7-4) is applied to each sublayer and the cumulative settlement is computed using the following equation.

$$S_c = \sum_{i=1}^{n} \frac{\Delta e_i H_{0i}}{1 + e_{0i}} \tag{7-5}$$

where, Δe_i —change in void ratio of the ith sublayer;

H_{0i} —initial thickness of the ith sublayer;

e_{0i} —initial void ratio of the ith sublayer.

The appropriate Δe_i for each sublayer within of the compressible soil must be determined. The initial vertical effective stress, σ'_{v0}, and the final vertical effective stress (after excess pore pressures is fully dissipated), σ'_{vf}, are needed. The distribution of σ'_{v0} with depth is usually obtained by subtracting the in situ pore pressure from the vertical total stress, σ_v. Vertical total stress at a given depth is calculated using the following equation.

$$\sigma_v = \sum_{j=1}^{m} \gamma_j z_j \tag{7-6}$$

where, γ_j—unit weight of the jth stratigraphic layer;

z_j—thickness of the jth stratigraphic layer;

m—number of layers above the depth of interest point.

It should not be assumed that in situ pore pressures are hydrostatic. Rather, a significant upward or downward groundwater flow may be present. For some important structures, the installation of piezometers to measure the in situ distribution of pore pressure is warranted. In addition, the piezometers will provide a valuable check on the estimated initial excess pore pressures and indicate when the consolidation process is completed.

The final vertical effective stress is equal to the initial vertical effective stress plus the change of vertical effective stress, $\Delta \sigma'_v$, due to loading.

$$\sigma'_{vf} = \sigma'_{v0} + \sigma'_v \tag{7-7}$$

For truly one-dimensional loading conditions, such as a wide fill, $\Delta \sigma'_v$ is constant with depth and equal to the change in total stress applied at the surface of the soil stratum. For situations in which the load is applied over a limited surface area, such as a strip foundation, $\Delta \sigma'_v$ will decrease with depth as the surface load is transmitted to deeper larger portions of the soil mass. In this case, the theory of elasticity can be used to estimate $\Delta \sigma'_v$ as a function of depth under the center of the loaded area, and the details can be found in Chapter 6.

Beside the calculation of ultimate consolidation settlement related to the compete dissipation of excess pore pressure, we also figure out the time needed for the consolidation settlement, i. e., how long it will take for the completion of settlement. As we know, at any time during the process of consolation, the amount of settlement is directly related to the excess pore pressure that has been dissipated. The consolidation theory is used to predict the dissipation of excess pressure as a function of time. Therefore, the theory is also used to predict the rate of consolidation settlement. The one-dimensional consolidation theory proposed by Karl Terzaghi in 1925, which is most commonly used for predication of consolidation settlement rate. Details will be discussed in section 7.4.

7.2.3 Secondary compression settlement

Secondary compression settlement results from the time-dependent rearrangement of soil particles under the constant effective stress conditions. For highly compressible soils, such as soft clays and/or peats, secondary compression is important whenever there is a net increase in σ'_v due to surface loading. Although structures would rarely be founded on these soils. Highways, for example, must cross areas of compressible soils that are either too deep or too extensive to excavate. Secondary compression settlement can be predicted using the secondary compression index, C_α, defined as the change of void ratio per time in logarithmic scale.

$$C_\alpha = -\frac{\Delta e}{\Delta \lg t} \tag{7-8}$$

Laboratory values for the secondary compression index should be measured at a stress level corresponding to that expected in field. Besides, the temperature, chemical factors and biological effects may lead to secondary compression settlement. The secondary compression settlement is very important for some soft clays, but it is relatively small and can be ignored for hard or overconsolidated soils. The phenomenon that the deformation of soil increases with time under the action of a constant stress level is called creep, which is the most common rheological phenomenon of soil. The secondary compression settlement is the result of soil creep.

Once a C_α value has been selected, secondary compression settlement S_s can be calculated using the following equation.

$$S_s = \frac{C_\alpha H_0}{1+e_0} \lg \frac{t_f}{t_p} \tag{7-9}$$

where, t_p—time at the end of primary consolidation;

t_f—time when secondary compression settlement is predicted (typically the design life of the structure).

7.3 Compressibility characteristics

Soils are often subjected to uniform loading over large areas, such as from wide foundations, fills or embankments. Under such conditions, the soil which is remote from the edges of the loaded area undergoes vertical strain, but there is no horizontal strain. Thus, the settlement occurs in one-dimension.

7.3.1 Consolidation test

To learn the fundamentals of compression and consolidation, it includes preparing and analyzing compression and consolidation curves. Compression curves show the change in void ratio with effective stress and consolidation curves show the vertical deformation with time. One will analyze the curves to calculate total settlements. and the time it takes for the settlements to occur in embankments with different soil stress conditions. Once the initial and final stress conditions have been established, it is necessary to determine the relationship between void ratio and vertical effective stress for the in-situ soil. The information is generally obtained from a laboratory consolidation test, also known as the oedometer test. The general consolidation test procedure per relevant codes is to place successive loads on an undisturbed soil specimen (typically with a 20mm in height and a

61.8mm in diameter). We can measure the void ratio corresponding to the end of consolidation at each load increment. The load increment ratio (*LIR*) is defined as the added load divided by the previous total load on the specimen. The load increment duration (*LID*) is the elapsed time permitted for each load increment. For the standard consolidation test, the load is doubled every day, giving $LIR=1$ and $LID=24$ hours.

Consolidation test subjects to a specimen (usually a clay or clayey silt) placed in a rigid wall oedometer (no lateral strain, $\varepsilon_a = \varepsilon_{vol}$), as shown in Figure 7-4, with increasing constant vertical load. The vertical load is applied faster than pore water can dissipate, causing an initial increase in pore water pressure within the sample. The initial increase in pore pressure is equal to the instantaneous change in applied total stress.

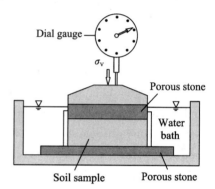

Figure 7-4 Schematic of oedometer test cell

When a soil is subject to rapid loading, the induced hydraulic gradient causes water to flow towards the drainage boundaries (upper and lower porous stones) and the sample consolidates (decreases in volume while reducing excess pore water pressure) at a rate that is depend on several factors including hydraulic conductivity, thickness, compressibility, degree of saturation (it is assumed the soil is fully saturated), pore fluid, initial void ratio, and soil structure.

7.3.2 One-dimensional compression curve

The consolidation test can include a series of load increments to create the void ratio to effective stress curve, known as the compression curve. Typically, the void ratio at the end of primary consolidation is taken from the curve because at this time the excess pore pressure is zero; thus the effective stress is equal to the applied total stress of that particular load increment. The compressibility of soils under one-dimensional compression can be described by the decrease in the void ratio with the increase of effective stress. The relation of void ratio (e) and effective vertical stress (σ'_v) can be depicted either in an arithmetic plot (Figure 7-5) or a semi-logarithmic plot (Figure 7-6).

In the arithmetic plot as shown in Figure 7-5, as the soil compresses, for the same increase of vertical effective stress $\Delta\sigma'_v$, the void ratio reduces by a smaller value from Δe_1 to Δe_2. It is to account of an increasingly denser packing of the soil particles as the pore water is squeezed out. In fine-grained soils, a much longer time is required for the pore water to escape compared to coarse-grained soils.

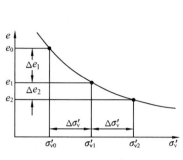
Figure 7-5 $e \sim \sigma_v'$ curve

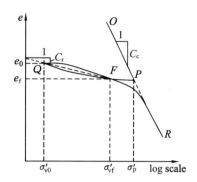
Figure 7-6 $e \sim \lg \sigma_v'$ curve

It can be said that the compressibility of a soil decreases as the effective stress increases. This can be presented by the slope of the void ratio to effective stress relation, which is called the coefficient of compressibility, a_v.

$$a_v = -\frac{\Delta e}{\Delta \sigma_v'} = \frac{e_1 - e_2}{\sigma_{v2}' - \sigma_{v1}'} \qquad (7\text{-}10)$$

The negative sign is introduced to make a_v a positive parameter. The value of a_v depends on the stress range over which it is calculated.

The coefficient of compressibility (a_v) is one of the most important indexes for characterizing the soil compression behavior. The higher the value of a_v is, the steeper is the $e \sim \sigma_v'$ curve, and the higher is the compressibility of soil; on the contrary, the smaller the value of a_v is, the smoother is the $e \sim \sigma_v'$ curve, and the lower is the compressibility of soil. Generally, the $e \sim \sigma_v'$ curve is not a straight line, so the value of a_v for the same soil is not a constant. Its value depends on the initial vertical effective stress σ_{v0}' and the effective stress increment $\Delta \sigma_v'$. In engineering practice, in order to facilitate unified comparison, it is customary to use the coefficient of compressibility a_{v1-2} in the range of 100kPa and 200kPa to measure the compressibility of soil per Chinese code GB 50007—2011 "Code for design of building foundation". When $a_{v1-2} < 0.1\text{MPa}^{-1}$, it belongs to a low compressibility soil; when $0.1\text{MPa}^{-1} \leqslant a_{v1-2} < 0.5\text{MPa}^{-1}$, it belongs to a medium compressible soil; when $a_{v1-2} \geqslant 0.5\text{MPa}^{-1}$, it belongs to a high compressibility soil.

If e_0 is the initial void ratio of the consolidating layer, another useful parameter is the coefficient of volume compressibility, m_v, which is defined as the ratio of change in volumetric strain (ε_v) over change in vertical effective stress (σ_v').

$$m_v = \frac{\Delta \varepsilon_v}{\Delta \sigma_v'} = \frac{a_v}{1 + e_0} \qquad (7\text{-}11)$$

It represents the compression of the soil per unit original thickness due to a unit increase of pressure. The unit of m_v is the inverse of pressure, i.e., m^2/kN or kPa^{-1}.

Figure 7-6 shows the relation of void ratio (e) and vertical effective stress (σ_v') of a clayey soil in a semi-logarithmic plot. OP corresponds to initial loading process of the soil. PQ corresponds to unloading of the soil. QFR corresponds to a reloading of the soil. Upon reloading beyond P point. The stress history continues along the path that would follow the loading trend from O to R. The preconsolidation stress (σ'_p) is defined as the maximum vertical effective stress experienced by the soil. This stress is identified in comparison with the effective stress in the present state. For soil at state Q or F, it would correspond to the effective stress at point P. If the current effective stress (σ'_v) is equal (note that it cannot be greater than) to the preconsolidation stress, then the deposit is normally consolidated (NC). If the current effective stress is less than the preconsolidation stress, then the soil is said to be over-consolidated (OC).

It may be seen that for the same increase in effective stress, the change in void ratio is much less with an overconsolidated soil (from e_0 to e_f) than that with a normally consolidated soil as in path OP. In unloading, the soil swells but the increase in volume is much less than the initial decrease in volume for the same amount of decrease in stress. The distance from that of the normal consolidation line has an important influence on soil behaviour. It is described numerically by the over-consolidation ratio (OCR), which is defined as the ratio of the preconsolidation stress to the current vertical effective stress.

$$OCR = \sigma'_p / \sigma'_v \tag{7-12}$$

Note that when the soil is normally consolidated, $OCR=1$, settlements will generally be much smaller for structures built on overconsolidated soils. Most soils are overconsolidated to some degree. It can be attributed to shrinking and swelling of the soil on drying and rewetting, changes in ground water levels, and unloading due to erosion of overlying strata. For NC clays, the plot of void ratio versus logarithm effective stress can be approximated to a straight line(like OP), the slope of which is indicated by a parameter termed as compression index, C_c.

$$C_c = \frac{e_1 - e_2}{\lg\sigma'_{v2} - \lg\sigma'_{v1}} = \frac{\Delta e}{\lg\left(\frac{\sigma'_{v1} + \Delta \sigma'_v}{\sigma'_{v1}}\right)} \tag{7-13}$$

For OC clays, in over-consolidated stress range, i. e., the current vertical effective stress is less than the preconsolidation stress, the plot of void ratio versus logarithmic vertical effective stress can also be approximated to be a straight line (like QF), the slope of which is indicated by another parameter called recompression index or expansion index, C_r.

$$C_r = \frac{-\Delta e}{\lg\Delta\sigma'_v} = \frac{e_0 - e_f}{\lg\sigma'_{vf} - \lg\sigma'_{v0}} \tag{7-14}$$

It is possible to empirically determine the preconsolidation stress σ'_p of soil. The soil sample is loaded in the laboratory to obtain the void ratio-effective stress relationship. Empirical procedures are used to estimate the preconsolidation stress, the most widely used method is Casagrande's method which is illustrated as follows:

The steps are shown below according to the Figure 7-7:

① Draw the graph using an appropriate scale.

② Determine the point of maximum curvature at point A.

③ At A, draw a tangent line AB of the curve.

④ At A, draw a horizontal line AC.

⑤ Draw an extension line ED of the straight portion of the curve.

⑥ Where the line ED cuts the bisector AF of angle CAB, that point corresponds to the preconsolidation stress of soil.

Figure 7-8 shows a typical compression $e \sim \lg\sigma'_v$ curve on a semi-logarithmic scale. The open points are viod ratios measured at the end of 24 hours for each load increment. It includes the immediate and secondary

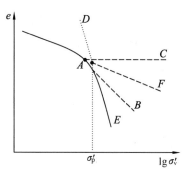

Figure 7-7 Schematic of Casagrande's method to determine the preconsolidation stress

compression settlements. The solid points represent the sum of changes in void ratio during consolidation, and are calculated by subtracting out the immediate and secondary compression from that of the previous load increments. As indicated in Figure 7-8, the laboratory compression curve is best drawn through the solid points. The reason for this procedure is that S_i, S_c, and S_s are computed separately and then summed to calculate total settlement S using Eq. (7-1). Therefore, the immediate and secondary compression settlements should likewise be removed from the laboratory compressibility curve to compute S_c.

For the compression curve in Figure 7-8, both the preconsolidation pressure σ'_p and the in situ initial vertical effective stress σ'_{v0} are indicated. The stress history of a soil layer is generally expressed by its over-consolidation ratio (OCR), as mentioned above. Normally consolidated soils have $OCR=1$, while soils with an $OCR>1$ are preconsolidated or over-consolidated. For the example shown in Figure 7-8, the soil is overconsolidated. In addition, a soil can be underconsolidated if excess pore pressures is within the deposit (i. e., the soil is still undergoing consolidation). With the exception of recently

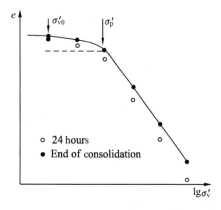

Figure 7-8 Typical laboratory compression curve on a semi-logarithm scale

deposited materials, soils in the field are often overconsolidated as a result of unloading, desiccation, secondary compression, and/or aging effects.

The preconsolidation pressure is the stress at which the soil begins to yield in volumetric compression, it therefore separates the region of small strains ($\sigma'_v < \sigma'_p$) from the region of large strains ($\sigma'_v > \sigma'_p$) on the $e \sim \lg\sigma'_v$ curve. As a result, for given initial and final stress condition, the total consolidation settlement of a compressible layer is highly depend on the value of the preconsolidation pressure. If a foundation applies a stress increment such that the final stress is less than σ'_p, the consolidation settlement will be relatively small. However, if the final stress is larger than σ'_p, much larger settlements will occur. Therefore, accurate determination of the preconsolidation pressure and its variation with depth is the most important step in settlement analysis. The determination of σ'_p is generally performed using the Casagrande graphical construction method.

For analysis purposes, the laboratory compressibility curve is usually approximated as linear line (in log scale) for both the overconsolidated and normally consolidated ranges. A typical example is shown in Figure 7-9. The slope of the overconsolidated range is the recompression index, C_r. Although this portion of a compression curve is generally not linear on the semilog plot, a constant C_r is usually fitted for simplicity. The slope of the normally consolidated portion in the compression curve is the compression index, C_c. It is often constant over typical stress ranges of engineering interest. C_c and C_r can be calculated using Eq. (7-13) and Eq. (7-14) respectively.

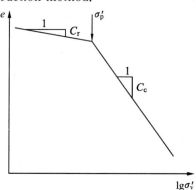

Figure 7-9 Simplified approximation of a laboratory compressibility curve

In many clay deposits, σ'_p, C_c and C_r vary considerably with depth and the practice of performing one or two consolidation tests to evaluate the entire profile is seldom satisfactory. The following procedure is recommended for performing and interpreting the consolidation test to well obtain these parameters:

1) Obtain the high quality and undisturbed specimens, each representative of one sublayer in the profile, and testing on the same day if possible. The value of σ'_p determined in the laboratory is very sensitive to sample disturbance and will generally be underestimated by using poor-quality soil samples.

2) Perform a consolidation test in which each load increment is placed on the specimen at the end of consolidation after the previous increment. The final void ratio obtained from each load will then directly provide the solid points in Figure 7-8. The end of consolidation can be determined from pore pressure measurements or, less accurately, from graphical procedures such as the Casagrande log time or Taylor square root of time methods. $LIR=1$ is satisfactory; however, to more accurately measure the value of σ'_p, it may be advantageous to run another test with smaller load increments in the vicinity of the preconsolidation pressure. If a standard consolidation test is performed using $LID=24$ hours, plot the cumulative void ratio reduction during consolidation for each increment as shown by the solid dots in Figure 7-8.

3) Obtain C_r by unloading from σ'_p to σ'_{v0} and then reloading. This path is indicated by the dashed line in Figure 7-9. Using the initial reloading curve in Figure 7-5 will yield a large value of C_r. When the value of C_r is critical to a particular design, a backpressured oedometer should be used for testing. The C_r value obtained by unloading and reloading in

a conventional oedometer (without backpressure) is about twice the value obtained in a backpressured oedometer due to diffusion of gas within the pore water.

4) Reconstruct the in situ compression curve using the methods of Schmertmann.

Once the in-situ compression curve is established for a given sublayer, the change of void ratio can be calculated knowing $\Delta\sigma'_v$. In normally consolidated conditions, the change of void ratio for the ith sublayer is

$$\Delta e_i = C_{ci} \lg\left(\frac{\sigma'_{vfi}}{\sigma'_{v0i}}\right) \tag{7-15}$$

where σ'_{v0i} and σ'_{vfi} are inital and final vertical effective stresses for the ith sublayer, respectively.

Substituting Eq. (7-15) into Eq. (7-5), the ultimate consolidation settlement for a normally consolidated soil is

$$S_c = \sum_{i=1}^{n} \frac{C_{ci} H_{0i}}{1+e_{0i}} \lg\left(\frac{\sigma'_{vfi}}{\sigma'_{v0i}}\right) \tag{7-16}$$

where the summation is performed over n^{th} sublayer.

In the case of over-consolidated clays, the change of void ratio for i^{th} sublayer at a given $\Delta\sigma'_{vi}$ ($\Delta\sigma'_{vi} = \sigma'_{vfi} - \sigma'_{v0i}$) can be obtained as

when $\sigma'_{vf} < \sigma'_p$,

$$\Delta e_i = C_{ri} \lg\left(\frac{\sigma'_{vfi}}{\sigma'_{v0i}}\right) \tag{7-17}$$

when $\sigma'_{vf} > \sigma'_p$,

$$\Delta e_i = C_{ri} \lg\left(\frac{\sigma'_{pi}}{\sigma'_{v0i}}\right) + C_{ci} \lg\left(\frac{\sigma'_{vfi}}{\sigma'_{v0i}}\right) \tag{7-18}$$

Substituting Eq. (7-17) and Eq. (7-18) into Eq. (7-5), the total consolidation settlement of an overconsolidated soil can be obtained as

when $\sigma'_{vf} < \sigma'_p$,

$$S_c = \sum_{i=1}^{n} \frac{C_{ci} H_{0i}}{1+e_{0i}} \lg\left(\frac{\sigma'_{vfi}}{\sigma'_{v0i}}\right) \tag{7-19}$$

when $\sigma'_{vf} > \sigma'_p$,

$$S_c = \sum_{i=1}^{n} \frac{C_{ci} H_{0i}}{1+e_{0i}} \lg\left(\frac{\sigma'_{pi}}{\sigma'_{v0i}}\right) + \frac{C_{ci} H_{0i}}{1+e_{0i}} \lg\left(\frac{\sigma'_{vfi}}{\sigma'_{pi}}\right) \tag{7-20}$$

As noted earlier, the discussion of consolidation settlement is limited to conditions of one-dimensional compression. In the cases where the thickness of the compressible strata is large relative to the dimensions of the loaded area, the three-dimensional condition of the problem may influence the rate of consolidation settlement.

7.4 Analysis of consolidation

A general theory for consolidation, incorporating three-dimensional flow is complicated and only applicable to a very limited range of problems in geotechnical engineering. For the vast majority of practical settlement problems, it is sufficient to consider that the seepage and strain in one direction only, as one-dimensional consolidation in the vertical direction.

The theory of consolidation was originally developed by Terzaghi (1925) in a study of the creep in the deformation caused by the slow expulsion of water through material of low permeability under compressive loading, e.g., a sample of clay. For the one-dimensional case, he developed the mathematical description of the phenomenon, on the basis of Darcy's Law for the flow through a porous medium, and the concept of the effective stress. He realized that in a soft soil, such as clay, the deformations are caused by the effective stresses, defined as the difference of the total stress and the pore pressure, where the latter must be considered over the entire surface of a cross section.

7.4.1 Piston-spring analogy

The process of consolidation is often explained with an idealized system composed of a spring, a container with a hole in its cover and water (Figure 7-10). In this system, the spring represents the soil with a compressibility, and the water which fills the container represents the pore water in the soil.

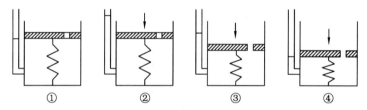

Figure 7-10 Schematics of piston-spring analogy of consolidation

① The container is completely filled with water, and the hole is blocked (fully saturated soil).

② A load is applied on the cover, while the hole is still blocked. At this stage, only the water resists the applied load (development of excess pore water pressure).

③ As soon as the hole is unblocked, water starts to flow out of the hole and the spring shortens (dissipation of excess pore water pressure).

④ After a certain time, there is no water drainage. The spring resists the applied load (full dissipation of excess pore water pressure at the end of consolidation).

7.4.2 Terzaghi's consolidation theory

The total stress increases when additional vertical load is applied. Instantaneously, the pore water pressure increases by exactly the same amount. Subsequently there will be flow from regions of higher excess pore pressure to that of lower excess pore pressure causing dissipation. The effective stress will change and the soil will consolidate with time. This process is shown schematically in Figure 7-11.

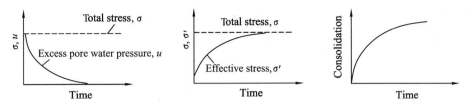

Figure 7-11 Schematics of Terzaghi consolidation theory

On the assumption that the excess pore water moves only vertically, an analytical procedure can be developed for computing the rate of consolidation. Consider a saturated soil element of sides dx, dy and dz.

The initial volume of soil element equals to $dxdydz$. If n is the porosity, the volume of water in the element equals to $ndxdydz$, the continuity equation for one-dimensional flow in the vertical direction is

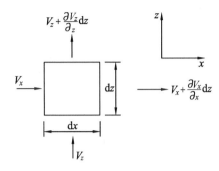

Figure 7-12 Soil element

$$\frac{\partial V_z}{\partial z}dxdydz = -\frac{\partial}{\partial t}(ndxdydz) \qquad (7\text{-}21)$$

Only the excess head (h) causes consolidation, which is related to the excess pore water pressure (u) by $h=u/\gamma_w$. The Darcy's Equation can be written as

$$V_z = -k_v\frac{\partial h}{\partial z} = -\frac{k_v}{\gamma_w}\frac{\partial u}{\partial z} \qquad (7\text{-}22)$$

where γ'_w is unit weight of water and k_v is soil permeability coefficient or hydraulic conductivity.

The Darcy's Equation can be substituted in the continuity equation, and the porosity n can be expressed in terms of void ratio e, thus obtain the flow equation as

$$\frac{k_v}{\gamma_w}\frac{\partial^2 u}{\partial z^2}dxdydz = \frac{\partial}{\partial t}\left(\frac{e}{1+e}dxdydz\right) \qquad (7\text{-}23)$$

The soil element can be represented schematically as shown in Figure 7-13.

If e_0 is the initial void ratio of the consolidating layer, the initial volume of solids in the element is $(dxdydz)/(1+e_0)$, which remains constant. The change in water volume can be represented by small changes ∂e in the void ratio e. The flow equation can be written as

$$\frac{k_v}{\gamma_w}\frac{\partial^2 u}{\partial z^2}dxdydz = \frac{\partial e}{\partial t}\cdot\frac{dxdydz}{1+e_0} \qquad (7\text{-}24)$$

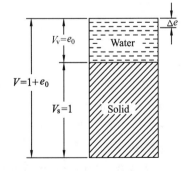

Figure 7-13 Schematic of soil element

or

$$\frac{k_v}{\gamma_w}\frac{\partial^2 u}{\partial z^2} = \frac{1}{1+e_0}\cdot\frac{\partial e}{\partial t} \qquad (7\text{-}25)$$

This is the hydrodynamic equation of one-dimensional consolidation. If a_v = coefficient of compressibility, the change in void ratio can be expressed as $\partial e = a_v(-\partial\sigma') = a_v(\partial u)$, since any increase in effective stress equals the decrease in excess pore water pressure. Thus,

$$\frac{\partial e}{\partial t} = a_v \cdot \frac{\partial u}{\partial t} \tag{7-26}$$

The flow equation can be expressed as

$$\frac{k_v}{\gamma_w} \frac{\partial^2 u}{\partial z^2} = \frac{a_v}{1+e_0} \cdot \frac{\partial u}{\partial t} \tag{7-27}$$

or

$$\frac{k_v}{a_v} \cdot \frac{1+e_0}{\gamma_w} \cdot \frac{\partial^2 u}{\partial z^2} = \frac{\partial u}{\partial t} \tag{7-28}$$

By introducing the coefficient of consolidation,

$$c_v = \frac{k_v(1+e_0)}{a_v \gamma_w} = \frac{k_v}{m_v \gamma_w} \tag{7-29}$$

where the coefficient of consolidation c_v indicates the rate at which a soil will undergo deformation in response to changes in effective stress induced by a rapidly applied load. This value can be determined by an oedometer test, which is used to plot the deformation versus time curve. Analysis of the curve by logarithm of time or square root of time curve-fitting methods suggests the time that the specimen experiences primary consolidation, secondary consolidation, and the expected deformations at each phase. Primary consolidation occurs from the time when the load increment is increased until the excess pore pressures dissipate. Meanwhile, secondary consolidation is analogous to creep; it describes the deformation in the soil under constant effective stress therefore occurs once the excess pore pressures dissipate. It is important to decide the duration of each load increment to study the consolidation behavior of soils. For example, organic or peat soils display very large deformations in the creep phase of consolidation over long periods of time.

Substituting Eq. (7-29) into Eq. (7-28), we have

$$c_v \cdot \frac{\partial^2 u}{\partial z^2} = \frac{\partial u}{\partial t} \tag{7-30}$$

This is Terzaghi's one-dimensional consolidation equation. The solution for a set of boundary conditions will describe how the excess pore water pressure u dissipates with time t and location z. When all the u is dissipated completely throughout the depth of the compressible soil layer, consolidation is complete and the transient flow situation does not exist any more.

7.4.3 Solution of Terzaghi's theory

During the consolidation process, the assumptions in Terzaghi's theory are as follows:
① Drainage and compression are one-dimensional.
② The compressible soil layer is homogenous and completely saturated.

③ The mineral grains and pore water are incompressible.

④ Darcy's Law governs the water outflow in the soil.

⑤ The applied load increment produces only small strains. Therefore, the thickness of the layer remains unchanged during the consolidation process.

⑥ The permeability coefficient k_v and the coefficient of volume compressibility m_v of the soil are constant.

⑦ The relationship between void ratio and vertical effective stress is linear. This assumption also implies that there is no secondary compression settlement.

⑧ Total stress remains constant throughout the consolidation process.

By accepting these assumptions, the fundamental governing equation for one-dimensional consolidation is expressed as shown in Eq. (7-30).

The parameter c_v is called the coefficient of consolidation and is mathematically analogous to the diffusion coefficient in Fick's Second Law. It indicates material properties that govern the process of consolidation and is with dimensions of area per time. In general, c_v is not constant because the component parameters vary during the consolidation process. However, in order to reduce Eq. (7-30) to a linear form, c_v is assumed constant for each load increment. The consolidation equation Eq. (7-30) can be solved analytically using the Fourier method. There are three variables shown in the consolidation equation of Eq. (7-30):

① The depth of the soil element in the layer (z).

② The excess pore water pressure (u).

③ The time elapsed since loading (t).

Three non-dimensional parameters are provided:

① Drainage path ratio, $Z = z/H$, where H equals to the drainage path which is the longest path taken by the pore water to reach a permeable sub-surface layer above or below.

② Time factor, T, which is a fuction of c_v.

③ Consolidation ratio at depth z, U_z, which is the ratio of dissipated pore pressure to the initial excess pore pressure. This represents the stage of consolidation at a certain point in the compressible layer.

The mathematical expressions of the three dimensionless parameters are as follows

$$Z = \frac{z}{H} \tag{7-31}$$

$$T = \frac{c_v t}{H^2} \tag{7-32}$$

$$U_z = 1 - \frac{u}{u_i} \tag{7-33}$$

where, z—depth below top of the compressible stratum;

H—length of the longest drainage path;

t—elapsed time of consolidation;

u—excess pore pressure at time t and depth z;

u_i—initial excess pore pressure at depth z.

Z is a measure of the dimensionless depth in the consolidating stratum, T is the time factor as a measure of dimensionless time, and U_z is the consolidation ratio, which is a function of Z and T, thus it varies throughout the consolidation process with time and depth in the layer. U_z expresses the progress of consolidation at a specific point in the consolidating layer. The value of H depends on the boundary drainage conditions of the layer. Figure 7-14 shows the two typical drainage conditions for the consolidation problem. A single-drained layer has an impervious and pervious boundary. Pore water can escape only through the previous boundary, giving $H=H_0$. A double-drained layer is bounded by two pervious strata. Pore water can escape to either boundary, therefore $H=H_0/2$.

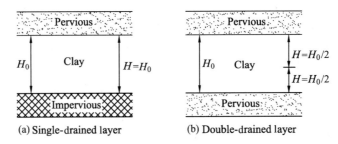

Figure 7-14   Boundary drainage conditions in the consolidation problem

Figure 7-15 shows the solution to Eq. (7-30) in terms of the above dimensionless parameters. For a double-drained layer, pore pressure dissipation is modeled using the entire figure. However, for a single drained layer, only the upper or lower half of figure is used. As expected, U_z is zero for Z at the beginning of the consolidation process ($T=0$). As time elapses and pore pressures dissipates, U_z gradually increases to 1.0 for and σ'_v increases accordingly. From Figure 7-6, it is possible to find the consolidation ratio (and u and σ'_v) at any time t and any point Z in the consolidating layer after the start of loading. The time factor T can be calculated from Eq. (7-32) with c_v for a particular deposit, the total thickness of the layer, and the boundary drainage conditions.

Figure 7-15 also provides some insight as to the progress of consolidation. The isochrones (curves of constant T) represent the percent consolidation at a given time throughout the compressible layer. For example, the percent consolidation at the midheight of a doube-drained layer at $T=0.2$ is approximately 23% (point A in Figure 7-15). However, at $Z=0.5$, $U_z=44\%$ with the same time factor. Similarly, near the drainage surfaces at $Z=0.1$, the clay is already 86% consolidated. It also means, at the same depth and time, 86% of the original excess pore pressure has dissipated and the effective stress has increased by a corresponding amount.

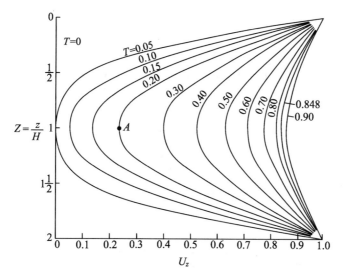

Figure 7-15 Consolidation ratio as a function of Z and T

In settlement analysis, the consolidation ratio U_z is not of immediate interest. Rather, geotechnical engineer needs to know the average degree of consolidation of a layer, U, defined as

$$U = \frac{s(t)}{S_c} \tag{7-34}$$

where, $s(t)$—the consolidation settlement at time t;

S_c—the total (ultimate) consolidation settlement.

There are useful approximations relating to the degree of consolidation U and the time factor T (Figure 7-16).

For $U < 0.60$,

$$T = \frac{\pi}{4} U^2 \tag{7-35}$$

and for $U > 0.60$,

$$T = 1.781 - 0.933 \lg[100(1-U)] \tag{7-36}$$

With S_c is known, Eq. (7-34), Eq. (7-35), and Eq. (7-36) can be used to predict the consolidation settlement as a function of time.

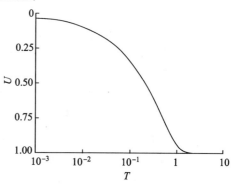

Figure 7-16 Plot of Terzaghi consolidation theory

7.4.4 Settlement and time

To estimate the amount of consolidation occurred and the time it would take, it is necessary to know:

① The boundary and drainage conditions;

② The loading conditions;

③ The relevant parameters of the soil including initial void ratio e_0, coefficient of compressibility a_v, coefficient of volume compressibility m_v, compression index C_c, and coefficient of consolidation c_v. They are obtained from consolidation tests on undisturbed samples from the compressible soil stratum.

Comparing the compressible soil layer with a soil element in this layer (Figure 7-17),

$$\frac{\text{Change in thickness}}{\text{Total thickness}} = \frac{\text{Change in volume}}{\text{Total volume}} \tag{7-37}$$

$$\frac{\Delta D}{D} = \frac{\Delta e}{1+e_0} \tag{7-38}$$

So

$$\Delta D = \frac{\Delta e}{1+e_0} D \tag{7-39}$$

Δe can be expressed in terms of a_v or C_c.

$$\Delta e = a_v \Delta \sigma'_v \tag{7-40}$$

or

$$\Delta e = C_c \lg(\sigma'_{v0} + \Delta \sigma'_v / \sigma'_{v0}) \tag{7-41}$$

The settlement is

$$\Delta D = \frac{a_v \cdot \Delta \sigma'_v}{1+e_0} D = m_v \Delta \sigma'_v D \tag{7-42}$$

or

$$\Delta D = \frac{C_c \lg\left(\frac{\sigma'_{v0} + \Delta \sigma'_v}{\sigma'_{v0}}\right) D}{1+e_0} \tag{7-43}$$

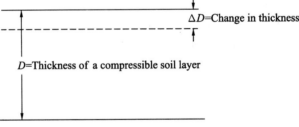

Figure 7-17 Schematic of settlement calculation

【Example 7-1】 A 3m-thick layer of saturated clay in the field under a surcharge loading will achieve 90% consolidation in 75 days in double drainage conditions. Find the coefficient of consolidation of the clay.

【Solution】

As the clay layer has two-way drainage, $H = H_0/2 = 1.5\text{m} = 150\text{cm}$, $t_{90} = 75$ days $= 75 \times 24 \times 60 \times 60$ seconds for 90% consolidation ($U = 90\%$).

$$T_{90} = C_v t_{90}/H^2 = 0.848$$

$$C_v = \frac{T_{90} H^2}{t_{90}} = \frac{0.848 \times (150)^2}{75 \times 24 \times 60 \times 60} = 2.94 \times 10^{-3} \text{cm}^2/\text{s}$$

【Example 7-2】 A 15m-thick layer of saturated clay in the filed with both above and bottom sides drainage. Now we take a sample (2cm with double drained condition) from the center of the clay layer for the consolidation test. Under a certain load the sample will achieve 70% consolidation in 8 minutes. Calculate the time of the soil (with thickness of

15m) layer to achieve the 70% consolidation under the same load. If the soil layer is in single drainage condition, find the time that is needed to achieve the 70% consolidation.

[Solution]

We know that the thickness of clay layer $H_1=15$m, and the thickness of consolidation test sample $H_2=2$cm, and time to achieve 70% consolidation is $t_2=8$ minutes.

Since the soil have the same degree of consolidation, the time factor should be the same, $T_v = \dfrac{C_v t}{H^2}$, $T_{v1} = T_{v2}$.

For the same soil type $C_{v1}=C_{v2}$, hence we get

$$\frac{t_1}{\left(\dfrac{H_1}{2}\right)^2} = \frac{t_2}{\left(\dfrac{H_2}{2}\right)^2}$$

so $t_1 = \dfrac{H_1^2}{H_2^2} t_2 = \dfrac{1500^2}{2^2} \times 8 = 4500000$ minutes $=8.56$ years.

If it is single drainage, we have

$$\frac{t_1}{\left(\dfrac{H_1}{2}\right)^2} = \frac{t_3}{H_1^2}$$

so $t_3 = 4t_1 = 4 \times 8.56 = 34.24$ years.

We can conclude that for the same soil, it will take 4 times longer under single drainage than that under double drainage to achieve the same degree of consolidation.

Knowledge expansion

Introduction of Soil Compressibility Parameters

Some useful parameters of soil compressibility mentioned in this chapter include the coefficient of compressibility (a_v), coefficient of volume compressibility m_v, compression index (C_c), recompression or expansion index (C_r). Besides, the modulus of compression E_s and modulus of deformation E are often used to describe compressibility of soil.

The modulus of compression E_s is defined as the ratio of vertical stress to vertical strain of soil without lateral deformation, $E_s = \sigma_z / \varepsilon_z$. It is equal to $1/m_v$. The magnitude of E_s reflects the ability of soil to resist compression deformation under one-dimensional compression.

For linear elastic body, the relationship between the stress and strain is linear. According to the generalized Hooke's law, the strain in x, y and z coordinates can be expressed as

$$\left.\begin{array}{l}\varepsilon_x=\dfrac{\sigma_x}{E}-\dfrac{v}{E}(\sigma_y+\sigma_z)\\[6pt] \varepsilon_y=\dfrac{\sigma_y}{E}-\dfrac{v}{E}(\sigma_x+\sigma_z)\\[6pt] \varepsilon_z=\dfrac{\sigma_z}{E}-\dfrac{v}{E}(\sigma_x+\sigma_y)\end{array}\right\} \quad (7\text{-}44)$$

where E is Young's modulus (in kPa) and v is Poisson's ratio.

Young's modulus E reflects the ability of material to resist deformation. It represents the ratio of stress to strain of linear elastic material under unconfined condition. Because soil is not an ideal elastic body, E is called deformation modulus for soil. The range of Poisson's ratio v of soil is small, generally between 0.3 and 0.4. The Poisson's ratio of saturated clay under undrained condition may be close to 0.5. The deformation modulus of soil E is often used to estimate the immediate settlement. It can be determined by the laboratory triaxial test or field test (such as in-situ plate load test PLT and pressuremeter test PMT), or by the soil compression modulus E_s. The relationship between E and E_s is deduced as follows:

Assuming that the vertical effective stress is σ_z and the lateral effective stress is σ_x and σ_y, there are $\sigma_x=\sigma_y=\sigma_z\kappa_0$ and $\varepsilon_x=\varepsilon_y=0$ under the confined condition with no lateral deformation. Therefore, we can get

$$\sigma_x-v(\sigma_x+\sigma_z)=0 \quad (7\text{-}45)$$

and
$$\dfrac{\sigma_x}{\sigma_z}=\dfrac{v}{1-v} \quad (7\text{-}46)$$

So
$$\kappa_0=\dfrac{v}{1-v} \quad (7\text{-}47)$$

Under the condition of no lateral deformation, there is $\sigma_x=\sigma_y=\sigma_z K_0$. From Eq. (7-44), we can get

$$E=\dfrac{\sigma_z}{\varepsilon_z}\left(1-\dfrac{2v^2}{1-v}\right)=E_s\left(1-\dfrac{2v^2}{1-v}\right) \quad (7\text{-}48)$$

This is the theoretical relationship between deformation modulus E and compression modulus E_s. Since the Poisson's ratio v of soil is less than or equal to 0.5, the deformation modulus of soil E is always less than the compression modulus E_s. It should be noted that Eq. (7-44) is derived from the generalized Hooke's law in the theory of elasticity. However, the soil mass is not an ideal elastic body, its deformation properties is not fully satisfied with Hooke's law. Therefore, Eq. (7-44) is only an approximate expression for estimating the deformation modulus E. Table 7-2 lists the empirical values of deformation modulus E for different soil types.

Table 7-2 Empirical values of deformation modulus E

Soil type	E/kPa	Soil type	E/kPa
Peat	100~500	Loose sand	10000~20000
Soft plastic clay	500~4000	Dense sand	50000~80000
Hard plastic clay	4000~8000	Dense gravel	100000~200000
Hard clay	8000~15000		

Exercises

[7-1] What are the main reasons that causes the compression of soil?

[7-2] What are the difference between coefficient of compressibility and coefficient of volume compressibility?

[7-3] What is the influences of change in groundwater level (rising or lowering) on the settlement of building foundation?

[7-4] A ground soil improvement method is called surcharge preloading. It was achieved by applying surcharge load on the soil before constructing a building foundation. The load is removed after a period of time and the building is built on the pretreated ground. Please find the mechanism of this method in view of settlement controling.

[7-5] How does it change between pore water pressure and effective stress in the process of soil consolidation? Please clarify the relationship between them?

[7-6] Can the coefficient of consolidation reflect the soil compressibility?

[7-7] What is the difference of a soil mass in normally consolidated or over-consolidated? Tell the reason for the difference.

[7-8] The foundation of a culvert gate is 6m in width and 18m in length (along the flow direction), and it is subjected to a central vertical load $F_v = 10800$kN. The ground soil below the foundation is homogeneous cohesive soil, and the groundwater table is 3m below the ground surface. The wet unit weight of the soil above the groundwater table is $\gamma = 19.1$kN/m^3, and the saturated unit weight of the soil below the groundwater table is $\gamma_{sat} = 21$kN/m^3. The embedment depth of the foundation is 1.5m, and the compression curve of cohesive soil is shown in Figure 7-18. Try to calculate the settlement of the center point of the foundation.

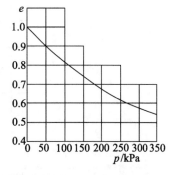

Figure 7-18 Schematic diagram of Exercise [7-8]

[7-9] The profile of a homogeneous earth dam and ground soil is shown in Figure 7-19, in which the average coefficient of compressibility of clay layer is $a_v = 2.4 \times 10^{-4}$kPa^{-1}. The initial

void ratio $e_0 = 0.97$, and the permeability coefficient is $k = 2.0$ cm/a. The induced stress distribution in the clay layer along the axis of the dam is shown in the shaded part of Figure 7-19. If the dam body is impermeable. Try to find: (1) the final settlement of the clay layer; and (2) the time required when the settlement of the clay layer reaches 12cm.

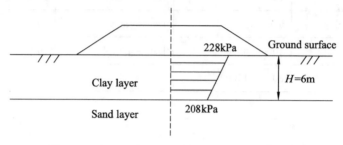

Figure 7-19　Schematic diagram of Exercise [7-9]

[7-10] There is a clay layer with a thickness of 5m between two sand layers. Now a sample is taken from the center of the clay layer for the laboratory consolidation test (the thickness of the specimen is 2cm and both top and bottom are placed with permeable sand stone). It is measured that it takes 8 min when the degree of consolidation reaches 60%. How long does it take when the degree of consolidation of the natural clay layer reaches 80% (assuming that the induced stress in the clay layer is distributed in a straight line)?

[7-11] Figure 7-20 shows a profile of soil strata. A line is the original ground surface. In modern artificial activities, 2m has been excavated, that is, the current ground surface is B line. After excavation, the soil below the ground surface is allowed to swell fully. If a large area of loading is applied on the current ground surface, the distributed load pressure is 150kPa. How much compression settlement will the clay layer produce (initial void ratio of clay layer is $e_0 = 1.00$, $C_c = 0.36$ and $C_r = 0.06$)?

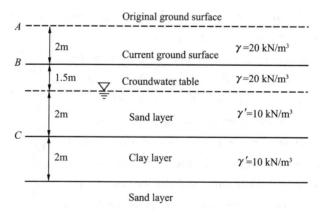

Figure 7-20　Schematic diagram of Exercise [7-11]

Chapter 8 Shear Strength of Soil

8.1 Introductory case

The Transcona Grain Elevator project was built in 1913 at North Transcona in Canada by the Canadian Pacific Railway Company. The structure consisted of a reinforced-concrete work-house, and an adjoining bin-house, which contained five rows of 13 bins, each 28m in height and 4.4m in diameter. The bins were based on a concrete structure containing belt conveyors supported by a reinforced-concrete shallow raft foundation. After the structure was

Figure 8-1 The collapse of the Transcona Grain Elevator in Canada

completed, the filling was begun and grain was distributed uniformly between the bins. On October 18,1913, after the elevator was loaded to 87.5% of its capacity, settlement of the bin-house was noted. Within an hour, the settlement had increased uniformly to about 30cm following by a tilt towards the west (Figure 8-1), which continued for almost 24 hours until it reached an inclination of almost 27 degrees.

Although the western end of the grain elevator sinks 7.3m and the east end rises 1.5m, the upper reinforced concrete silo structure is still intact with few surface cracks. Based on the test results of foundation excavation of adjacent structures, the calculated bearing capacity of foundation soil was up to 352kPa, which was applied to the grain elevator project. According to the site investigation in 1952, the actual bearing capacity calculated for the grain elevator foundation soil was ranging from 193.8kPa to 276.6kPa, which is far less than the pressure of 329.4kPa when the collapse of the grain elevator. Therefore, a general shear failure occurred when the shear stress along the failure surface exceeds the shear strength of soil (Figure 8-1).

8.2 Shear strength and Mohr-Coulomb failure criterion

8.2.1 Coulomb's law of shear strength

Under the external load, any point in the soil will yield a stress, which can be decomposed into a normal stress that is perpendicular to the cross section and a shear stress that is parallel to the cross section. The effect of normal stress is to make the soil mass compaction or compression, while the shear stress mainly result in shear deformation of soil mass. When the shear stress at a cross section reaches the soil shear strength, the soil mass on both sides of the cross section will have a larger relative sliding along the direction of shear stress, i. e., shear failure occurs at this point. Engineering practice and laboratory tests have shown that the failure of soil is mainly caused by the shear deformation. Shear failure is the typical failure mode of soil, which can be determined by the soil properties and its mechanical characteristics.

Figure 8-2 shows an embankment on soft soil ground. As the weight of fill gradually increased, the shear stress and normal stress in ground soil under the fill load will increase gradually. When the shear stress increases to the peak value (shear strength) that the soil mass can resist, shear failure will occur along a sliding surface, as shown in Figure 8-2. The surface of shear failure often exhibits approximately a slip arc, or a logarithmic spiral. For a small unit of soil mass, its failure surface can be regarded as a plane.

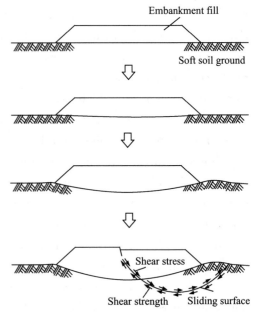

Figure 8-2 Failure process of an enbankment on soft soil ground

The shear failure of soil can be simulated by the direct shear test in the laboratory with the direct shear apparatus, as shown in Figure 8-3a. Following relevant testing code, a cylindrical soil sample with 61.8mm in diameter and 20mm in height was put in the shear box under a vertical load P. When apply a shear force S, the upper and lower shear box will have a relative movement along the shear surface. The area (A) of the shear plane is 30cm². The normal stress is $\sigma = P/A$ and the shear stress is $\tau = S/A$. When shear stress τ reaches a certain value, shear failure occurs along the contact surface of upper and lower shear box. τ_f is the shear strength, which refers to the shear stress at failure. Appling three or more times with different values of normal stress σ on replicate soil samples, the relationship between τ_f and σ can be obtained, which is approximate a straight line through the testing data points, i. e., (σ_1, τ_{f1}), (σ_2, τ_{f2}) and (σ_3, τ_{f3}), as shown in Figure 8-3b.

$$\tau_f = c + \sigma \cdot \tan\phi \tag{8-1}$$

where, τ_f—shear strength of soil, kPa;

σ—normal stress on the shear plane, kPa;

c—cohesion of soil, kPa;

ϕ—internal friction angle of soil, (°).

Eq. (8-1) is the shear strength equation proposed by Coulomb in 1773, known as the Coulomb's law of shear strength. It shows that the normal stress is proportional to the shear resistance and the shear strength of soil is a linear function of normal stress on the surface of the sliding plane. Eq. (8-1) also illustrates that the shear strength of soil mass contains two component: one is the friction component ($\sigma\tan\phi$) caused by slipping among soil particles and the other is the cohesion component (c) produced by the bonding force among particles.

Consider a small soil unit on the shear surface, the show stress τ is caused by external force and its weight. When $\tau < \tau_f$, one can know that the soil unit is in a stable state; when $\tau = \tau_f$, the soil unit is in a limit equilibrium state; when $\tau > \tau_f$ the soil unit is at failure. Note that the value of τ cannot be longer than τ_f in the actual soil unit.

If the shear strength of soil is expressed using the effective stress, we can get the following equation.

$$\tau_f = c' + \sigma' \cdot \tan\phi' \tag{8-2}$$

where, σ'—effective normal stress on the shear plane, kPa;

c'—effective cohesion of soil, kPa;

ϕ'—effective friction angle of soil, (°).

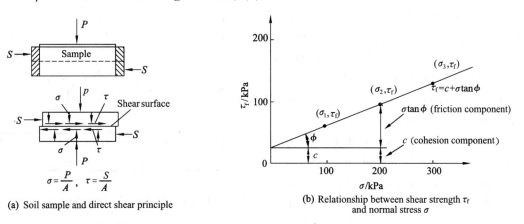

(a) Soil sample and direct shear principle

(b) Relationship between shear strength τ_f and normal stress σ

Figure 8-3 Direct shear of soil and Coulomb's law of shear strength

8.2.2 Mohr's circle

Consider a small soil unit at depth z in a homogeneous steady-state soil slope, as shown in Figure 8-4a. Assuming that it is a plane stress problem, the stress distribution on the small soil unit is shown in Figure 8-4b. It can be seen that the major principle stress σ_1 and minor principal stress σ_3 with known directions are acting on the unit. We can analyze any given plane with an angle α to the horizontal plane. We have the normal stress σ_α and shear stress τ_a on this plane. The following equations can be obtained from the force equilibrium condition for the unit abc shown in Figure 8-4c.

$$\sigma_a \overline{ab} = \sigma_1 \overline{ac} \cos \alpha + \sigma_3 \overline{bc} \sin \alpha \tag{8-3}$$

$$\tau_a \overline{ab} = \sigma_1 \overline{ac} \sin \alpha - \sigma_3 \overline{bc} \cos \alpha \tag{8-4}$$

Since $\cos\alpha = \overline{ac}/\overline{ab}$, $\sin\alpha = \overline{bc}/\overline{ab}$, $\cos 2\alpha = 2\cos^2\alpha - 1 = 1 - 2\sin^2\alpha$, $\sin 2\alpha = 2\sin\alpha\cos\alpha$, we can obtain

$$\sigma_a = \sigma_1 \cos^2\alpha + \sigma_3 \sin^2\alpha = \frac{\sigma_1 + \sigma_3}{2} + \frac{\sigma_1 - \sigma_3}{2}\cos 2\alpha \tag{8-5}$$

$$\tau_a = \sigma_1 \sin\alpha\cos\alpha - \sigma_3 \cos\alpha\sin\alpha = \frac{\sigma_1 - \sigma_3}{2}\sin 2\alpha \tag{8-6}$$

By $\sin^2 2\alpha + \cos^2 2\alpha = 1$, we can obtain

$$\left(\sigma_a - \frac{\sigma_1 + \sigma_3}{2}\right)^2 + \tau_a^2 = \left(\frac{\sigma_1 - \sigma_3}{2}\right)^2 \tag{8-7}$$

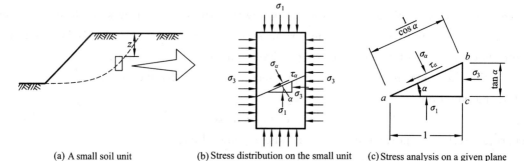

(a) A small soil unit (b) Stress distribution on the small unit (c) Stress analysis on a given plane

Figure 8-4 Schematic diagram of stress analysis for a small soil unit in a slope

Eq. (8-7) can form a circle where the center of the circle is $(\sigma_1 + \sigma_3)/2$ with $\tau = 0$ and the radius is $(\sigma_1 - \sigma_3)/2$, as shown in Figure 8-5. This circle is called the Mohr's circle. Therefore, Mohr's circle can be regarded as the locus of normal and shear stresses (σ_a, τ_a) on any inclined plane of soil element under the action of major and minor principal stresses (σ_1, σ_3). Once the Mohr's circle for a soil element is determined, the stress state i. e., the value of σ_a and τ_a on any inclined surface of the element can be calculated by Eq. (8-5) and Eq. (8-6), respectively.

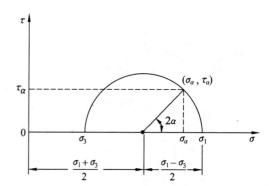

Figure 8-5 Mohr's circle for a soil unit

8.2.3 Mohr-Coulomb failure criterion

A distributed load is applied on the top of a soil slope shown in Figure 8-6a. A small soil unit at depth z is under the major principal stress σ_1 and the lateral minor principle stress σ_3. If the load is increased, the value of σ_3 will basically remain unchanged while the value of σ_1 will increase obviously. When the load increases to a certain value, shear failure of soil unit may occur, in which the major principle stress at failure is σ_{1f}, as shown in Figure 8-6b. Since Coulomb's law of shear strength shows the relationship between normal stress and shear stress of soil at shear failure, the Mohr's circle at failure must be tangent to the Coulomb's strength line DG, which is called the Mohr-Coulomb failure envelope, as shown in Figure 8-7. The tangent point C represents the stress state on the failure surface. Assume that the angle is α_f between the direction of minor principle stress σ_3 and the failure plane shown in Figure 8-6b, then angle of CAB is α_f and the angle of $CO'B$ is $2\alpha_f$, as shown in Figure 8-7. We can determine α_f as

$$(180°-2\alpha_f)+\phi=90°, \quad \alpha_f=45°+\frac{\phi}{2} \tag{8-8}$$

where, ϕ—internal friction angle of soil, (°).

As shown in Figure 8-7, we have the major principal stress σ_{1f} and the minor principle stress σ_{3f} at failure. The subscript f indicates soil is at shear failure or limit equilibrium state. For a soil unit at failure plane, the relationship between the major and minor principal stresses is called Mohr-Coulomb failure criterion at limit equilibrium condition.

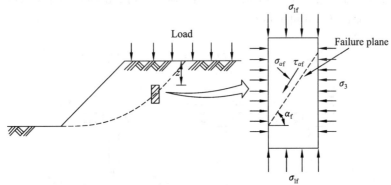

(a) A soil slope under distributed load   (b) Stress analysis on the sliding surface

Figure 8-6 Schematic diagram of stress analysis for a soil unit at failure under vertical load

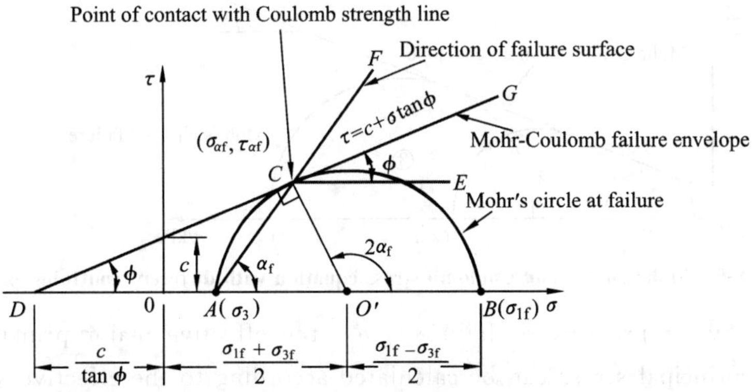

Figure 8-7 Mohr's circle and Mohr-Coulomb failure envelope

According to the graphical relation in Figure 8-7,

$$\sin\phi=\frac{(\sigma_{1f}-\sigma_{3f})/2}{(\sigma_{1f}+\sigma_{3f})/2+c\cdot\cot\phi} \tag{8-9}$$

By transformation, it can be rewritten in the following forms.

$$\sigma_{3f} = \sigma_{1f} \tan^2\left(45° - \frac{\phi}{2}\right) - 2c \cdot \tan\left(45° - \frac{\phi}{2}\right) \quad (8\text{-}10)$$

or
$$\sigma_{1f} = \sigma_{3f} \tan^2\left(45° + \frac{\phi}{2}\right) + 2c \cdot \tan\left(45° + \frac{\phi}{2}\right) \quad (8\text{-}11)$$

Eq. (8-9), Eq. (8-10) and Eq. (8-11) are the soil limit equilibrium equations. When the major principal stress σ_1 and minor principal stress σ_3 at a given point satisfy the above equations if the strength parameters c and ϕ are known, the soil is at failure, i. e., $\sigma_1 = \sigma_{1f}$ and $\sigma_3 = \sigma_{3f}$. The strength parameters c and ϕ of soil can be obtained by triaxial compression test.

In the test, a cylindrical specimen was applied a confining pressure σ_3, and the axial stress at failure is σ_{1f}. Figure 8-8 shows three replicate specimens in triaxial test with different confining pressures and corresponding vertical stresses at failure, i. e., $\sigma_{3①} < \sigma_{3②} < \sigma_{3③}$ and $\sigma_{1f①} < \sigma_{1f②} < \sigma_{1f③}$. Figure 8-9 shows the different Mohr's circle for three specimens at failure, which can be plotted according to different values of σ_3 and σ_{1f}. The common tangent line of the Mohr's circle is called Mohr-Coulomb failure envelope, which represents the limit stress state of soil element.

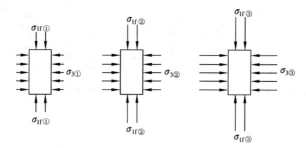

Figure 8-8 Schematics of replicate specimens at failure with different confining pressures and vertical stresses

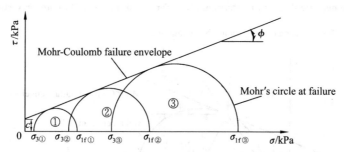

Figure 8-9 Mohr circles and Coulomb stress equation with different confining pressure

If the pore water pressure at failure is u_f, the effective major principal stress and effective minor principal stress can be calculated according to the effective stress principle by Eq. (8-12) and Eq. (8-13), respectively.

$$\sigma'_{1f} = \sigma_{1f} - u_f \quad (8\text{-}12)$$
$$\sigma'_{3f} = \sigma_{3f} - u_f \quad (8\text{-}13)$$

where, σ'_{1f} —effective major principal stress;

σ'_{3f} —effective minor principal stress;

u_f —pore water pressure at failure.

Similar equations can be obtained under the effective stress limit equilibrium condition. By replacing $\sigma_{1f}, \sigma_{3f}, c$ and ϕ with $\sigma'_{1f}, \sigma'_{3f}, c'$ and ϕ' in Eq. (8-10) and Eq. (8-11), we can obtain

$$\sigma'_{3f} = \sigma'_{1f} \tan^2\left(45° - \frac{\phi'}{2}\right) - 2c' \cdot \tan\left(45° - \frac{\phi'}{2}\right) \qquad (8\text{-}14)$$

or

$$\sigma'_{1f} = \sigma'_{3f} \tan^2\left(45° + \frac{\phi'}{2}\right) + 2c' \cdot \tan\left(45° + \frac{\phi'}{2}\right) \qquad (8\text{-}15)$$

【Example 8-1】 The major principal stress of a sandy soil ground is 300kPa, and the minor principal stress is 150kPa. The internal friction angel of the sandy soil is 25° and the cohesion is 0. What state is the soil in?

【Solution】 Given that $\sigma_1 = 300\text{kPa}, \sigma_3 = 150\text{kPa}, \phi = 25°$, and $c = 0$.

According to Eq. (8-10),

$$\sigma_{3f} = \sigma_{1f} \tan^2\left(45° - \frac{\phi}{2}\right) - 2c \cdot \tan\left(45° - \frac{\phi}{2}\right)$$

$$= 300 \times \tan^2\left(45° - \frac{25°}{2}\right) = 122\text{kPa}$$

$\sigma_3 = 150\text{kPa} > \sigma_{3f}$, indicating that the current Mohr's circle is within the Mohr's circle at failure. Thus the sandy soil is in a stable state.

【Example 8-2】 The major principal stress of a soil element is 480kPa, and the minor principal stress is 210kPa. The shear strength parameters measured in the triaxial test are: $c = 20\text{kPa}$ and $\phi = 18°$. What state is the soil element in?

【Solution】

Given that $\sigma_1 = 480\text{kPa}, \sigma_3 = 210\text{kPa}, c = 20\text{kPa}$ and $\phi = 18°$, so $\alpha_f = 45° + \frac{\phi}{2} = 54°$.

(1) According to the relationship between τ and τ_f:

By Eq. (8-5) and Eq. (8-6), we can calculate the normal stress σ and shear stress τ on the failure plane:

$$\sigma = \frac{1}{2}(\sigma_1 + \sigma_3) + \frac{1}{2}(\sigma_1 - \sigma_3)\cos 2\alpha_f$$

$$= \frac{1}{2}(480 + 210) + \frac{1}{2}(480 - 210)\cos 108° = 303\text{kPa}$$

$$\tau = \frac{1}{2}(\sigma_1 - \sigma_3)\sin 2\alpha_f$$

$$= \frac{1}{2}(480 - 210)\sin 108° = 128\text{kPa}$$

By Eq. (8-1), we can obtain τ_f on the failure plane:

$$\tau_f = c + \sigma \cdot \tan\phi$$

$$= 20 + 303\tan 18° = 118\text{kPa}$$

Since $\tau > \tau_f$, the soil element has already failed.

(2) Judge by the limit equilibrium condition:

① By Eq. (8-10), to reach the limit equilibrium condition, we can calculate the minor principal stress σ_{3f}. Since the major principal stress $\sigma_1 = 480$kPa, substitute σ_1, c and ϕ into Eq. (8-10), we have

$$\sigma_{3f} = \sigma_1 \cdot \tan^2\left(45° - \frac{\phi}{2}\right) - 2c \cdot \tan\left(45° - \frac{\phi}{2}\right)$$

$$= 480 \times \tan^2\left(45° - \frac{18°}{2}\right) - 2 \times 20 \times \tan\left(45° - \frac{18°}{2}\right)$$

$$= 480 \times \tan^2 36° - 40 \times \tan 36° = 224 \text{kPa}$$

Since $\sigma_{3f} > \sigma_3 = 210$kPa, the Mohr's circle at failure is within the given Mohr's circle. Thus the soil element has already failed.

② By Eq. (8-11), to reach the limit equilibrium condition, we can calculate the major principal stress σ_{1f}. Since the minor principal stress $\sigma_3 = 210$kPa, substitute σ_3, c and ϕ into Eq. (8-11), we have

$$\sigma_{1f} = \sigma_3 \cdot \tan^2\left(45° + \frac{\phi}{2}\right) + 2c \cdot \tan\left(45° + \frac{\phi}{2}\right)$$

$$= 210 \times \tan^2\left(45° + \frac{18°}{2}\right) + 2 \times 20 \times \tan\left(45° + \frac{18°}{2}\right)$$

$$= 210 \times \tan^2 54° + 40 \times \tan 54° = 453 \text{kPa}$$

Since $\sigma_1 > \sigma_{1f}$, the Mohr's circle at failure is within the given Mohr's circle. Thus the soil element has already failed.

8.3 Laboratory tests for soil shear strength

8.3.1 Type of shear tests and drainage conditions

The test for determining the shear strength of soils is called the shear strength test, or shear test. Tests can be conducted in the laboratory and/or in the field (in-situ test). Common shear tests include direct shear test, triaxial compression test, unconfined (uniaxial) compression test and in-situ vane shear test. The vane shear test can be done in the field and the other three tests are usually done in the labs. Table 8-1 illustrates the principle, the determination of strength parameters, the testing method and the basic characteristics for the common shear test. Using the shear tests in Table 8-1 to determine the strength parameters of c and ϕ, the sample in the process of consolidation and drainage conditions should be considered.

Table 8-2 shows the comparison of shear tests under different drainage conditions with the corresponding loading conditions in the field. For example, according to the drainage condition, triaxial compression test can be categorized into: unconsolidated undrained (UU) test, consolidated undrained (CU) test and consolidated drained (CD) test. The three kinds of strength parameters can be obtained accordingly, i.e., c_u, ϕ_u; c_{cu}, ϕ_{cu} and c_d, ϕ_d. Besides, according to the drainage conditions, direct shear test can be categorized into unconsolidated quick shear test without drainage, consolidated quick shear test without drainage, and consolidated slow shear test with drainage. The three kinds of strength parameters can be obtained accordingly, i. e., c_q, ϕ_q; c_{cq}, ϕ_{cq} and c_s, ϕ_s.

Table 8-1 Type of shear tests

Items	Direct shear	Triaxial compression	Unconfined compression	In-situ vane shear
Principle				
Method	Sample is put in the shear box with an upper and lower parts under the action of vertical normal stress σ and then a horizontal shear force S is applied. The maximum shear stress τ_f is measured under three or more different normal stresses.	A cylindrical sample with rubber membrane under the confining pressure of σ_3, σ_1 is gradually increased till the sample failure. Three or more samples needed to the determine the vertical normal stress at failure σ_{1f}. Pore water pressure can also be obtained to get effective stress parameters.	Cylindrical sample without rubber membrane is in unconfined condition; the vertical normal stress is gradually increased till the sample failure to determine the unconfined compressive strength q_u; the undrained strength c_u can be calculated by $q_u/2$.	The vane shear device is inserted into the soil layer and the upper torsion moment is applied to make the soil inside the vane and surrounding soil have relative torsional shear till the failure. The corresponding maximum torsional moment M_{max} can be measured as well as the height H and diameter D of the vane.
Parameters (c, ϕ)				Calculate in-situ undrained strength by $\tau_f = \dfrac{M_{max}}{\dfrac{\pi D^2}{2}\left(H+\dfrac{D}{3}\right)}$
Characteristics	It is suitable for all kinds of soil; the soil sample is easily prepared and the test is easy to operate; the sample is constrained with a fixed shear plane; the drainage condition is difficult to control.	It is suitable for all kinds of soil; the drainage condition can be controlled, but the operation of test is complicated.	It is only suitable for cohesive soil; the operation of test is the simplest	It is applicable to cohesive soil (especially saturated soft clay); the test is directly done in the field.

Table 8-2 Comparison of field loading condition and the drainage condition of shear tests

	Construction just completed	Graded loading	Long time settlement after construction
Field loading condition	For rapid construction, the ground soil is not under consolidation with external load all turns into pore water pressure in soil. The shear strength of the ground remains unchanged. This is the most dangerous condition.	The external load applied onto the clay ground with multi-stage loading. The ground soil is fully consolidated under the first stage loading before applying the second stage loading.	After a long time of consolidation for clay ground, the vertical stress increase under loading has turned into effective stress of soil skeleton.
Triaxial test condition	Drain valve is closed during consolidation and shearing process, which is a unconsolidated undrained (UU) test.	Drain valve is opened during consolidation and closed in the shearing process. This is a consolidated undrained (CU) test.	Drain valve is opened during consolidation and shearing process, which is a consolidated drained (CD) test.
Direct shear test condition	Quick shear test is conducted without consolidation and drainage of soil.	Consolidated quick shear test is conducted with drainage during consolidation and no drainage in the shearing process.	Slow shear test is conducted with consolidation and drainage of soil.

8.3.2 Direct shear test

The direct shear test apparatus is shown in Figure 8-10. It is the simplest and most commonly used test for determining the shear strength of soil. The direct shear apparatus consists of an upper and a lower metal shearing boxes that can be shifted from each other. The lower box is usually free to move and the upper box is fixed to a load ring. During the test, the soil sample with 2cm in height and 30cm^2 in area is put in the shear box, a normal lead P is applied, and the box was under the gradually increased horizontal load S at a given rate, the upper box has a relative displacement (shear deformation) to the lower box until the sample reaches shear failure. Failure plane is forced to be horizontal. The shear stress ($\tau = S/A$) and normal stress ($\sigma = P/A$) can be obtained on the failure plane respectively.

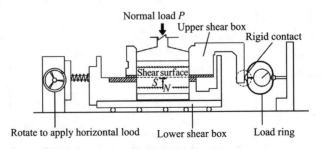

Figure 8-10 Direct shear apparatus

After the horizontal load S is applied, the displacement of the soil specimen is measured and the corresponding shear stress is calculated. The relationship between shear stress and shear displacement is plotted as shown in Figure 8-11a. The peak shear stress on the curve is taken as the shear strength τ_f of the soil under this normal pressure σ.

If there is no peak shear stress, the shear stress is selected as shear strength τ_f corresponded to the specified displacement such as 4mm per Chinese code GB/T 50123—2019 "Standard for geotechnical testing methed". Generally no less than 4 tests are conducted to get the relation between normal stress σ and corresponding shear strength τ_f, as shown in Figure 8-11b. The interception is cohesion c, and the angle between the connection line of data points and horizontal axis is the internal friction angle ϕ. There are two loading phases in a direct shear test, i.e., the application of normal load P and application of horizontal load S. Since the pore water pressure can not be measured in direct shear test, it can only represent the total stress of soil.

However, to consider the consolidation degree and the effect of drainage condition, fine-grained soil with the permeability coefficient less than 10^{-6} cm/s is selected. Base on the loading rate of the two stages of loading, the direct shear test can be categorized into quick shear test, consolidated quick shear and slow shear test. ① quick shear test: apply τ immediately after the application of σ, until the sample failure to get c_q and ϕ_q; ② consolidated quick shear test: after the application of σ to consolidate the sample, τ is rapidly applied until the sample failure to get c_{cq} and ϕ_{cq}; ③ slow shear test: after the application of σ to consolidate the sample, τ is applied slowly until the sample failure to get c_s and ϕ_s. Specific test requirements can be found in relevant codes such as Chinese code GB/T 50123—2019. For cohesive soil, generally we have $c_q > c_{cq} > c_s$, and $\phi_q < \phi_{cq} < \phi_s$.

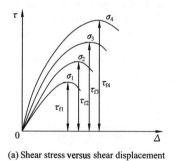

(a) Shear stress versus shear displacement (b) Shear stress at failure versus noral stress

Figure 8-11 Results of direct shear test

8.3.3 Triaxial compression test

The triaxial compression test can directly measure the compressive strength of soil specimen under constant confining pressure, and then indirectly determine the shear strength of soil using the Mohr-Coulomb failure criterion. Figure 8-12 shows the schematic of triaxial compression apparatus. Triaxial compression apparatus is mainly composed of pressure chamber, loading system and measurement system.

Testing procedures can be found in the relevant code such as Chinese code GB/T

50123—2019 for saturated soil. The soil specimen is first consolidated or unconsolidated and then sheared in drained or undrained condition. Table 8-3 lists the details of the three-type triaxial compression tests: ① unconsolidated undrained (UU) test; ② consolidated undrained (CU) test; and ③ consolidated drained (CD) test. The consolidation of the specimen is accomplished by increasing the confining pressure (usually water pressure) σ_3 in the chamber, where the drain valve is open and the confining pressure is controlled by the pressure system. The volume change of the specimen during the consolidation is determined by the amount of water drained from the specimen, which can be measured through the upper and/or lower drainage that connected to the burette. The confining pressure σ_3 keeps unchanged before the consolidation is completed, and the axial stress is gradually increased through the loading piston until the shear failure of the specimen.

Table 8-3 Drainage condition and the testing method of the triaxial compression test

Drainage condition			Unconsolidated undrained (UU) test	Consolidated undrained (CU) test	Consolidated drained (CD) test
Testing method	Isotropic consolidation		Close the drain valve	Open the drain valve	Open the drain valve
			No drainage is kept during the consolidation and shearing processes.	Drainage is allowed until the end of consolidation	Drainag is allowed until the end of consolidation.
	Shearing process		Close the drain valve	No drainage is kept during the shearing process. The excess pore water pressure is measured during shear.	Drainage is allowed during the shearing process. The deviator stress is applied slowly so that no excess pore water pressure is developed during the shear.
				Close the drain valve	Open the drain valve

During the shearing process, the drain valve stays open for drained tests and closed for undrained tests. For undrained test, the pore water pressure (u) is measured and the effective major principal stress σ_1' and the effective minor principal stress σ_3' can be calculated from the measured pore water pressure.

As shown in Figure 8-12, the axial stress increment, know as the deviator stress q ($q = \sigma_1 - \sigma_3$), is applied by the loading positon. In the shearing process, the axial displacement needs to be recorded to calculate the axial strain ε. From Figure 8-12, the major principal stress σ_1 is equal to the confining pressure σ_3 (minor principal stress) plus the deviator stress q.

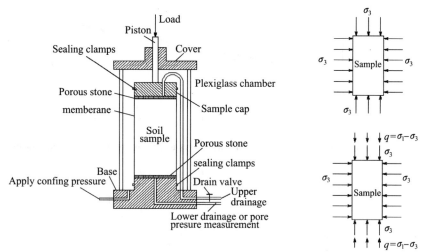

Figure 8-12　Schematic diagram of triaxial apparatus

With σ_{1f}, σ_{3f}, σ'_{1f}, and σ'_{3f}, the Mohr's circle of total stress and Mohr's circle of effective stress can be plotted, as shown in Figure 8-13. The diameter of Mohr's circle $(\sigma'_{1f}-\sigma'_{3f})$. is equal to $(\sigma_{1f}-\sigma_{3f})$. indicating that the effective stress circle and the total stress circle are of same size. It should be noted that when the pore water pressure at failure (u_f) is positive, as shown in Figure 8-13, the effective stress Mohr's circle moves left to the total stress Mohr's circle. However, if the if u_f is negative, the effective stress Mohr's circle will move to the right.

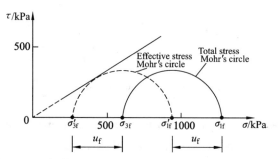

Figure 8-13　Mohr circle of effective stress and Mohr circle of total stress for a soil sample

There are three ways to calculate the strength parameters (c, ϕ) by conducting triaxial compression test. Once the pore water pressure is determined, the same methed can be used to obtain effective cohesion c' and effective internal friction angle φ'.

① Connect the common tangent points of Mohr's circles at failure (failure envelope) to determine the strength parameters (Figure 8-14);

② Determine the strength parameters using the relationship between the difference of principal stresses at failure $(\sigma_1-\sigma_3)_f$ and confining pressure σ_3 (Figure 8-15);

③ Use the relationship between $\left(\dfrac{\sigma_1-\sigma_3}{2}\right)_f$ and $\left(\dfrac{\sigma_1+\sigma_3}{2}\right)_f$ to determine the strength parameters (Figure 8-16).

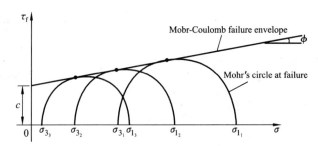

Figure 8-14　Determination of strength parameters using Mohr-Coulomb failure envelope

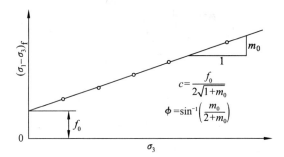

Figure 8-15　Determination of strength parameters using $(\sigma_1-\sigma_3)_f \sim \sigma_3$ curve

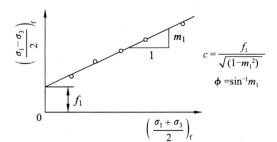

Figure 8-16　Determination of strength parameters using $\left(\dfrac{\sigma_1-\sigma_3}{2}\right)_f \sim \left(\dfrac{\sigma_1+\sigma_3}{2}\right)_f$ curve

8.3.4　Pore pressure coefficients in triaxial compression test

Skempton (1954) proposed the concept of the pore pressure coefficient A and B through laboratory triaxial test. In unsaturated soil, there are air and water. In this case, the pore air pressure u_a and pore water pressure u_w is not equal to each other due to the surface tension of water and the presence of meniscus, which makes $u_a > u_w$. When the degree of saturation is more than 95%, the surface tension is small, and u_a is roughly equal to u_w. The following discussion would not distinguish u_a and u_w, we just use the pore pressure u.

In conventional triaxial compression test, the sample was firstly consolidated under overburden pressure σ_c to simulate the soil in-situ stress state. After consolidation, the initial pore pressure u_0 dissipates to zero. $\Delta\sigma_3$ and $\Delta\sigma_1$ are applied to simulate the additional load, They can be applied by two steps in triaxial compression test. First, the pore pressure generated by the applied confining pressure increment $\Delta\sigma_3$ is Δu_1; second, under the constant confining pressure $\sigma_3 = \sigma_c + \Delta\sigma_3$, the axial principal stress increment ($\Delta\sigma_1 - \Delta\sigma_3$) (i.e., the deviator stress q) is applied on the sample. As a result, pore pressure increment Δu_2 is produced. As shown in Figure 8-17, if the test is under undrained condition, the pore pressure increment Δu induced by $\Delta\sigma_3$ and ($\Delta\sigma_1 - \Delta\sigma_3$) is the sum of pore pressure increments Δu_1 and Δu_2,

$$\Delta u = \Delta u_1 + \Delta u_2 \tag{8-16}$$

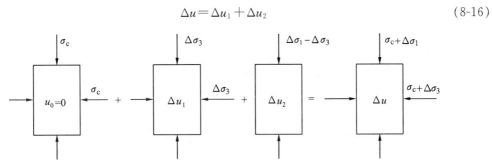

Figure 8-17 Pore pressure during triaxial consolidated undrained test

I. Pore pressure coefficient B

The ratio of Δu_1 and coressponding confining pressure increment $\Delta\sigma_3$ is defined as the pore pressure coefficient B, namely

$$B = \frac{\Delta u_1}{\Delta\sigma_3} \tag{8-17}$$

where B is the pore pressure coefficient under the condition of isotropic consolidation in the triaxial test. It is an indicator of the pore pressure increase when the soil is subjected to the isotropically consolidated pressure. For saturated soil, soil particles and water are considered to be incompressible, thus the action of $\Delta\sigma_3$ can neither change sample volume nor the shape. Therefore, $\Delta\sigma_3$ will be fully taken by the pore water, where $\Delta u_1 = \Delta\sigma_3$, so $B = 1$. For dry soil, due to the compressibility of air is much higher than the compressibility of soil skeleton, the action of $\Delta\sigma_3$ will be fully taken by the soil skeleton, thus do not produce any pore pressure and $\Delta u_1 = 0$, namely $B = 0$. For unsaturated soil, the action of $\Delta\sigma_3$ will make both the fluid in pores and soil skeleton being compressed. Therefore, we have $0 < \Delta u_1 < \Delta\sigma_3$, where B is between 0 and 1. The larger the degree of saturation for a soil sample, the value of B would be closer to 1. Thus we can say that B is also an indicator reflecting the saturation degree of soil.

II. Pore pressure coefficients A and $\bar{B}$

As the specimen is subjected to the deviator stress $q = \Delta\sigma_1 - \Delta\sigma_3$, the pore pressure generated is Δu_2. We can define another pore pressure coefficient A below

$$A = \frac{\Delta u_2}{B(\Delta\sigma_1 - \Delta\sigma_3)} \tag{8-18}$$

In the process of shearing, the deviator stress $q = \Delta\sigma_1 - \Delta\sigma_3$ is applied onto the sample. The pore pressure coefficient A can be calculated by measuring the Δu_2 and B. The magnitude of A depends on the types of soil, dry density, saturation degree and stress history, etc. For normally consolidated saturated clay or saturated loose sand, we have $0 < A < 1$; for over-consolidated saturated clay and saturated dense sand, we generally have $A < 0$.

In the process of shearing in triaxial compression test, the total pore pressure increment Δu generated by $\Delta\sigma_3$ and $\Delta\sigma_1 - \Delta\sigma_3$ is

$$\Delta u = \Delta u_1 + \Delta u_2 = B\Delta\sigma_3 + BA(\Delta\sigma_1 - \Delta\sigma_3) \tag{8-19}$$

or
$$\Delta u = B[\Delta\sigma_3 + A(\Delta\sigma_1 - \Delta\sigma_3)] \tag{8-20}$$

Eq. (8-20) can also be written as

$$\Delta u = B[\Delta\sigma_1 - (1-A)(\Delta\sigma_1 - \Delta\sigma_3)]$$
$$= B\Delta\sigma_1 \left[1 - (1-A)\left(1 - \frac{\Delta\sigma_3}{\Delta\sigma_1}\right)\right] \tag{8-21}$$

Define
$$\bar{B} = \frac{\Delta u}{\Delta\sigma_1} = B\left[1 - (1-A)\left(1 - \frac{\Delta\sigma_3}{\Delta\sigma_1}\right)\right] \tag{8-22}$$

where $\bar{B}$ is also a pore pressure coefficient. It is an indicator that reflects the pore pressure increment Δu generated by the increase of major principal stress $\Delta\sigma_1 = \Delta\sigma_3 + q$. $\bar{B}$ is a very important parameter in the embankment stability analysis, where it can be used to estimate the initial pore pressure during the construction of embankment.

In unconsolidated undrained (UU) test, we have

$$\Delta u = B\Delta\sigma_3 + A(\Delta\sigma_1 - \Delta\sigma_3) \tag{8-23}$$

In consolidated undrained (CU) test, since $\Delta u_1 = 0$ and $B = 1$, we have

$$\Delta u = \Delta u_2 = A(\Delta\sigma_1 - \Delta\sigma_3) \tag{8-24}$$

In consolidated drained (CD) test, we have $\Delta u = 0$ due to the dissipation of excess pore pressure in drained condition.

【Example 8-3】 CU test was conducted on a saturated clay sample. The values of confining pressure σ_3 applied onto the three soil specimens are set 60, 100 and 150kPa, respectively. The values of major principal stress σ_1 at failure are 143, 220 and 313kPa, and the measured values of pore water pressure at failure are 23, 40 and 67kPa. Determine the total stress strength parameters c_{cu}, ϕ_{cu}, and effective stress strength parameters c' and ϕ'.

【Solution】

(1) Plot the Mohr's circle for each specimen with respect to total stress on $\tau \sim \sigma$ coordinate, where $\left(\frac{\sigma_1+\sigma_3}{2}, \frac{\sigma_1-\sigma_3}{2}\right)$ is the center of each Mohr's circle, as shown in solid line in Figure 8-18. Connecting the common tangent points to get the total stress strength envelope, thus c_{cu} is 10kPa and ϕ_{cu} is 18°.

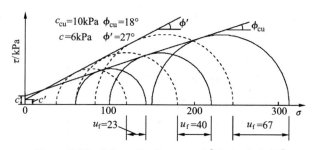

Figure 8-18 Schematic diagram of 【Example 8-23】

(2) Since the pore water pressure at failure is known for each specimen, we can plot the effective stress Mohr's circle, as shown in dash line in Figure 8-18. Connecting the common tangent points to get the effective stress strength envelope, thus c' is 6kPa and ϕ' is 27°.

【Example 8-4】 A saturated clay had a preconsolidation stress of 800kPa. Table 8-4 lists the value of pore water pressure at failure (u_f) from consolidated undrained (CU) test. Please determine the pore water pressure coefficient at failure (A_f) for specimens under different values of σ_3 and over consolidation ratio (OCR).

【Solution】 In consolidated undrained (CU) test, we have $\Delta u_1 = 0$ and $B = 1$. According to the Eq. (8-24), the pore pressure coefficient at failure $A_f = \Delta u / (\Delta \sigma_1 - \Delta \sigma_3) = u_f / (\sigma_1 - \sigma_3)_f$. The results of A_f are listed in Table 8-5.

Table 8-4 Results of pore water pressure at failure from CU test

σ_3 /kPa	$\sigma_1 - \sigma_3$ /kPa	u_f /kPa
100	42	−66
200	530	−10
400	730	82
600	1000	183

Table 8-5 Results of A_f at different values of σ_3 and OCR

σ_3 /kPa	OCR	A_f /kPa
100	8	−1.571
200	4	−0.019
400	2	0.112
1600	1	0.183

8.3.5 Unconfined compression test

I. Unconfined compressive strength

Unconfined compression test is used to determine the unconfined compressive strength of cohesive soil. It is by far the most common, cheap and fastest method for soil shearing test. The test may be carried out in the laboratory or at the field. This test is inappropriate for dry sand or crumbly clays. This is essentially a special case of the triaxial compression test where the minor principal stress $\sigma_3 = 0$. Figure 8-19 shows the apparatus for conducting the unconfined compression test.

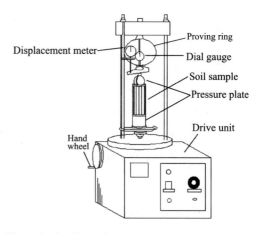

Figure 8-19 Unconfined compression test apparatus

The detailed testing procedures can be found in the relevant code, such as Chinese code GB/T 50123−2019 and American code ASTM D2166. Plot the axial compressive stress $q = \sigma_1 - \sigma_3 = \sigma_1$ versus axial strain ε and obtain the maximum stress which gives the unconfined compressive strength q_u, as shown in Figure 8-20. In case no pronounced peak is observed, take the strength corresponding to 20% strain as the unconfined compressive strength.

For saturated clay, triaxial unconsolidated undrained (UU) test result indicates that the internal friction angle $\phi_u = 0$. Unconfined compression test is conducted in undrained condition, thus q_u can be used to calculate the undrained strength c_u of saturated soil, where a Mohr's circle is plotted by q_u from the origin ($\sigma_{3f} = 0$) and then a horizontal strength envelope with $\phi_u = 0$ can be obtained, as shown in Figure 8-21. The cohesion c_u of saturated clay, also known as the undrained strength, can be obtained

$$c_u = \frac{q_u}{2} \tag{8-25}$$

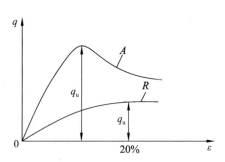

Figure 8-20 Typical stress and strain curve from unconfined compression test

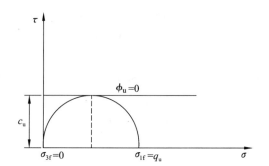

Figure 8-21 Mohr's circle of saturated clay from unconfined compression test

II. Sensitivity and thixotropy of clay

Unconfined compressive strength is commonly used to calculate the sensitivity of saturated clays. In the process of shearing, the original structure of clay is disturbed. It not only changes the arrangement (fabric) of soil particles, but also destroys the connection (bonding) between soil particles. As a result, the strength decreases and the compressibility increases.

Under the condition of the same water content, the clay with thoroughly disturbed structure is called remolded or reconstituted clay. The susceptibility of clay to structural disturbance can be expressed in terms of sensitivity. Sensitivity is defined as the ratio of unconfined compressive strength of undisturbed specimen to unconfined compressive strength of remolded specimen at same water content.

$$S_t = \frac{q_u}{q'_u} \tag{8-26}$$

where, S_t—sensitivity of clay;

q_u—unconfined compressive strength of undisturbed clay specimen;

q'_u—unconfined compressive strength of remolded clay specimen.

According to the sensitivity, clays can be classified by different types as listed in Table 8-6.

Table 8-6 Classification of clays by sensitivity

S_t	1	1~2	2~4	4~8	8~16	16~32	32~64	>64
Types of clay	Insensitive clays	Slightly sensitive clays	Medium sensitive clays	Very sensitive clays	slightly quick clays	Medium quick clays	Very quick clays	Extra quick clays

Therefore, engineers should pay attention to avoid disturbing the structure of clayey soils, especially for those with high sensitivity, since disturbance will reduce the strength of the soil.

It is observed that cohesive soils may lose some of their strength upon remolding. The loss of the strength of a soil upon remolding is partly due to change in the soil structure and partly due to disturbance caused to water molecules in the adsorbed layer.

Some of these changes are reversible. If a remolded soil is left alone undisturbed at the same water content for some time, it may regain part of its lost strength. This gain of strength in the soil with the passage of time after remolding is called thixotropy. This regain in strength is mainly due to gradual reorientation of molecules of water in the adsorbed water layer and due to re-establishment of chemical equilibrium. This recovery of the strength is not hundred percent. Soil cannot achieve its original strength which was before remolding because strength loss due the destruction of the soil structure cannot be recovered with time. Only strength loss due to disturbance of water molecules can be regained. In practical engineering, engineers can make use of soil thixotropy.

8.4 Shear behavior of sand

8.4.1 Internal friction angle of sand

Since the high permeability of sand, the triaxial test can be regarded as in consolidated and drained condition. In a CD test, we have $\Delta u = 0$, thus $\sigma = \sigma'$. The strength envelope obtained is usually a straight line through the origin of $\tau_f \sim \sigma$ coordinate, as shown in Figure 8-22.

$$\tau_f = \sigma' \cdot \tan\phi_d \tag{8-27}$$

where, ϕ_d is related to compactness of sandy soil; ϕ_d is higher for dense sand than that for loose sand; the cohesion c is 0, which indicates that there is no cohesive forces among soil particles.

ϕ_d can be decomposed into the following two parts.

$$\phi_d = \phi_r + \Delta\phi_d \tag{8-28}$$

where ϕ_r is related to the sand particle shape, surface roughness and gradation, etc. For a sandy soil with a constant volume, the value of ϕ_r is equal to the angle of repose, as shown in Figure 8-23. $\Delta\phi_d$ corresponds to the shear strength upon the volume change during shearing. The denser the sand, the greater is the $\Delta\phi_d$.

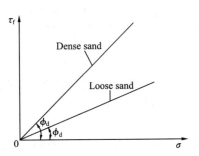

Figure 8-22　Strength envelopes for dense and loose sand

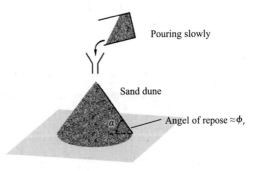

Figure 8-23　Angle of repose for a sand dune

8.4.2 Stress strain characteristics of sand

The stress strain behavior of sand is mainly influenced by the initial compaction state of sand. Extensive tests have shown that when the loose sand is under shear, the particles will fill into the voids thus the soil particles are arranged more closely, as shown in Figure 8-24a, so its volume decreases (void ratio decreases). This phenomenon of volume decrease due to shear is called shear contraction. Conversely, when the dense sand is under shear, the particles need to rearrange thus the particles would be able to slide over each other, causing the volume to expand (void ratio increases), as shown in Figure 8-24b. This phenomenon of volume expansion due to shear is called shear dilation. Figure 8-24c shows the comparison of the volume change of loose and dense sand during shear. For loose sand, a decrease trend in volume can be found showing the shear contraction. while for dense sand, a slight decrease in volume can be found due to the adjustment of the initial soil particles, then a expansion of volume is found showing the shear dilation. However, the contraction of loose sand and dilation of dense sand gradually level off with the increase of axial strain.

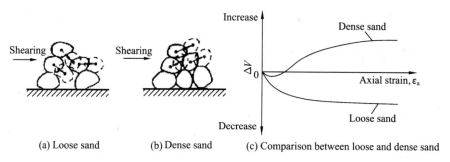

Figure 8-24 Volume change ΔV of loose and dense sands during shearing

Figure 8-25 shows the comparison of deviator stress q and axial strain ε_a between loose and dense sand obtianed from CD test. A gradual increase of q can be found with the increase of ε_a for loose sand (curve A). For dense sand, q increases first to a peak value and then decreases with increasing ε_a (curve B). q of dense sand finally tends to close to that of loose sand, as shown in Figure 8-25.

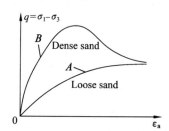

Figure 8-25 Comparison of $q \sim \varepsilon_a$ curve between loose and dense sand under drained condition

If saturated loose sand or dense sand is sheared in undrained condition (no volume change is allowed), the pore water pressure generated during shear is different. Fore loose sand, there is a tendency that the pore water in the pores is squeezed out due to shear contraction. Since the sample tries to maintain the same volume, the excess pore water pressure u will increase (and effective stress σ' will decrease); In contrast, for dense sand, there is a trend of volume increasing during shear. Pore water in the pores is under the suction. Since the sample tries to keep the volume unchanged, the excess pore water pressure u will decrease and sometimes even become negative, thus the effective stress σ' will increase.

8.4.3 Sand liquefaction

If some dry loose sand is put into a box, we can imagine that the application of vibration load can make the sand compacted, i.e., the void ratio of the sand decreases. If the sand is fully saturated, as mentioned above, the decrease in void ratio will make the pore water pressure increase. In general, pore water pressure can dissipate rapidly because of the high permeability coefficient of sand. However, it was difficult to dissipate during an earthquake. A large area of loose sand ground during an earthquake vibration can be considered as in undrained shear condition. Thus the pore water pressure at the same time would increase rapidly, and cannot be dissipated for a short time period. The sudden increase of pore water pressure makes the effective stress and shear strength of sand a sudden drop even to zero, so the sand behaves like a liquid, known as the sand liquifaction.

The liquefaction of sand is one of the primary factors leading to the damage of the structures during earthquakes. Figure 8-26 shows the schematic of liquefaction of sand due to earthquake. The liquefaction normally occurs due to the generation of excess pore pressure under undrained loading in loose sand. While for dense sand, due to the shear dilation potential, negative pore pressure will generate during shear under undrained condition, resulting in the increase of effective stress and there is little possibility of liquefaction occurred on dense sand. Therefore, the compaction degree of sand is often used as the criterion to judge liquefaction in engineering applications.

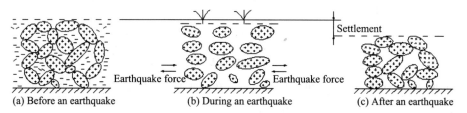

Figure 8-26 Schematic diagram of sand liquefaction due to an earthquake

8.5 Shear behavior of clay

The shear strength of saturated clay would be affected by the degree of consolidation, drainage condition and stress history. In triaxial compression test, the consolidation stress σ_c is applied to simulate the soil preconsolidation stress, and then different values of confining pressure σ_3 can be applied to consolidate the soil sample before shearing. The method is used to simulate the over-consolidated and normally consolidatied soil properties. A specimen is normally consolidated where the confining pressure σ_3 is equal to or greater than the preconsolidation pressure. In contrast, a specimen is over-consolidated, when the confining pressure σ_3 is smaller than its preconsolidation pressure. For a certain type of soil, the void ratio of over-consolidated soil is smaller than that of normally consolidated soil. The higher the over-consolidation ratio (OCR), the smaller is the void ratio.

8.5.1 Shear strength of normally consolidated clay

I. Unconsolidated undrained (UU) test

As shown in Figure 8-27, a set of saturated clay samples (generally 3~4) are used to conduct UU test to obtain the Mohr's circle for each sample (only two sample results are plotted). First, the consolidation pressure σ_c is applied onto each sample until the pore water pressure u_0 dissipated to zero. At this time, the Mohr's circle is only a point, where the specimen has not been sheared. Second, taking one sample with $\Delta\sigma_3 = 0$, the vertical deviator stress q is applied and increased gradually until the shear failure of sample in undrained condition. In this case, the minor principal stress $\sigma_3 = \sigma_c$, $\Delta u_1 = 0$, and the major principal stress $\sigma_1 = \sigma_c + q$, the pore water pressure generated during shear is Δu_2. Therefore, the total pore water pressure increment $\Delta u = \Delta u_2$. At failure, the effective stress Mohr's circle is on the left of the two total stress Mohr's circle, as shown in Figure 8-27. Third, take another sample with the confining pressure $\sigma_3 = \sigma_c + \Delta\sigma_3$, we have $\Delta u_1 = \Delta\sigma_3$. Then the deviator stress q is applied and increased until the failure of specimen. The major principal stress $\sigma_1 = \sigma_3 + q$. The total pore water pressure increment is $\Delta u = \Delta u_1 + \Delta u_2$. The total stress Mohr's circle for this case is on the right side of the two total stress Mohr's circles, as shown in Figure 8-27.

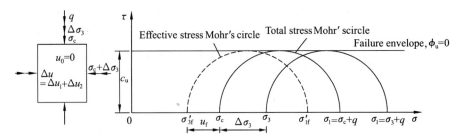

Figure 8-27 Mohr's circle and failure envelope for normally consolidated clay under UU test

In UU test, since samples are subjected to the increased cell pressure $\Delta\sigma_3$ under undrained conditions, and the pore water press coefficient B for saturated soil is equal to 1, the application of $\Delta\sigma_3$ will cause an equivalent increase in pore water pressure, i. e., $\Delta u_1 = \Delta\sigma_3$. Therefore, the effective consolidation pressure before shearing have not changed. They have the same void ratio before shearing and remain unchanged during the undrained test. It is expected that soil specimens will have the same undrained shear strength c_u, which is equal to the radius of the effective stress Mohr's circle.

$$c_u = \frac{\sigma_{1f} - \sigma_{3f}}{2} \tag{8-29}$$

where c_u is the undrained shear strength.

For saturated normally consolidated clay in UU test, the total strength failure envelope is a horizontal line, where $\phi_u = 0$. If the pore water pressure at failure u_f is measured, we can obtain the effective major and minor principal stresses at failure; i. e., $\sigma'_{1f}(=\sigma_{1f} - u_f)$ and $\sigma'_{3f}(=\sigma_{3f} - u_f)$. Therefore, we can obtain $(\sigma'_{1f} - \sigma'_{3f}) = (\sigma_{1f} - \sigma_{3f})$, which indicates that the diameter of the effective stress Mohr's circle is the same as that of the total stress Mohr's circle. Since the pore water pressure caused by the normally consolidated clay is positive, the effective stress Mohr's circle is on the left of the total stress Mohr's circle, as shown by the dotted circle in Figure 8-27. If the pore pressure coefficients A_f of soil specimens at failure are identical, the specimens under the different confining pressures should have the same major and minor effective principal stress.

II. Consolidated undrained (CU) test

The procedures of consolidated undrained test for saturated clay can be found in the relevant code. First, the specimen is consolidated with drain valve open under a given isotropic consolidation stress σ_3. After the consolidation process is finished, a deviator stress $q = \sigma_1 - \sigma_3$ is applied onto the specimen until its failure in undrained condition with drain valve closed.

As shown in Figure 8-28, a set of saturated clay samples (generally 3~4) are used to conduct CU test to obtain the Mohr's circle for each sample (only two sample results are plotted). Different values of confined pressure σ_3 are respectively applied on replicate samples. It can be expected that as σ_3 increases, the void ratio of sample at the end of consolidation will reduce and the shear strength will increase accordingly.

The failure envelope by connecting the common tangent points of Mohr's circle in total stress for normally consolidated clay is a straight line through the origin. The total stress strength parameters obtained in CU test include internal friction angle ϕ_{cu} and cohesion $c_{cu} = 0$, as shown in Figure 8-28.

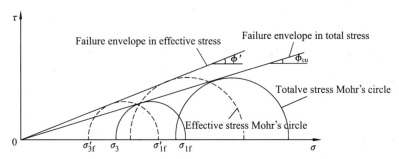

Figure 8-28 Mohr's circle and failure envelope for normally consolidated clay under CU test

If the pore water pressure at failure u_f is measured in CU test, the effective major principal stress σ'_{1f} and effective minor principal stress σ'_{3f} can be obtained using the effective stress principle. The failure envelope in effective stress can be drawn accordingly. When the pore water pressure at failure is positive, the effective stress Mohr's circle should be on the left of the total stress Mohr's circle, as shown by the dotted circle in Figure 8-28. The strength envelope in effective stress is also a straight line through the origin, and the corresponding effective stress strength parameters are $\phi'(>\phi_{cu})$ and $c'(=0)$.

III. Consolidated drained (CD) test

The procedures of consolidated drained test for saturated clay can be found in relevant code. First, the specimen is isotropically consolidated under a given confining pressure σ_3 in drained condition. Keep the drain valves open, a deviator stress $q = \sigma_1 - \sigma_3$ is then applied slowly onto the soil specimen in drained condition. No excess pore water pressure develops during the CD test.

As shown in Figure 8-29, a set of saturated clay samples (generally 3~4) are used to conduct CD test to obtain the Mohr's circle for each sample (three sample results are plotted). Different values of σ_3 are respectively applied onto replicate samples. It can be expected that as the σ_3 increases, the void ratio of sample at the end of consolidation will reduce and the shear strength will increase accordingly. The failure envelope for normally consolidated clay is a straight line through the origin, as shown in Figure 8-29.

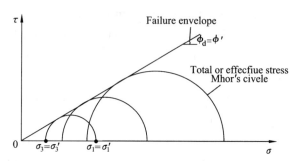

Figure 8-29 Mohr's circle and failure envelope for normally consolidated clay under CD test

Since the excess pore water pressure is kept zero through the whole test in drained condition, the effective stress is equal to the total stress. Thus total and effective Mohr's circle coincide. The failure envelope in total or effective stress is the same straight line through the origin. The shear strength parameters obtained from CD test are c_d or c' and ϕ_d or ϕ'. For normally consolidated clay, we have $c_d = c' = 0$ and $\phi_d = \phi'$.

8.5.2 Shear strength of overconsolidated clay

The overconsolidated clay has a present overburden pressure that is less than the maximum effective overburden pressure in the past. While a normally consolidated clay has a present overburden pressure that is equal to or larger than the maximum effective overburden pressure in its history. It can be seen from the $e \sim \lg\sigma'$ relation in Figure 8-30a that under the same consolidation stress, the overconsolidated clay has a lower void ratio than that of normally consolidated clay. Therefore, the overconsolidated clay should have a greater shear strength than that of normally consolidated clay, as shown in Figure 8-30b. It should be noted that for overconsolidated clay, three sets of triaxial tests including UU, CU and CD test also can be preformed to get shear strength parameters. The test procedures are the same with that for normally consolidated clay.

Chapter 8 Shear Strength of Soil

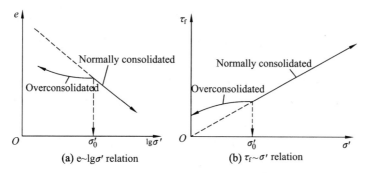

Figure 8-30 Comparison of normally consolidated and overconsolidated clay

8.5.3 Stress strain behavior of normally and overconsolidated clay

Figure 8-31 shows the typical relations of deviator stress $(\sigma_1 - \sigma_3)$ and axial strain (ε_a), pore water pressure increment (Δu) and ε_a obtained by consolidated undrained (CU) tests on normally consolidated (NC) and overconsolidated (OC) clays, respectively. It can be seen from Figure 8-31a that the deviator stress $(\sigma_1 - \sigma_3)$ increases with the increase of axial strain ε_a for NC clay. The phenomenon that $(\sigma_1 - \sigma_3)$ increases with increasing axial strain ε_a is called strain hardening behavior. According to the relevant testing code, the deviator stress at failure $(\sigma_1 - \sigma_3)_f$ in this case can be obtained at $\varepsilon_a = 15\%$. Meanwhile, the pore water pressure increment (Δu) shows an generally increase trend with ε_a, as shown in Figure 8-31c. This is because the normally consolidated clay has a volume contraction tendency during shear, the shear contraction potential under undrained conditions will increase the pore water pressure.

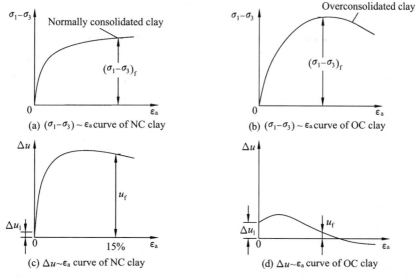

Figure 8-31 $(\sigma_1 - \sigma_3) \sim \varepsilon_a$ and $\Delta u \sim \varepsilon_a$ curves of NC and OC clays in CU test

As shown in Figure 8-31b, the stress strain behavior of overconsolidated (OC) clay is different from that of NC clay. The deviator stress $(\sigma_1 - \sigma_3)$ increases first with increasing axial strain ε_a and then decreases after $(\sigma_1 - \sigma_3)$ reaches the peak value, i.e., the deviator stress at failure $(\sigma_1-\sigma_3)_f$. This stress strain relation is called strain softening behavior. Meanwhile, the pore water pressure increment (Δu) shows an similar first increase and then decrease trend with the increase of ε_a, as shown in Figure 8-31d. This is because the overconsolidated clay is denser than the normally consolidated (NC) clay, the particles of OC clay is more difficult to move and the soil produces a trend of shear dilation, which causes the pore water pressure decreased or even negative. For OC clay, the dense structure will become loose during undrained shear, thus the peak value of $(\sigma_1-\sigma_3)$ will continue to decrease with increasing axial strain.

Figure 8-32 shows the relations of deviator stress $(\sigma_1-\sigma_3)$ and axial strain (ε_a), volumetric strain (ε_v) and ε_a obtained from consolidated drained (CD) test on normally consolidated (NC) and overconsolidated (OC) days, respectively. It can be seen from Figure 8-32a that $(\sigma_1-\sigma_3)$ increases with increasing axial strain ε_a, i.e., a strain hardening behavior is found for NC clay. Meanwhile, the volumetric strain ε_v reduces with the increase of ε_a, which reflects a shear contraction of NC clay. It is because the NC of clay has a volume contraction tendency during shear. In drained condition, the contraction potential is characterized by a reduction of void ratio.

As shown in Figure 8-32b, the stress strain behavior of overconsolidated (OC) clay is different from that of NC clay. The deviator stress $(\sigma_1-\sigma_3)$ increases first with increasing axial strain ε_a and then decreases after $(\sigma_1-\sigma_3)$ reaches the peak value, i.e., a strain softening behavior is found for OC clay. Moreover, the volumetric strain ε_v shows a slightly decrease with the increase of ε_a followed by an increase trend. This is because OC clay is denser than NC clay. When the axial strain is small, the soil particles show a position adjustment, thus the soil mass is slightly compressed at first. Then soil mass exhibits a dilation trend in drained condition. The dilation potential of soil is characterized by the increase in pore volume. It should be noted that the value of deviator stress $(\sigma_1-\sigma_3)$ in drained condition is the same with that in undrained condition for NC or OC clay since $\sigma_1'-\sigma_3'=(\sigma_1-u)-(\sigma_3-u)=\sigma_1-\sigma_3$.

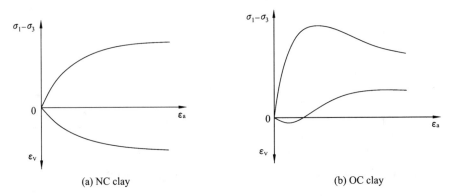

Figure 8-32 $(\sigma_1-\sigma_3) \sim \varepsilon_a$ and $\varepsilon_v \sim \varepsilon_a$ curves of NC and OC clays in CD test

8.5.4 Discussion on shear strength problems

Figure 8-33 shows the strength envelope for the same saturated clay in overconsolidated

(OC) and normally consolidated (NC) state from UU, CU and CD tests. It can be seen that the shear strength obtained for overconsolidated clay is higher than that for the normally consolidated clay. This feature is widely used in ground improvement and various practical engineering projects. In Figure 8-33, σ_p is the preconsolidation stress of clay sample. It can be seen that the strength envelopes are different when determining the shear strength by different testing methods. Strength envelope obtained from a UU test is a horizontal line compared with that from CU or CD test with a straight line through the origin. The relationships of the strength parameters obtained from different tests include: $c_u > c_{cu} = c_d = 0$ for NC clay, $c_u > c_{cu} > c_d > 0$ for OC clay, and $\phi_d > \phi_{cu} > \phi_u = 0$ for both NC and OC clays. In CD test, we have $\sigma = \sigma'(u_f = 0)$, the total stresses are exactly the same with the effective stresses, thus $c_d = c'$, $\phi_d = \phi'$. So the strength envelopes in total stress and effective stress are the same.

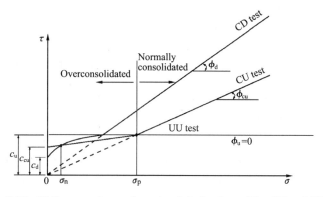

Figure 8-33 Failure envelopes of a saturated clay from CD, CU and UU tests

For overconsolidated clay, the void ratio is smaller than that of normally consolidated clay, and there is a peak value of stress on its stress strain curve, known as the strain-softening behavior. After the yielding point, the stress gradually levels off with the increasing strain to the residual value and the corresponding shear stress is called the residual strength of clay. When a soil is likely to have a large deformation under the long-term loading, the stability checking should be considered whether the residual strength would be used. Beside, the direct shear test can also be used to determine the peak strength and residual strength.

Figure 8-34a shows the shear stress strain curves obtained from CD tests for overconsolidated, normally consolidated and remolded clay specimens at a given normal stress σ. Figure 8-34b shows the strength envelopes of clay specimens and the residual strength is expressed as

$$\tau_r = \sigma \cdot \tan\phi_r \qquad (8-30)$$

where, τ_r—residual strength of clay;

σ—normal stress on the shear surface;

ϕ_r—residual internal friction angle.

Besides, if the strength parameters obtained from UU, CU and CD tests are substituted into the Eq. (8-8), different values of shear failure angle θ_f can be obtained. In fact, when the soil sample is sheared to fail, it can only have one failure angle regardless of the testing method. Test results show that the same strength envelopes and effective stress parameters can be obtained when we use effective stress parameters for UU, CU and CD test. The testing data show that the measured shear failure angle θ_f is close to $(45°+\phi'/2)$. Thus the actual shear failure angle can be calculated with the following equation

$$\theta_f = 45° + \phi'/2 \tag{8-31}$$

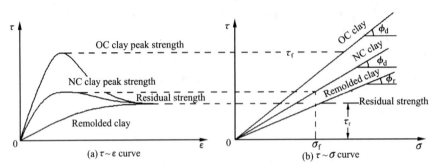

Figure 8-34 Peak and residual strength of NC and OC clay from CD test

Eq. (8-31) also indicates that the failure angle obtained from UU, CU and CD test should be the same when we adopt effective stress analysis. Besides, For UU test under unsaturated state, the air in soil pores will be soluble in pore water as a result of increasing total stress. Finally the soil specimen can be said in saturation state since air dissolves in water. In this case, the friction angel ϕ_u is equal to 0, but the undrained strength c_u in unsaturated state is probably slightly larger than that in saturated state.

8.5.5 Creep of clay during shear

Creep of clay is the phenomenon describing the increase of shear strain and volumetric strain under a constant shear stress, where its rate is controlled by soil structure and cohesive resistance. Figure 8-35a shows the change in axial strain (ε_a) over time under constant deviator stress ($\sigma_1 - \sigma_3$) with different values, which is called the creep curve. From the figure, we can see that the axial strain increases significantly at low elapsed time. When the deviator stress is small, the creep curve is almost a horizontal line, where creep rupture will not occur. When the deviator stress increases, the axial strain increases with elapsed time. When the deviator stress increases to a certain value, the axial strain increases remarkably with elapsed time and ultimately leads to a creep rupture, as shown in Figure 8-35b. The creep of clay will develop as follows:

① Instantaneous elastic strain: it corresponds to the *OA* segment in Figure 8-35b. In this stage, the axial strain increment is small.

② Primary creep: the *AB* segment corresponds to this stage, which is a non-stable creep. In this stage, the strain rate decreases with time. If the deviator stress is removed, the instantaneous elastic strain and the unstable creep will recover.

③ Secondary creep: the *BC* segment corresponds to this stage, which is a steady creep. The strain rate at this stage is a constant. When the deviator stress is removed, the soil will have a permanent creep deformation.

④ Tertiary creep: the CD segment corresponds to this stage, where the strain rate increases rapidly until the soil sample is ruptured.

If the deviator stress exceeds a certain value, the long-term creep deformation developed will cause the soil rupture, and the long-term soil strength may be much lower than the strength determined by lab test. The creep of soil is one of the important reasons that atributes to the lateral movement of retaining structures and the failure of soil slopes. Some engineering projects need considering long-term deformation and stability. We should study the creep characteristics of soil and take the creep subceptibility into account.

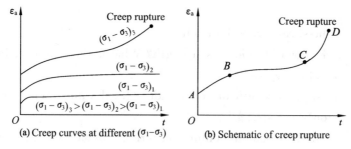

Figure 8-35 Creep behaviors of clayey soil

Knowledge expansion

How to Choose a Proper Testing Method for Shear Strength in Practical Engineering

There are three kinds of the triaxial compression tests: unconsolidated undrained (UU) test, consolidated undrained (CU) test and consolidated drained (CD) test. In these three tests, the "consolidated" is used to simulate the consolidation of natural soil under its own weight by applying the confining stress and keeping the drain valve open in the test. However, the stress state of the natural soil has changed due to the stress release after unloading when sampling. Therefore, the applied confining stress should reach its geostatic stress or self-weight stress of sample obtained from a certain depth to simulate the real stress states.

Additionally, the "consolidated" means the soil layer has been consolidated under additional loading. CD test is used when the construction speed is low, otherwise, CU or UU test can be adopted.

According to the natural stress state of soil, the starting point can be found on the shear strength envelope. If consolidated undrained condition is satisfied in the practical engineering, the strength increase caused by the external loading will develop along the envelope corresponding to CU test; and if both consolidation and drainage are not allowed, the strength will develop along the horizontal line passing through the starting point obtained from a UU test.

Besides, there are two methods, i. e., total stress and effective stress method that can be used to deal with the triaxial compression testing data. However, the concept of total stress method is only an approximate method, which is not reflecting the actual stress state of soil. The more accurate quantitative calculation can be done by the effective stress method.

Exercises

[8-1] What is the shear strength of soil?

[8-2] What is the law of Coulomb's strength? What is the difference in shear strength between sandy soil and clayey soil?

[8-3] Why the shear strength of soil is not a fixed value? What are the effects of drainage conditions on shear strength?

[8-4] What is Mohr-Coulomb failure criterion? What is the limit equilibrium condition?

[8-5] Is the plane with maximum shear strength the same with the plane of shear failure?

[8-6] What are the methods for determining the shear strength of soil? Try to discuss their advantages and disadvantages.

[8-7] What are soil sensitivity and thixotropy?

[8-8] What are the factors influencing the shear strength of sandy soil?

[8-9] What is the liquefaction of sand? How it will behave?

[8-10] Describe the strength behaviors of saturated normally consolidated clay in UU, CU and CD tests.

[8-11] Describe the strength behaviors of saturated overconsolidated clays in UU, CU and CD tests.

[8-12] Explain the relationship between the total stress envelope and effective stress envelope for normally consolidated soil and overconsolidated soil.

[8-13] The strength parameters of a non-cohesive soil obtained from a triaxial compression test are known as $c=0$, and $\phi=30°$:

(1) If the major and minor principal stresses are 200kPa and 120kPa, respectively, will the sample fail?

(2) If the minor principal stress is kept unchanged and the major principal stress is increased. Do you think the major principal stress can be increased to 400kPa? Why?

[8-14] The major principal stress at a certain point in the ground soil is 450kPa and the minor principal stress is 200kPa. The internal friction angle of soil is 20° and the

cohesion is 50kPa. What is the state of this point?

[8-15] The major principal stress at a given point in the ground soil is 460kPa; the minor principal stress is 150kPa; the pore water pressure is 50kPa and the effective strength parameters of soil are $\phi'=30°$, $c'=0$. What is the state of this point?

[8-16] The results for a CU test for a soil are shown in Table 8-7. Try using graphical methods to get total stress strength parameters c_{cu} and ϕ_{cu}, and effective stress strength parameters c' and ϕ'.

Table 8-7 Results of a CU test for Exercise [8-16]

σ_3/kPa	$(\sigma_1-\sigma_3)_f$/kPa	u_f/kPa
100	200	35
200	320	70
300	460	75

[8-17] The internal friction angle $\phi'=\phi_d=30°$ was measured by the triaxial CD test for a normally consolidated clay. If we use the same soil to run CU test and apply $\sigma_3=200$kPa. The deviator stress q is 175kPa at failure. Calculate the strength parameters c_{cu}, ϕ_{cu} and the pore water pressure at failure u_f.

[8-18] The maximum torsional moment of the vane shear test at different depths of ground soil are shown in Table 8-8. The height of the vane is 10cm and the width is 5cm. Try to determine the shear strength of soil at different depths.

Table 8-8 Vane shear test data for Exercise [8-18]

Depth/m	5	10	15
Torsional moment/(kN·m)	120	160	190

[8-19] A saturated normally consolidated clay specimen was consolidated at $\sigma_3=150$kPa and then sheared to failure in undrained condition. The measured undrained strength $c_u=60$kPa, and the measured angle between the shear failure surface and the major principal stress acting surface is $\theta_f=57°$. Try to determine the pore pressure coefficient A_f and internal friction angle ϕ_{cu} at failure.

[8-20] Below an engineering foundation is a soft clay ground. The average water content $w=70.5\%$, $\gamma=15.9$kN/m³, $c_u=5.0$kPa, $\phi_u=1.8°$, $c_{cu}=1.0$kPa and $\phi_{cu}=15°$. A large area of surcharge (=30kPa) is applied on the ground. When the consolidation degree reaches 80%, what is soil strength at depth of 5m? What is the net increase in strength?

[8-21] A saturated normally consolidated clay has $c'=0$ and $\phi'=30°$. When the confining pressure $\sigma_3=100$kPa, try to determine c_{cu} and ϕ_{cu} (assuming $A_f=1$).

[8-22] A circular foundation under a concentrated load is shown in Figure 8-36. Below the foundation is a deep layer of clay with $\gamma=18.0$kN/m³ and $\gamma_{sat}=21.0$kN/m³. The groundwater table is 3m below the surface. Before loading, point M at depth of 5m has a

inital water head of 2m. Upon loading, the water head in the pressure tube is 7m above ground surface. If the vertical additional stress $\Delta\sigma_1 = 140$kPa and horizontal additional stress $\Delta\sigma_3 = 60$kPa at point M, try to solve:

(1) The vertical effective stress and pore pressure coefficients A and B at point M upon loading;

(2) If the ground soil is a normally consolidated soil before loading, the effective internal friction angle $\phi' = 30°$, and the earth pressure coefficient $K_0 = 0.6$, will the point M fail?

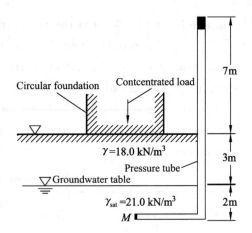

Figure 8-36 Schematic diagram of Exercise [8-22]

Chapter 9 Earth Pressure on Retaining Structure

9.1 Introductory case

Earth pressure is the force per unit area exerted by soil. Earth pressure is an important knowledge for the design and stability checking of retaining structures. The major objective of this chapter is to present the distribution of different types of earth pressures on retaining structures and corresponding calculation method.

Retaining structures are commonly used in port engineering, road and bridge engineering, building foundation engineering, etc. There are many common retaining structures in engineering projects, such as wharf wall (Figure 9-1a), side walls of tunnel (Figure 9-1b), retaining walls of various slopes (Figure 9-1c), bridge abutment (Figure 9-1d), basement wall (Figure 9-1e). The role of the retaining structure is to support the side soil mass and bear the lateral earth pressure force (E).

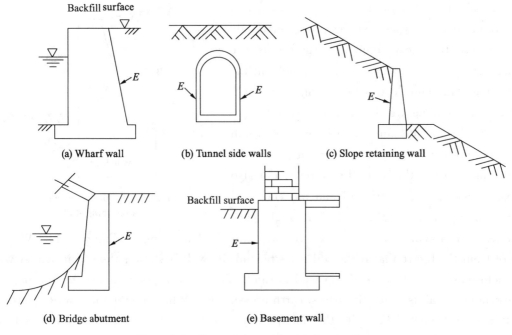

Figure 9-1 Common types of retaining structures

Therefore, it is of great importance to have a thorough knowledge of the lateral forces interacting between the retaining structures and the soil mass retained. If the retaining structure can not bear the weight and the surcharge load, serious safety problems will occur. Figure 9-2 shows the photograph of the retaining structure collapse accident happened on the construction site of No. 1 line of Fengqing Avenue Subway in Hangzhou city, Zhejiang Province in China. The accident occurred in the afternoon of November 15, 2008, in which 21 people died and 24 were injured. How to ensure the safety and stability of the retaining structure is an important issue, where the knowledge of earth pressure on retaining structures is needed.

Figure 9-2 Retaining structure collapse accident

Retaining structures can be classified according to the types, applications, materials etc, where the most common structure is the retaining wall. There are vertical and incline retaining walls applied in engineering. The main parts of a retaining are shown in Figure 9-3.

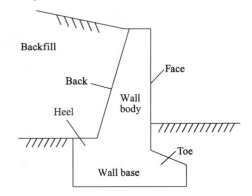

Figure 9-3 Basic elements of a retaining wall

The magnitude and distribution of earth pressure are not only related to the height of the retaining wall and the backfill materials, but also closely related to the stiffness and movement of the retaining wall. In 1929, Karl Terzaghi gave the relationship curve of earth pressure with the wall movement (Figure 9-4). We can learn from the figure that if the soil mass behind the wall is stable, the soil is at rest with no deformation, and the retaining wall is rigid. The earth pressure acting on the back of the retaining wall is called the at-rest earth pressure E_0. When the retaining wall rotates or moves away from the backfill, the unit pressure on the wall gets gradually reduced and after a particular displacement of the wall at the top, the pressure reaches a constant value. This pressure is the minimum possible value when the wall comes to the limit equilibrium state. The minimum earth pressure acting on the back of the retaining wall is called the active earth pressure E_a. The pressure is called as active since the weight of the backfill causes the movement of the wall. When the retaining wall rotates or moves towards the backfill, the unit pressure on the wall increases from the value of at-rest condition to the maximum possible value. The maximum earth pressure acting on the back of the retaining wall is called the passive earth pressure E_p. The pressure is called as passive since the weight of the backfill opposes the move of the wall.

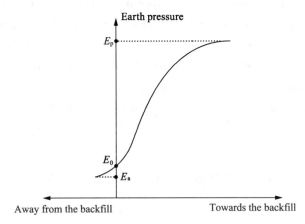

Figure 9-4   Variation of lateral earth pressure with the movement of retaining wall

Generally, neither rotation nor displacement of wall may lead the backfill to the equilibrium state. Therefore, the earth pressure may be a value between active earth pressure and at-rest earth pressure or between at-rest earth pressure and passive earth pressure. The magnitude of earth pressure is related to movement characteristic and the value of displacement of wall.

9.2 Calculation of earth pressure at rest

The earth pressure at rest only occurs when the retaining wall is rigid and does not have any displacement. In the practical engineering, the earth pressure acting on the retaining walls of deep foundation or the U-shaped abutment can be approximated regarded as at-rest earth pressure.

At-rest earth pressure calculation is relatively simple. Since the wall does not move, the stress state of soil behind the wall is similar to free state which can be determined by the calculation equation of horizontal stress in semi-infinite ground, that is

$$e_0 = K_0 \sigma_z = K_0 \gamma z \tag{9-1}$$

where, e_0—at-rest earth pressure per unit area;

K_0—coefficient of lateral earth pressure at rest;

σ_z—vertical effective stress at depth z;

γ—unit weight for single soil layer;

z—depth of calculation point.

Eq. (9-1) shows that the at-rest earth pressure varies linearly along the depth. The distribution of at-rest earth pressure for single soil layer is shown in Figure 9-5a. The total at-rest earth pressure force per unit length E_0 is

$$E_0 = \frac{1}{2}\gamma H^2 K_0 \quad (\text{kN/m}) \tag{9-2}$$

where, H—the height of wall, and the action point of horizontal resultant force E_0 is on the 1/3 wall height from the wall base.

If the soil behind the retaining wall is layered, we can use the unit weight of each soil layer to calculate the vertical effective stress for each layer and add them together to obtain the total effective stress at depth z. If there is groundwater in the backfill behind the wall, the unit weight γ of the soil should be replaced by the buoyant unit weight γ' ($\gamma' = \gamma - \gamma_w$) when we calculate the at-rest earth pressure. γ_w is the unit weight of water ($\gamma_w = 9.8 \text{kN/m}^3$). The distribution of at-rest earth pressure for this case is shown in Figure 9-5b. Thus the total at-rest earth pressure force (E_0) is equal to the area of the pressure distribution.

$$E_0 = \frac{1}{2}\gamma H_1^2 K_0 + \gamma H_1 H_2 K_0 + \frac{1}{2}\gamma' H_2^2 K_0 \tag{9-3}$$

The action point of E_0 is at the centroid position of the lateral earth pressure distribution area.

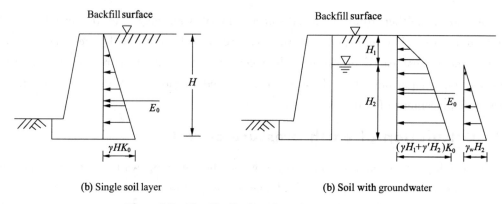

(b) Single soil layer   (b) Soil with groundwater

Figure 9-5   The distribution of earth pressure at rest

Meanwhile, the water pressure action on the wall should be taken into account when we analyze the total force on retaining wall. The total water pressure force (P_w) on the back of the wall is

$$P_w = \frac{1}{2}\gamma_w H_2^2 \tag{9-4}$$

The action point of P_w is on the $1/3 H_2$ from the wall base. The total pressure force acting on the wall is the vector resultant force of the earth pressure and the water pressure.

Determining the at-rest earth pressure coefficient K_0 is important to calculate at-rest earth pressure, which can be determined by lab or field at-rest earth pressure test. For sand or normally consolidated clay, K_0 can be estimated by ϕ'.

$$K_0 = 1 - \sin \phi' \tag{9-5}$$

where, ϕ'—effective internal friction angle of backfill.

The value of K_0 is usually less than 1. For over-consolidated clay and compacted soils, it may be greater than 1.

[Example 9-1] Calculate the distribution of the at-rest earth pressure and its resultant force on the retaining wall (Figure 9-6), where q is the uniformly distributed surcharge.

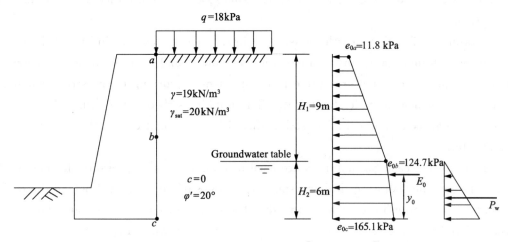

Figure 9-6 Schematic diagram of [Example 9-1]

[Solution] The at-rest earth pressure coefficient is

$$K_0 = 1 - \sin \phi' = 1 - \sin 20° = 0.66$$

The at-rest earth pressures at different points in the soil are

Point a: $e_{0a} = K_0 q = 0.66 \times 18 = 11.8 \text{kPa}$

Point b: $e_{0b} = K_0(q + \gamma H_1) = 0.66 \times (18 + 19 \times 9) = 124.7 \text{kPa}$

Point c: $e_{0c} = K_0(q + \gamma H_1 + \gamma' H_2) = 0.66 \times [18 + 19 \times 9 + (20 - 9.8) \times 6] = 165.1 \text{kPa}$

The resultant force of the at-rest earth pressure E_0 is

$$\begin{aligned}
E_0 &= \frac{1}{2}(e_{0a} + e_{0b})H_1 + \frac{1}{2}(e_{0b} + e_{0c})H_2 \\
&= \frac{1}{2}(11.8 + 124.7) \times 9 + \frac{1}{2}(124.7 + 165.1) \times 6 = 1484.3 \text{kN/m}
\end{aligned}$$

The distance y_0 between action point of E_0 and the bottom of the wall is

$$\begin{aligned}
y_0 &= \frac{1}{E_0}\left[e_{0a}H_1\left(\frac{H_1}{2} + H_2\right) + \frac{1}{2}(e_{0b} - e_{0a})H_1\left(H_2 + \frac{H_1}{3}\right) + e_{0b} \times \frac{H_2^2}{2} + \frac{1}{2}(e_{0c} - e_{0b})\frac{H_2^2}{3}\right] \\
&= \frac{1}{1484.3}\left[9 \times 11.8 \times 10.5 + \frac{1}{2} \times (124.7 - 11.8) \times 9 \times (6 + \frac{9}{3}) + 124.7 \times \frac{6^2}{2} + \right. \\
&\quad \left. \frac{1}{2}(165.1 - 124.74) \times \frac{6^2}{3}\right] = 5.51 \text{m}
\end{aligned}$$

The resultant force of hydrostatic water pressure P_w on the wall is

$$P_w = \frac{1}{2}\gamma_w H_2^2 = \frac{1}{2} \times 9.81 \times 6^2 = 176.6 \text{kN/m}$$

The distributions of at-rest earth pressure and water pressure are shown in Figure 9-6.

9.3 Rankine's earth pressure theory

9.3.1 Basic assumptions and principles

Rankine's earth pressure theory and Coulomb's earth pressure theory are two basic theories to study the active earth pressure and the passive earth pressure, which were presented in 1857 and 1776, respectively. Rankine's earth pressure theory is introduced in this section.

According to Rankine's earth pressure theory, when the soil behind the wall reaches the limit equilibrium state, any point in the soil unit is under the limit state. Then the calculation equation of earth pressure can be established according to the stress conditions when the soil element is under limit state. The Rankine's theory was originally proposed for dry homogeneous non-cohesive soils, then it was extended to cohesive soils with groundwater.

The following basic assumptions are used in Rankine's earth pressure theory: ① the back surface of the wall is vertical and smooth; ② the surface of backfill behind the wall is horizontal and semi-infinite, where the directions of principal stresses at depth z are horizontal and vertical, respectively. The stress state is shown in Figure 9-7.

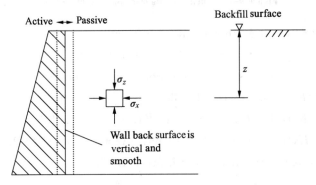

Figure 9-7 Stress state of soil unit for Rankine's theory

When the retaining wall does not move, the earth pressure on the back of the wall is the at-rest earth pressure. The stress state of the soil behind the wall can be represented by the Mohr's circle in Figure 9-8, where $\sigma_1 = \sigma_z$, $\sigma_3 = \sigma_x = K_0 \sigma_z$. When the wall moves away from the backfill, the vertical stress σ_z will remains constant and be equal to the major principal stress σ_1. With the increase of displacement, the horizontal stress σ_x (or σ_3) gradually decreases, and the radius of Mohr's circle gradually increases. When the increased Mohr's circle reaches to the failure envelope, soil will reach the active limit equilibrium state. The earth pressure acting on the wall equals to the minimum horizontal stress $\sigma_{x\min}$ (minor principal stress σ_{3f}), which is the active earth pressure; when the wall is approaching to the backfill, the vertical stress σ_z remains the same. However, with the increase of the displacement, the horizontal stress gradually increases beyond the vertical stress, thus the direction of major and minor principal stress changes, where $\sigma_z = \sigma_1$ becomes σ_3, and $\sigma_x = \sigma_3$ becomes σ_1. When the Mohr's circle reaches to the failure envelope, the soil reaches the passive limit equilibrium state, and the earth pressure acting on the wall is equal to the maximum horizontal stress $\sigma_{x\max}$ (major principal stress σ_{1f}), which is the passive earth pressure.

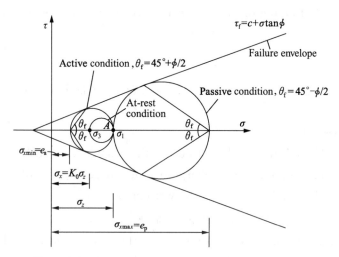

Figure 9-8 Active and passive limit equilibrium state of soil

According to the theory of shear strength, when the soil reaches the limit equilibrium state, the angle between the failure surface and the major principal stress acting surface is $\theta_f = 45° + \phi/2$. Therefore, when the soil reaches the active limit equilibrium state, the major principal stress acting surface is the horizontal plane, thus the angle between the failure surface and the horizontal plane is $\theta_f = 45° + \phi/2$. When the soil reaches the passive equilibrium state, the major principal stress acting surface is the vertical plane, the angle between the failure surface and the horizontal plane is $\theta_f = 45° - \phi/2$.

9.3.2 Calculation of Rankine's active earth pressure

When the soil behind the wall is in the active state, the vertical stress of soil is the major principal stress and the horizontal stress of soil unit at depth z is the minor principal stress.

$$\sigma_1 = \sigma_z, \quad \sigma_3 = \sigma_x = e_a \tag{9-6}$$

When the soil unit is in the limit equilibrium state, the Rankine's active earth pressure equation is

$$e_a = \sigma_{3f} = \sigma_{1f}\tan^2(45° - \frac{\phi}{2}) - 2c \cdot \tan(45° - \frac{\phi}{2}) = \sigma_z K_a - 2c\sqrt{K_a} \tag{9-7}$$

where $K_a = \tan^2(45° - \phi/2)$ is the Rankine's active earth pressure coefficient.

Eq. (9-7) is the basic equation of Rankine's theory for active earth pressure calculation.

I. For non-cohesive soil

Assuming that behind wall is filled with homogeneous non-cohesive soil without groundwater, and there is no surcharge on the surface, we have $\sigma_z = \gamma z$ and $c = 0$. The active earth pressure at depth z can be obtained by Eq. (9-7).

$$e_a = \gamma z K_a \tag{9-8}$$

From Eq. (9-8), the active earth pressure is linearly distributed along the depth z, which can be seen from Figure 9-9. If the height of the retaining wall is H, the total active earth pressure force (E_a) acting on the wall can be calculated by the area of the triangular distribution.

$$E_a = \frac{1}{2}\gamma H^2 K_a \tag{9-9}$$

The action point of E_a is acting on the centroid of the triangular area with $H/3$ from the wall base.

II. For cohesive soil

Assuming that behind the wall is filled with homogeneous cohesive soil without groundwater ($c \neq 0$) and there is no surcharge on the surface of backfill, the active earth pressure value at depth z by Eq. (9-7) can be written as

$$e_a = \gamma z K_a - 2c\sqrt{K_a} \tag{9-10}$$

It can be seen that the active earth pressure of the cohesive soil also linearly changes along the depth z. However, it is different from that of the non-cohesive soil. It consists of two parts: one is caused by soil weight, which is a triangular distribution along the height; another part is caused by the cohesion of the soil, which is independent of depth and shows a rectangular distribution along the height. Due to the presence of cohesion of soil, the active earth pressure at a certain depth will be a negative value. It means that there will be a certain tensile area, where the depth z_0 can be estimated at $e_a = 0$ from Eq. (9-10).

$$z_0 = \frac{2c}{\gamma \sqrt{K_a}} \tag{9-11}$$

Since the soil cannot bear tensile force, the distribution of the active earth pressure acting on the wall should be the lower triangle abc shown in Figure 9-10. The total active earth pressure force can be calculated according to the area of the triangle abc.

$$E_a = \frac{1}{2}\gamma(H - z_0)^2 K_a \tag{9-12}$$

The action point of E_a is on the centroid of the triangle abc with the distance $(H - z_0)/3$ from the bottom of the wall.

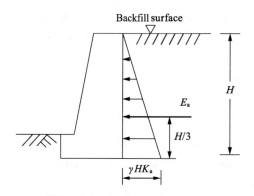

Figure 9-9 Active earth pressure distribution of non-cohesive soil

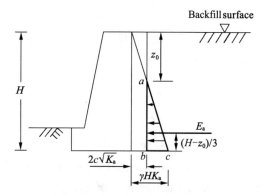

Figure 9-10 Active earth pressure distribution of cohesive soil

Ⅲ. For soil with a uniform surcharge load

As shown in Figure 9-11, if the backfill soil is non-cohesive soil with a uniform surcharge q, the vertical stress of the soil unit at depth z is $\sigma_z = \gamma z + q$. We can substitute it and $c = 0$ into Eq. (9-7) to obtain the active earth pressure acting on the wall at depth z.

$$e_a = (\gamma z + q) K_a = \gamma z K_a + q K_a \quad (9\text{-}13)$$

It can be seen that the active earth pressure is composed of two parts. One is caused by soil weight with a triangular distribution along the wall, which is proportional to the depth; another part is caused by the uniform surcharge with a rectangular distribution along the wall, which is independent of the depth. The total active earth pressure force acting on the back of the wall can be calculated by the area of the trapezoidal distribution as below

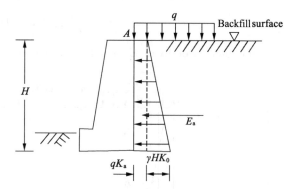

Figure 9-11 Active earth pressure of non-cohesive soil with a uniform surcharge load

$$E_a = \frac{1}{2} \gamma H^2 K_a + q H K_a \quad (9\text{-}14)$$

The action point of E_a is on the centroid of trapezoidal distribution area.

If the backfill soil is a cohesive soil with a uniform surcharge q, the vertical stress of the soil unit at depth z is $\sigma_z = \gamma z + q$, and the active earth pressure is

$$e_a = \sigma_z K_a - 2c\sqrt{K_a} = \gamma z K_a + q K_a - 2c\sqrt{K_a} \quad (9\text{-}15)$$

It can be seen that the active earth pressure consists of three parts: the first part is caused by soil weight with a triangular distribution along the wall, which is proportional to depth; the second part is caused by the uniform surcharge load with a rectangular distribution along the wall height, which is independent of the depth; the third part is caused by the tension due to cohesion with a rectangular distribution along the wall, which is independent of the depth.

When the uniform surcharge is acting on the wall, tension may occur in the area of active earth pressure. If $e_a = 0$, we can obtain the maximum depth of tensile stress zone from Eq. (9-15).

$$z_0 = \frac{2c}{\gamma\sqrt{K_a}} - \frac{q}{\gamma} \tag{9-16}$$

If the value of z_0 is greater than 0, tensile stress zone will exist and the distribution of active earth pressure will be a triangle, as shown in Figure 9-12a. The total active earth pressure force can be calculated by Eq. (9-12), which is acting on the centroid of the triangle with the distance of $(H - z_0)/3$ from the wall base.

If the value of z_0 is less than 0, tensile zone will not exist and the distribution of active earth pressure is a trapezium, as shown in Figure 9-12b. The total active earth pressure force can be calculated by the area of the trapezium.

$$E_a = \frac{1}{2}\gamma H^2 K_a + qHK_a - 2cH\sqrt{K_a} \tag{9-17}$$

The action point of E_a is acting on the centroid of trapezium.

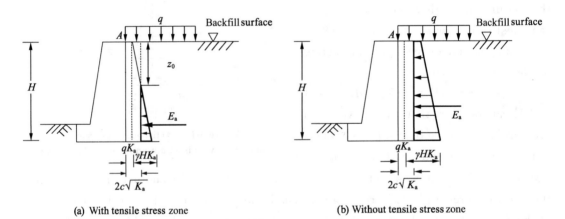

(a) With tensile stress zone (b) Without tensile stress zone

Figure 9-12 Active earth pressure of cohesive soil under a uniform surcharge load

IV. For layered soils

If the backfill soil behind the wall is composed of different soil layers, the effect of each soil property (unit weight, cohesion and internal friction angle) on the earth pressure should be considered. Two aspects should be taken into consideration: first, the distribution of vertical stress will have an abrupt turning point at the interface of different layers; second, due to the different cohesion and angle of internal friction, it is necessary to use the corresponding cohesion and internal friction angle of the soil when calculating the active or passive earth pressure coefficient, where abrupt change of soil pressure at the interface may occur. It should be pointed out that the sudden change only exist in the theoretical calculation, whereas the actual earth pressure at the interface of the soil layer is continuous in the vicinity of the interface. The following is an example of the active earth pressure distribution with two layers of non-cohesive soil shown in Figure 9-13.

Chapter 9 Earth Pressure on Retaining Structure

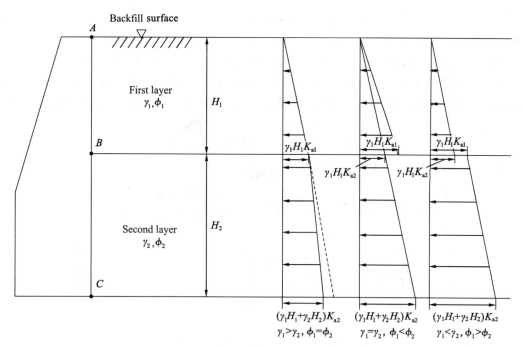

Figure 9-13 Active earth pressure of layered backfill

The active earth pressure above interface point B is

$$(e_a)_{B\text{-above}} = (\sigma_z)_B K_{a1} = \gamma H_1 K_{a1} \tag{9-18}$$

The active earth pressure below interface point B is

$$(e_a)_{B\text{-below}} = (\sigma_z)_B K_{a2} = \gamma H_1 K_{a2} \tag{9-19}$$

The active earth pressure on point C is

$$(e_a)_C = (\sigma_z)_C K_{a2} = (\gamma_1 H_1 + \gamma_2 H_2) K_{a2} \tag{9-20}$$

where, K_{a1}, K_{a2} —the coefficient of active earth pressure in the first layer and second layer, respectively.

The total earth pressure force can be determined by the area of distributed pressure in Figure 9-13.

$$E_a = \frac{1}{2} \gamma_1 H_1^2 K_{a1} + \gamma_1 H_1 H_2 K_{a2} + \frac{1}{2} \gamma_2 H_2^2 K_{a2} \tag{9-21}$$

The action point of E_a is acting on the centroid of the earth pressure distribution area.

V. For soil with groundwater

When there is groundwater in backfill soil behind the wall, the soil can be regarded as two layers separated by the groundwater table. We can use the unit weight (γ) to calculate the earth pressure of soil layer above water table. Below the groundwater table, we can use the buoyant unit weight (γ') to calculate the earth pressure and use the unit weight of water (γ_w) to calculate the water pressure.

Here comes an example of calculation of active earth pressure with non-cohesive soil as shown in Figure 9-14. The active earth pressure above the groundwater table (AB section) is the same as that without groundwater. The active earth pressure at point B is

$$(e_a)_B = (\sigma_z)_B K_a = \gamma H_1 K_a \tag{9-22}$$

where, H_1—height of backfill above the groundwater table.

For the active earth pressure below the groundwater table (BC section), the soil unit weight (γ) should be replaced by the buoyant unit weight (γ') in the part below groundwater (section BC in the figure). Assuming that the K_a value below water table is constant, the active earth pressure at point C is

$$(e_a)_C = (\sigma_z)_C K_a = \gamma H_1 K_a + \gamma' H_2 K_a \tag{9-23}$$

where, H_2—height of the backfill below the groundwater table.

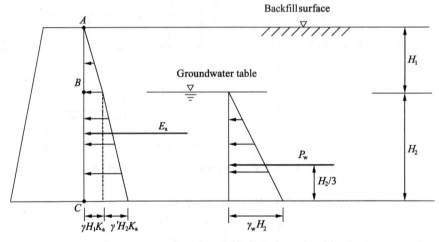

Figure 9-14 Active earth pressure of non-cohesive soil with groundwater

The total active earth pressure force can be determined by the distributed earth pressure area,

$$E_a = \frac{1}{2}\gamma H_1^2 K_a + \gamma H_1 H_2 K_a + \frac{1}{2}\gamma' H_2^2 K_a \tag{9-24}$$

The action point of E_a is acting on the centroid of the area of active earth pressure.

The water pressure P_w acting on the wall is shown in Eq. (9-4). So the total pressure force acting on the wall is

$$P = E_a + P_w = \frac{1}{2}\gamma H_1^2 K_a + \gamma H_1 H_2 K_a + \frac{1}{2}\gamma' H_2^2 K_a + \frac{1}{2}\gamma_w H_2^2 \tag{9-25}$$

【Example 9-2】 As shown in Figure 9-15, there is a 8m height retaining wall with two layers of backfill soil behind it and a uniform surcharge load of 18kPa. Please determine the value of active earth pressure force and its acting point on the wall.

【Solution】 The coefficient of active earth pressure in the first layer is

$$K_{a1} = \tan^2(45° - \frac{\phi_1}{2}) = \tan^2(45° - \frac{25°}{2}) = 0.406$$

The coefficient of active earth pressure in the second layer is

$$K_{a2} = \tan^2(45° - \frac{\phi_2}{2}) = \tan^2(45° - \frac{20°}{2}) = 0.490$$

Since c_1 is greater than 0, it is necessary to determine whether there is tensile earth pressure in the first layer. From Eq. (9-16), the depth corresponding to the zero earth pressure is calculated as

$$z_{01} = \frac{2c_1}{\gamma_1 \sqrt{K_{a1}}} - \frac{q}{\gamma_1} = \frac{2 \times 10}{15 \times \sqrt{0.406}} - \frac{18}{15} = 0.89 \text{m}$$

Since $z_{01} > 0$, it indicates that a tensile stress area within depth of 0.89m is in the first layer.

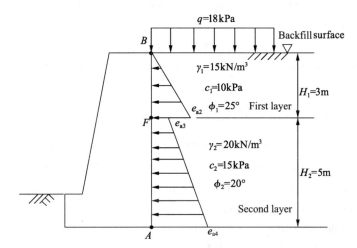

Figure 9-15 Schematic diagram of [Example 9-2]

The depth where the earth pressure is zero in the second layer is

$$z_{02} = \frac{2c_2}{\gamma_2 \sqrt{K_{a2}}} - \frac{q + \gamma_1 H_1}{\gamma_2} = \frac{2 \times 15}{20 \times \sqrt{0.490}} - \frac{18 + 15 \times 3}{20} = -1.01 \text{m}$$

Since $z_{02} < 0$, there is no tensile stress area in the second layer.

The active earth pressure at a given depth in the soil can be calculated by Eq. (9-15), where the active earth pressure above the interface through point F is

$$e_{a2} = qK_{a1} + \gamma_1 H_1 K_{a1} - 2c_1 \sqrt{K_{a1}} = 18 \times 0.406 + 15 \times 3 \times 0.406 - 2 \times 10 \times \sqrt{0.406} = 12.8 \text{kPa}$$

The active earth pressure below the interface through point F is

$$e_{a3} = qK_{a2} + \gamma_1 H_1 K_{a2} - 2c_2 \sqrt{K_{a2}} = 18 \times 0.490 + 15 \times 3 \times 0.490 - 2 \times 15 \times \sqrt{0.490} = 9.9 \text{kPa}$$

The active earth pressure at point A is

$$e_{a4} = qK_{a2} + (\gamma_1 H_1 + \gamma_2 H_2) K_{a2} - 2c_2 \sqrt{K_{a2}}$$
$$= 18 \times 0.490 + (15 \times 3 + 20 \times 5) \times 0.490 - 2 \times 15 \times \sqrt{0.490} = 58.9 \text{kPa}$$

The active earth pressure force in the first layer is

$$E_{a1} = \frac{1}{2} \gamma_1 (H_1 - z_{01})^2 K_{a1} = \frac{1}{2} \times 15 \times (3 - 0.89)^2 \times 0.406 = 13.6 \text{kN/m}$$

The active earth pressure force in the second layer is

$$E_{a2} = \frac{e_{a3} + e_{a4}}{2} \times H_2 = \frac{9.9 + 58.9}{2} \times 5 = 172 \text{kN/m}$$

The total active pressure force on the wall is

$$E_a = E_{a1} + E_{a2} = 13.6 + 172 = 185.6 \text{kN/m}$$

The distance between the action point of E_a and the point A is

$$y_0 = \frac{13.6 \times (5 + \frac{3-0.89}{3}) + 172 \times (\frac{2 \times 9.9 + 589}{3 \times 9.9 + 3 \times 58.9} \times 5)}{185.6} = 2.18 \text{m}$$

9.3.3 Calculation of Rankine's passive earth pressure

When the backfill behind the wall is in the passive limit equilibrium state, the vertical stress of the soil unit at depth z is the minor principal stress and the horizontal stress is the major principal stress.

$$\sigma_3 = \sigma_z, \quad \sigma_1 = \sigma_x = e_p \tag{9-26}$$

When the soil is in the limit equilibrium state, the Rankine's passive earth pressure is expressed as:

$$e_p = \sigma_{1f} = \sigma_{3f} \tan^2(45° + \frac{\phi}{2}) + 2c\tan(45° + \frac{\phi}{2}) = \sigma_z K_p + 2c\sqrt{K_p} \tag{9-27}$$

where $K_p = \tan^2(45° + \phi/2)$ is called the passive earth pressure coefficient of Rankine's theory.

Eq. (9-27) is the basic equation for calculating the Rankine's passive earth pressure. It can be seen that the values of two parts in Eq. (9-27) are positive, indicating there is no tensile area for passive earth pressure.

I. For non-cohesive soil

Assuming that behind the wall is filled with homogeneous non-cohesive soil, and there is no surcharge on the surface, we have $\sigma_z = \gamma z$ and $c = 0$. The passive earth pressure at depth z can be obtained by Eq. (9-27).

$$e_p = \gamma z K_p \tag{9-28}$$

From the Eq. (9-28), we can find that the distribution of passive earth pressure is in form of triangular which varies linearly along the depth z, as shown in Figure 9-16.

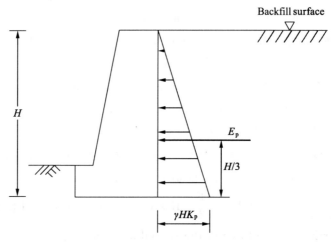

Figure 9-16 Passive earth pressure distribution of non-cohesive soil

The height of the retaining wall is H, the total passive earth pressure force acting on the wall can be obtained by calculating the area of the triangular distribution.

$$E_p = \frac{1}{2} \gamma H^2 K_p \tag{9-29}$$

The action point of E_p is located at the centroid of the triangular area with $H/3$ from base of the wall.

II. For cohesive soil

If the backfill behind the wall is homogeneous cohesive soil without groundwater, we have cohesion $c \neq 0$ and the passive earth pressure at depth z is:

$$e_p = \gamma z K_p + 2c\sqrt{K_p} \tag{9-30}$$

It can be seen that the passive earth pressure is linearly distributed along the depth z, which consists of two parts: one is in the form of triangular distribution along the height caused by the soil weight; another is in the form of rectangular distribution along the height caused by the cohesion, which is independent with depth. The distribution is shown in Figure 9-17. The total passive earth pressure force acting on the wall can be obtained by computing the distributed trapezoidal area.

$$E_p = \frac{1}{2}\gamma H^2 K_p + 2cH\sqrt{K_p} \tag{9-31}$$

The action point of E_p is on the centroid of the trapezoidal area.

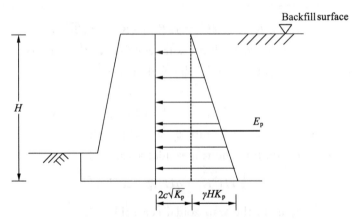

Figure 9-17 Passive earth pressure distribution of cohesive soil

III. For soil with a uniform surcharge load

When there is a uniform surcharge load q acting on the backfill surface and the backfill is non-cohesive soil, as shown in Figure 9-18. The vertical stress at depth z is $\sigma_z = \gamma z + q$, so the passive earth pressure is

$$e_p = (\gamma z + q)K_p = \gamma z K_p + qK_p \tag{9-32}$$

It can be seen that the passive earth pressure consists of two parts: one is in the form of rectangular distribution along the height caused by the uniform surcharge load, which is indepedent of the depth; another is in the form of triangular distribution along the height caused by the soil weight, which is proportional to the depth. So the total passive earth pressure force acting on the back of the wall can be obtained by computing the area of the trapezoidal distribution.

$$E_p = \frac{1}{2}\gamma H^2 K_p + qHK_p \tag{9-33}$$

The action point of E_p is on the centroid of the trapezoidal area.

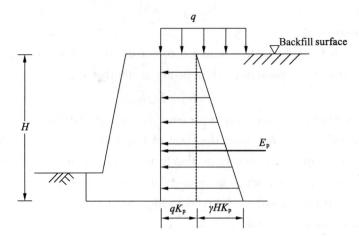

Figure 9-18   Passive earth pressure of non-cohesive soil with a uniform surcharge load

If the backfill is cohesive soil with a uniform surcharge load q, as shown in Figure 9-19, the passive earth pressure is

$$e_p = \sigma_z K_p + 2c\sqrt{K_p} = \gamma z K_p + q K_p + 2c\sqrt{K_p} \qquad (9\text{-}34)$$

It can be seen that the passive earth pressure consists of three parts: the first part is in the form of triangular distribution along the wall caused by the soil weight, which is proportional to the depth; the other two parts are respectively caused by the uniform load and the cohesion in the form of rectangular distribution, which is independent on the depth. The distribution is shown in Figure 9-19, where the total passive earth pressure force can be obtained by computing the trapezoidal area.

$$E_p = \frac{1}{2}\gamma H^2 K_p + q H K_p + 2c H\sqrt{K_p} \qquad (9\text{-}35)$$

The action point of E_p is on the trapezoidal centroid.

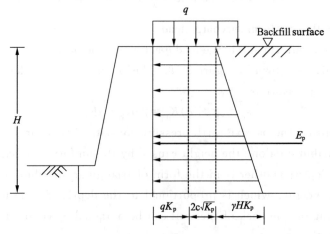

Figure 9-19   Passive earth pressure of cohesive soil with a uniform surcharge load

Besides, for the backfill behind the wall with groundwater or layered soil, the calculation procedures of passive earth pressure are similar to that of active earth pressure.

9.4 Coulomb's earth pressure theory

9.4.1 Basic assumptions and principles

Coulomb's earth pressure theory is based on the rigid body limit equilibrium theory. The earth pressure is determined by the balance of forces on the sliding soil body. In the retaining wall model test, the backfill behind the wall is non-cohesive soil without groundwater(Figure 9-20). When a displacement occurs away from backfill, there will be a nearly linear active sliding surface AD generating between wall surface AB and AC; on the contrary, if the wall is approaching to the backfill, a passive sliding surface AE will appear between the AC surface and the backfill surface. As long as the shape and location of failure surface are determined, the active or passive earth pressure can be determined based on the static equilibrium conditions of soil according to the assumption that the sliding soil is a rigid body.

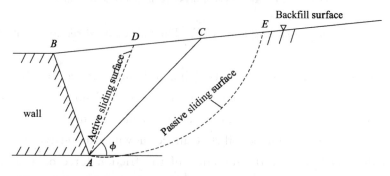

Figure 9-20 Sliding surfaces of non-cohesive backfill soil behind the wall

Thus, the basic assumptions of the Coulomb's theory is: ① the sliding soil is a rigid body; ② the sliding surface is a plane (or a predetermined sliding surface). Based on the assumptions, Coulomb established the equation of active earth pressure and passive earth pressure for non-cohesive soils. The theory is thereafter extended to be available in cohesive soil and saturated condition.

The key of the Coulomb's earth pressure theory is to determine the shape and location of the failure surface. In order to simplify the problem, it is generally assumed that the failure surface is a plane. This assumption often leads to the less accurate calculation, especially in the calculation of passive earth pressure. Since the actual passive failure surface is close to a logarithmic spiral line, it may cause errors that cannot be ignored.

9.4.2 Calculation of Coulomb's active earth pressure

I. For non-cohesive soil

As shown in Figure 9-21, the angle between the wall surface AB and the vertical line is ε. The backfill surface BC is a plane. The angle between the BC surface and horizontal plane is α. Assume the internal friction angle between the wall and the backfill is ϕ_0. For the wall moves away from the backfill under the action of earth pressure, when the soil reaches the limit equilibrium state (active state), two sliding surfaces AB and AC through point A appear in the soil. The angle between the AC plane surface and the horizontal plane is θ. Taking unit length of retaining wall, we can analyze the static force equilibrium condition for the sliding soil ABC. The forces acting on the sliding soil ABC are as follows:

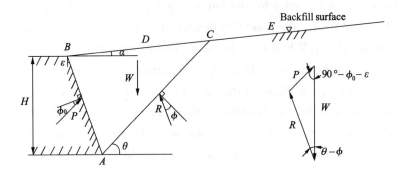

Figure 9-21   Active earth pressure of non-cohesive soil

1) The weight W of the sliding soil ABC. The value and direction of W are known if the value of θ is known.

2) The reaction force R of soil mass is acting on failure surface AC. It is the resultant force of the friction force and the normal reaction force on AC surface. The angle between the direction of R and normal direction of the failure surface is equal to the internal friction angle ϕ of the soil. Since the sliding soil ABC moves downwards relative to the soil on the right of the sliding surface AC, the direction of the friction force on the AC surface is inclined upward. The direction of R is known, and its value is unknown.

3) The reaction force P (the value is equal to active earth pressure force) of wall is acting on soil sliding surface AB. P is the resultant force of friction force on surface AB and normal reaction force, and the angle between P and the normal direction of wall surface AB is ϕ_0. Since the sliding soil ABC slides down relative to the wall, the direction of the friction force acting on AB surface is inclined upward, and the value is unknown.

According to the static force equilibrium condition, the directions of W, R, and P should intersect at one point then draw the triangle force vector acting on the sliding soil ABC, as shown in Figure 9-21. According to law of sines,

$$P = \frac{W \sin(\theta - \phi)}{\sin(90° + \phi + \phi_0 + \varepsilon - \theta)} \tag{9-36}$$

From Figure 9-21,

$$W = \gamma \cdot S_{ABC} = \gamma \cdot \frac{1}{2} \cdot AB^2 \cdot \frac{\sin(90° + \alpha - \varepsilon) \cdot \sin(90° + \varepsilon - \theta)}{\sin(\theta - \alpha)}$$
$$= \frac{1}{2} \gamma H^2 \cdot \frac{\sin(90° + \alpha - \varepsilon) \cdot \sin(90° + \varepsilon - \theta)}{\cos^2 \varepsilon \cdot \sin(\theta - \alpha)} \tag{9-37}$$

Substituting W into Eq. (9-36),

Chapter 9 Earth Pressure on Retaining Structure

$$P=\frac{1}{2}\gamma H^2 \cdot \frac{\cos(\varepsilon-\alpha)\cdot\cos(\varepsilon-\theta)\sin(\theta-\phi)}{\cos^2\varepsilon \cdot \sin(\theta-\alpha)\cos(\theta-\phi-\phi_0-\varepsilon)} \quad (9\text{-}38)$$

From the above equation, when the inclined angle of the wall ε, the angle of the backfill surface α, the backfill properties γ, ϕ, ϕ_0 and the height H of the wall are given, the value of the earth pressure P only depends on the inclined angle θ of the failure surface. Since the inclined angle θ in Figure 9-21 is hypothetical, the AC surface may not be the actual failure surface.

Once the exact failure surface is determined, the active earth pressure can be obtained. From Eq. (9-38), when $\theta=\phi, P=0$; when $\theta=90°-\varepsilon, P=0$. Therefore, when the value of θ is between ϕ and $90°-\varepsilon$, P will have a maximum value, which is the required active earth pressure force E_a. In fact, P can also be regarded as the force acting on the wall by the sliding body under the action of weight against the friction force on sliding surface AC. As the value of P increases, the sliding body has more tendency to slide downward. The actual sliding surface is the surface with the maximum P. Since the active earth pressure is assumed to be the maximum earth pressure calculated by a series of failure surfaces, we take the derivative of Eq. (9-38) with respect to θ and make it 0.

$$\frac{dP}{d\theta}=0 \quad (9\text{-}39)$$

After we obtain the limit value of θ, substituting it into Eq. (9-38), then the equation of total Coulomb's active earth pressure force E_a acting on the wall (in the opposite direction to P) is obtained.

$$E_a=\frac{1}{2}\gamma H^2 K_a \quad (9\text{-}40)$$

$$K_a=\frac{\cos^2(\phi-\varepsilon)}{\cos^2\varepsilon \cos(\varepsilon+\phi_0)\left[1+\sqrt{\frac{\sin(\phi+\phi_0)\sin(\phi-\alpha)}{\cos(\varepsilon+\phi_0)\cos(\varepsilon-\alpha)}}\right]^2} \quad (9\text{-}41)$$

where, γ, ϕ—unit weight and internal friction angle of the backfill, respectively;

H—height of wall;

ε—angle between the wall surface and the vertical plane. When the wall is inclined backwards, the symbol is positive (Figure 9-21). Otherwise the symbol is negative;

α—the inclined angle between the horizontal plane and the backfill surface. When the angle is anti-clockwise from horizontal line, it is positive (Figure 9-21). Otherwise the symbol is negative;

ϕ_0—the friction angle between wall and backfill, which depends on the roughness of wall surface, backfill properties, and tilted shape of the wall.

K_a in Eq. (9-41) is called the active earth pressure coefficient of the Coulomb's theory, which is related to ε, α, ϕ, ϕ_0 and can be found in the relevant design manual. Eq. (9-40) can be found in Chinese code GB 50007—2011 "Code for design of building foundation". It can be seen from Eq. (9-40) that the value of E_a is proportional to the square of the wall height H. Thus it can be estimated that the active earth pressure is in the form of rectangular distribution along the wall, and the pressure applied on the wall is $\gamma z K_a$, as shown in Figure 9-22b. The pressure along wall back is $\gamma z K_a \cos\varepsilon$, as shown in Figure 9-22a. The total active earth pressure force E_a is equal to the calculated area of the pressure distribution. The direction of E_a is inclined downward. The angle between the direction of E_a and the normal direction of the wall back is ϕ_0. The action point of E_a is at 1/3 height of the wall base.

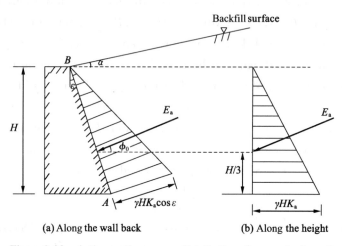

(a) Along the wall back   (b) Along the height

Figure 9-22 Active earth pressure distribution of non-cohesive soil

How to reduce the active earth pressure is a common problem in the design of a retaining wall. According to Eq. (9-41), the larger the angle ϕ is, the smaller is the K_a; the smaller the angle ϕ_0 is, the larger is the K_a; if the angle ε is positive (Figure 9-22a), the larger the value of ε is, the larger is the K_a value; if the angle ε is negative, the larger the value of ε is, the smaller is the K_a value; the larger the angle α is, the larger is the K_a value. Thus we can use a wall with negative ε, or reduce the backfill surface inclination angle α, or enhance the strength of backfill soil with high ϕ to reduce the active earth pressure on the wall. It should be noted that when $\alpha > \phi$, there will be an imaginary square root in Eq. (9-41). Therefore, the applicable conditions for Eq. (9-41) is $\alpha \leqslant \phi$.

The Coulomb's earth pressure theory is based on the limit equilibrium condition of the sliding soil body. Compared with the Rankine's earth pressure theory, the advantages of Coulomb's theory are that the backfill surface can be inclined, and the wall back surface can be rough or inclined. If the backfill surface is horizontal and the wall is vertical and smooth, then $\varepsilon = 0$, $\alpha = 0$, and $\phi_0 = 0$, so $K_a = \tan^2(45° - \phi/2)$ can be obtained. Therefore, Eq. (9-41) can be simplified to the same expression shown in Eq. (9-7) per Rankine's earth pressure theory. Thus, under certain conditions, the Rankine's earth pressure theory is a special case of the Coulomb's earth pressure theory.

Ⅱ. **Culmann's graphical method**

From Figure 9-21, we know that W, R, and P acting on the sliding soil can form a force vector triangle (wedge). The direction of the weight W is known as vertically downward. The direction of wall reaction force P (value is equal to the active earth pressure force) is known to form the angle with the vertical direction of $90° - \phi_0 - \varepsilon$. The magnitude of the sliding soil weight W and the direction of soil reaction force R below failure surface are related to the location of the failure surface. According to the determination process of the active earth pressure, if a series of sliding planes are assumed and the corresponding force vector triangles can be obtained. P can be determined directly by drawing the graphs. Then the maximum value of P is equal to the total active earth pressure force E_a (Figure 9-23). This is known as the Culmann's graphical method proposed in 1875. The detailed steps of Culmann's method are as follows:

① Drawn ϕ-line AF at an angle ϕ to the horizontal line. ϕ is the internal friction angle of soil. Then draw line AL and make angle between the AF and AL as $90° - \phi_0 - \varepsilon$.

② Assume a failure surface (rupture line) AC_1, calculate the weight W_1 of the sliding soil wedge ABC_1 and lay off on AF a distance An_1 to a suitable scale to represent the weight of wedge ABC_1. Draw the parallel line of AL from point n_1 to intersect assumed rupture line AC_1 at point m_1. The length of m_1n_1 represents the earth pressure E_1 when the failure surface is AC_1.

③ Assume that other failure surfaces are AC_2, AC_3, AC_4, etc. Repeat the step ②, we can obtain related points of n_2, n_3, n_4 and m_2, m_3, m_4, etc.

④ Connect points m_1, m_2, m_3, m_4, etc. by a smooth curve which is the pressure locus.

⑤ Select point m on the pressure locus such that the tangent to the curve at this point is parallel to the line AF.

⑥ Draw mn parallel to the pressure line AL. The magnitude of mn in its natural units gives the active pressure force E_a.

⑦ Connect Am and produce to meet the surface of the backfill at C. AC is the rupture line (failure surface).

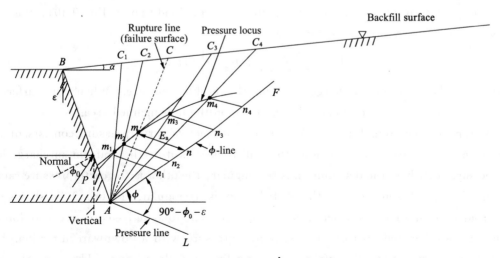

Figure 9-23 Active pressure by Culmann's method for non-cohesive soil

Compared Figure 9-23 with Figure 9-21, we can find that Figure 9-23 is the combination of two graphs in Figure 9-21 and rotating the force vector triangle clockwise with an angle of $(90°-\phi)$. This is the reason to draw the line AF and line AL.

The sliding surface AC and the total active earth pressure force E_a can be determined by the Culmann's graphical method. The direction of the total active earth pressure force is downward with an angle of ϕ_0 to the normal line of the wall, as shown in Figure 9-24. The action point can be determined by the following approximate method. In Figure 9-24, draw OO' parallel to AC, where point O is the gravity center of the sliding body ABC and point O' is the point of application of E_a.

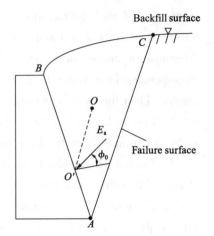

Figure 9-24　An approximate method to determine the action point of total active earth pressure force

Ⅲ. For soil with a uniform surcharge load

If a uniform surcharge load q is acting on the backfill surface, as shown in Figure 9-25a, the active earth pressure can also be determined according to the Coulomb's earth pressure theory. Compared with no surcharge condition, this condition with surcharge will exert a downward force on the sliding soil. The value is equal to q times the length of the sliding surface,

$$Q = q\,\overline{BC} = \frac{qH\cos(\theta-\varepsilon)}{\cos\varepsilon\,\sin(\theta-\alpha)} \tag{9-42}$$

Thus, the static equilibrium condition of the sliding soil is

$$P = \frac{(W+Q)\sin(\theta-\phi)}{\sin(90°+\phi+\phi_0+\varepsilon-\theta)} \tag{9-43}$$

Substitute Q and W into Eq. (9-43),

$$P = \left(\frac{1}{2}\gamma H^2 + qH\frac{\cos\varepsilon}{\cos(\varepsilon-\alpha)}\right) \cdot \frac{\cos(\varepsilon-\alpha)\cdot\cos(\varepsilon-\theta)\sin(\theta-\phi)}{\cos^2\varepsilon\cdot\sin(\theta-\alpha)\cos(\theta-\phi-\phi_0-\varepsilon)} \tag{9-44}$$

Comparing Eq. (9-44) with that of no surcharge condition in Eq. (9-40), the total active earth pressure force is calculated as

$$E_a = \frac{\cos\varepsilon}{\cos(\varepsilon-\alpha)}qHK_a + \frac{1}{2}\gamma H^2 K_a \tag{9-45}$$

where, q—the uniform surcharge load on the backfill surface. When the surface is inclined, it is calculated by the force on the unit inclined area.

It can be seen from Eq. (9-45) that the total active earth pressure consists of two parts: one is the earth pressure along the wall back caused by the uniform surcharge load on the surface, where the distribution is rectangular; the other is the earth pressure caused by the weight of soil, where the distribution is triangular. Figure 9-25 shows the trapezoidal distribution of the active earth pressure. The total active earth pressure force is equal to the calculated area of the distributed pressure with a downward direction. The angle between action direction of E_a and normal line of wall back is ϕ_0. The point of action is at the trapezoidal centroid.

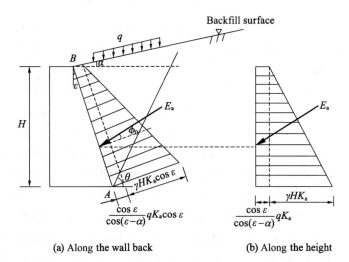

(a) Along the wall back   (b) Along the height

Figure 9-25   Distribution of active earth pressure with a uniform surcharge load

【Example 9-3】 As shown in Figure 9-26, behind the retaining wall is filled with non-cohesive soil. Its internal friction angle $\phi = 20°$ and the unit weight of soil $\gamma = 15\text{kN}/\text{m}^3$. The friction angle between the wall back surface and the backfill is $\phi_0 = 15°$. Other parameters are show in Figure 9-26. Try to calculate the active earth pressure force and its acting point on the wall.

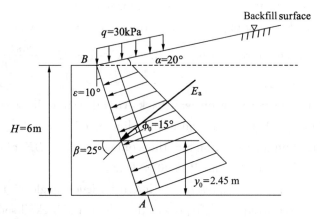

Figure 9-26   Schematic diagram of [Example 9-3]

【Solution】 The Coulomb's active earth pressure coefficient is 1.1 determined by Eq. (9-41). According to Eq. (9-45), the total active earth pressure force acting on the wall back is

$$E_a = \frac{\cos \varepsilon}{\cos(\varepsilon - \alpha)} q H K_a + \frac{1}{2} \gamma H^2 K_a$$

$$= \frac{\cos 10°}{\cos(10° - 20°)} \times 30 \times 6 \times 1.1 + \frac{1}{2} \times 15 \times 6^2 \times 1.1 = 225 + 270 = 495 \text{kN}/\text{m}$$

The distance from the bottom to the point of action E_a is

$$y_0 = \frac{225 \times 3 + 270 \times 6/3}{495} = 2.45 \text{m}$$

The angle between normal line of the wall back and the direction of E_a is 15°, i. e., the angle β between the wall and horizontal line is $\beta = \varepsilon + \phi_0 = 10° + 15° = 25°$.

IV. For cohesive soil

Compared with non-cohesive soil, the shear strength parameters include both cohesion c and internal friction angle ϕ. Therefore, when we determine the force by Coulomb's earth pressure theory on the sliding body (Figure 9-27a), in addition to the force P, W, R, there will be total cohesion forces $\overline{C}_w$ and $\overline{C}$ acting on the wall back surface and sliding surface, respectively. When the backfill behind the wall reaches the limit equilibrium state, the five forces (P, W, R, $\overline{C}$ and $\overline{C}_w$) will form a closed force vector polygon, as shown in Figure 9-27b. Then the earth pressure can be determined by the above principle. Here is an example for determination of the active earth pressure of the cohesive soil.

According to the Rankine's theory, for cohesive soil, when the backfill is at active limit equilibrium state, a tensile crack may occur in a certain range of depth. So there is no earth pressure on the failure surface and the wall back surface. The depth z_0 of tensile zone can be estimated by Rankine's theory, i. e., $z_0 = \dfrac{2c}{\gamma \sqrt{K_a}}$. We can assume a sliding surface, where the forces acting on the sliding body (Figure 9-27a) are:

① Gravity $W = W_1 + W_2 + W_3$ (direction is vertically downward), including the weight W_1, W_2, and the resultant force W_3 from the uniform surcharge load q;

② Reaction force R acting on surface BD (the angle between R and W is $\theta - \phi$);

③ The total cohesion force acting on BD is $\overline{C} = c \cdot H \dfrac{\cos(\varepsilon - \alpha)}{\cos \varepsilon \sin(\theta - \alpha)}$ (the angle between $\overline{C}$ and R is $90° + \phi$);

④ The reaction force P acting on wall back AB (the angle between P and W is $\phi = 90° - \phi_0 - \varepsilon$, and the angle between P and normal line of BF is ϕ_0);

⑤ The total friction force acting on the contact surface BF of the wall is $\overline{C}_w = c_w H / \cos \varepsilon$ (the angle between $\overline{C}_w$ and P is $90° + \phi_0$).

Where $H = H_0 - z_0$; ϕ, ϕ_0 represent the internal friction angle of backfill soil and the external friction angle respectively; c, c_w represent the cohesion of backfill soil and unit friction at contact surface, respectively.

The direction of the five forces in Figure 9-27b are known and thus can draw the force vector as a closed polygon, where the vector FA is the reaction force P acting on the surface BF of the wall (the value is equal to the earth pressure). In order to obtain the value of P, draw the extensions of lines AF and BD to the intersect point I. P can be obtained by the geometric relationship in the following equation:

$$P = P_1 - P_2 \tag{9-46}$$

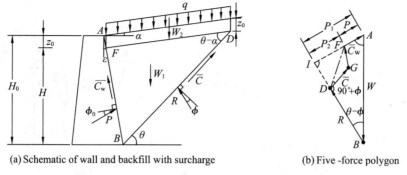

(a) Schematic of wall and backfill with surcharge (b) Five-force polygon

Figure 9-27 Active earth pressure determination of cohesive soil with a uniform surcharge

P_1 is the reaction force of wall back when both c and c_w are equal to 0. P_1 and P_2 can be expressed as a function of θ. According to the geometric relationship, we can obtain the detailed expression of P as follows

$$P = P_1 - P_2 = \frac{(W_1 + W_2 + W_3)\sin(\theta - \phi)}{\sin(\theta - \phi + \psi)} - \frac{cH\cos(\varepsilon - \alpha)\cos\phi + c_w H \sin(\theta - \phi - \varepsilon)\sin(\theta - \alpha)}{\cos\varepsilon \sin(\theta - \alpha)\cos(\theta - \phi - \varepsilon - \phi_0)}$$

(9-47)

According to Coulomb's earth pressure theory, the maximum value of P is equal to the active earth pressure force E_a. So we can get the critical angle θ_{cr} when $\dfrac{dP}{d\theta} = 0$. Then substitute it into equation Eq. (9-47) to obtain P_{max}. However, the explicit solution of θ_{cr} is complex. A common solution can be obtained based on the Culmann's graphical method.

V. For soil with groundwater

The poor drainage often leads to a higher level of water table in the backfill after construction. So the role of groundwater should be taken into consideration. The method of computing the earth pressure with groundwater is different from that without groundwater described above. It should be noted that the unit weight should be replaced by the buoyant unit weight γ' below the groundwater table, and the shear strength will reduce due to the presence of water.

Figure 9-28a shows the graphical method to determine active earth pressure without seepage for non-cohesive soil. The figure shows the polygon force vector when we assume that the failure surface is AC. We need to make a series of trails of sliding surface to find the maximum value of P, i. e., active earth pressure force E_a. It should be noted that when there is groundwater in the backfill, the water pressure should be taken into consideration.

While there is a stable seepage in backfill, we should consider the seepage force when we establish the static equilibrium condition of sliding soil body. Assuming that the seepage line is parallel to the backfill surface (the angle between it and horizontal line is α), the unit seepage force acting on the soil is $j = \gamma_w \sin\alpha$, and the seepage force on the sliding soil can be described as $J = \sum j$. Figure 9-28b shows the graphical method of active earth pressure for non-cohesive soil with seepage. We also need to make a series of trails of sliding surface to find the maximum value of P, i. e., active earth pressure E_a.

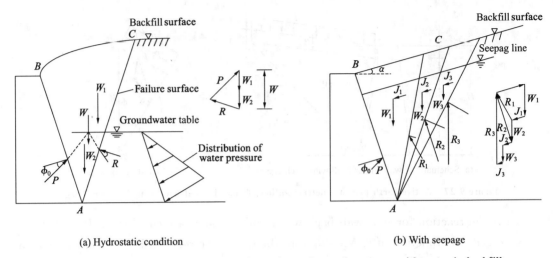

(a) Hydrostatic condition   (b) With seepage

Figure 9-28   The graphical method to determine active earth pressure with water in backfill

9.4.3 Calculation of Coulomb's passive earth pressure

When the retaining wall moves towards the backfill to a limit state under the external force, it is the passive failure condition. Figure 9-29 shows the failure surface AC for non-cohesive soil. Therefore, according to static force equilibrium condition of the wedge ABC, the triangle force vector can be obtained in Figure 9-29. According to the law of sines,

$$P = \frac{W \sin(\theta + \phi)}{\sin(90° + \varepsilon - \phi_0 - \phi - \theta)} \tag{9-48}$$

The symbol in the equation is the same as that in the Eq. (9-36).

Substituting W into the Eq. (9-48),

$$P = \frac{1}{2}\gamma H^2 \cdot \frac{\cos(\varepsilon - \alpha) \cdot \cos(\varepsilon - \theta)\sin(\theta + \phi)}{\cos^2\varepsilon \cdot \sin(\theta - \alpha)\cos(\theta + \phi + \phi_0 - \varepsilon)} \tag{9-49}$$

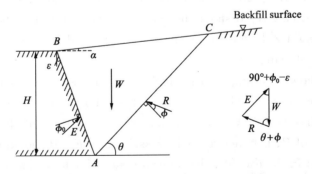

Figure 9-29   Passive earth pressure of non-cohesive soil

Similarly, P varies with the angle θ between the failure surface AC and horizontal line. When the retaining wall is approaching to the backfill, the P value is supposed to be minimum. This is because that the sliding force of the soil with the lowest value is easiest to be pushed upwards. The passive earth pressure E_p acting on the wall should be the minimum value P_{min} by making a series of trails for assumed failure surfaces. It is similar to that of the theory in active earth pressure. We can get the critical angle θ_{cr} by making $dP/d\theta = 0$, and then substitute it into Eq. (9-49) to obtain the Coulomb's passive earth pressure force acting on the wall

$$E_p = \frac{1}{2}\gamma H^2 K_p \qquad (9\text{-}50)$$

$$K_p = \frac{\cos^2(\phi+\varepsilon)}{\cos^2\varepsilon\cos(\varepsilon-\phi_0)\left[1-\sqrt{\dfrac{\sin(\phi+\phi_0)\sin(\phi+\alpha)}{\cos(\varepsilon-\phi_0)\cos(\varepsilon-\alpha)}}\right]^2} \qquad (9\text{-}51)$$

where, K_p—passive earth pressure coefficient of Coulomb's theory, which depends on the magnitudes of ε, α, ϕ and ϕ_0, the specific values can be found in the relevant design manual.

From Eq. (9-50), the value of the passive earth pressure force E_p is proportional to the square of the wall height H. Thus the passive earth pressure is distributed along the wall in a straight line. The direction of the total passive earth pressure is upward and the angle between it and normal line of the wall is ϕ_0. The action point of E_p is at a height of $H/3$ from the base of the wall.

Besides, for cohesive soil with or without a uniform surcharge and/or with or without groundwater, the determination of Coulomb's passive earth pressure is similar to that of active earth pressure.

9.5 Discussion on earth pressure

9.5.1 Influence of the wall movement

Existing research results have shown that the movement and deformation of retaining wall not only affect the earth pressure magnitude but also the distribution.

If the lower part of the retaining wall does not move, the upper part moves away from backfill. No matter how much displacement the wall would have, the pressure on the wall is distributed in form of a straight line, the total force is located at $H/3$ from the base of the wall, as shown in Figure 9-30a. With the movement of upper part of the wall, the backfill behind the wall will be in the active limit equilibrium state. The force acting on the wall is the active earth pressure.

If the upper part of the retaining wall does not move and the lower part moves away from backfill, soil cannot reach the active limit equilibrium no matter how much of the displacement. The distribution of pressure is a curve, and the total force is located at about $H/2$ from the base of the wall, as shown in Figure 9-30b.

If both the upper and lower parts of the retaining wall move away from backfill, the earth pressure is also a curve distribution when the displacement does not reach the limit state and the total force is acting at about $H/2$ from the base of the wall. When the displacement increases to a certain value, the backfill behind the wall will reach the active limit equilibrium state and then the pressure will be in the form of a linear distribution. The total active earth pressure force is acting at $H/3$ from the base of the wall, as shown in Figure 9-30c.

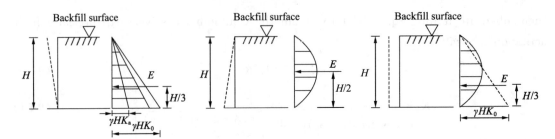

(a) Wall upper part movement (b) Wall lower part movement (c) Both upper and lower parts movement

Figure 9-30 Distribution of earth pressure for different types of wall movement

9.5.2 Differences between Rankine's and Coulomb's theory

Rankine's theory and Coulomb's theory are all based on some assumptions, so we should pay attention to the differences between the two theories.

① Rankine's theory on determination of the earth pressure is based on the stress state in elastic mechanics and the analysis of limit equilibrium theory, which can be used in both cohesive or non-cohesive soil. But the assumption about the wall is vertical and smooth, and the backfill surface behind the wall is horizontal, which limit the use of this theory. In contrast, Coulomb's earth pressure theory is based on the limit equilibrium state of soil wedge between the wall and the sliding surface. Under the static equilibrium condition, the calculation equation of the earth pressure is deduced, which considers the friction between the wall and the backfill, thus can be used for the inclined wall, or inclined backfill surface.

② The friction between the retaining wall and the soil are taken into account in Coulomb's theory, which can meet most engineering applications. Therefore, the direction of active earth pressure will change with soil properties according to Coulomb's theory, while Rankine's theory does not have such advantages.

③ In Coulomb's earth pressure theory, the earth pressure equation of non-cohesive soil can be applied to cohesive soil, but the related equation will be more complicated.

④ If the contact surface of the wall and backfill is smooth, and the back of the wall is vertical, the earth pressure obtained from Coulomb's theory will be the same with that obtained from Rankine's theory.

When we apply the theory of earth pressure, it is necessary to know the error between Rankine's theory and Coulomb's theory.

Rankine's theory assumes that the back of the wall is vertical and smooth, and the backfill surface is horizontal. However, the actual wall back is not smooth, so when we use Rankine's theory to calculate the earth pressure, we can find that the active earth pressure is larger and the passive earth pressure is lower compared with in actual conditions.

Coulomb's theory, although taking into account the friction between the wall and the soil, the assumptions in which are not as strict as those in Rankine's theory. It seems like the Coulomb's theory can be widly used. However, when the friction angle between the wall and the backfill is large, the sliding surface presented in the soil is often not linear but curved, which does not exactly conform to the assumption that the sliding surface is a plane. Thus it will produce a large error. Practical engineering applications have proved that if the inclined angle of the wall is not large (ε less than 15°) and the friction angle between the wall and the soil is small (ϕ_0 less than 15°), the failure surface is similar to a plane when the retaining wall moves away from the soil to achieve the active limit state. Therefore, the error of the active earth pressure calculated by the Coulomb's theory is often in the acceptable range, usually within an error of 2% to 10%. However, when the retaining wall is approaching to backfill, the failure surface is close to a logarithmic spiral surface which is very different from a plane. In this case, no matter use what theory, there is a great error. For simplicity, we often use Rankine's theory to calculate the passive earth pressure.

In the design of a retaining wall, how to choose a proper form of earth pressure, i. e., the at-rest earth pressure, the active earth pressure, and the passive earth pressure is depending on the relative displacement of the wall.

If the retaining wall is directly built on the rock foundation, the rigidity of the wall is very large, and the displacement of the wall is very small, which would not cause the active failure of the backfill, we can use the at-rest earth pressure, as shown in Figure 9-31. In addition, for the side wall of underground structure, if the wall rigidity is large and the wall displacement is very small, the pressure acting on the wall can be regarded as at-rest earth pressure. Since the static earth pressure coefficient K_0 is difficult to obtain, whereas the active earth pressure is easier to calculate, the active earth pressure is usually increased by 25% as the estimated value of the at-rest earth pressure is engineering applications.

If the retaining wall moves away from the backfill under the gravity, which leads the backfill behind the wall reach a limit equilibrium state, we can use the active earth pressure. Experimental studies have shown that when the displacement of the retaining wall reaches 0.1% $\sim$ 0.3% of the wall height, the backfill behind the retaining wall may fail. It is easier for common retaining wall to reach this condition. Therefore, the earth pressure on the retaining wall is usually calculated according to the active earth pressure, as shown of E_a on AB surface in Figure 9-32.

If the retaining wall moves towards the backfill under external force, which leads the backfill behind the wall reach a limit equilibrium state, it can be calculated by the passive earth pressure. But the existing experiment results show that the passive failure of backfill can be achieved when the retaining wall displacement is required to reach 2% to 5% of the wall height. This is not allowed in the practical projects. To study the stability of the retaining wall, we usually use 30% of the passive earth pressure instead of the entire passive earth pressure, as shown of E_p on EF surface in Figure 9-32.

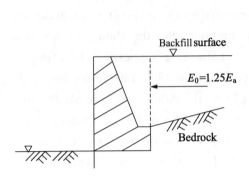

Figure 9-31 Earth pressure on retaining wall with bedrock

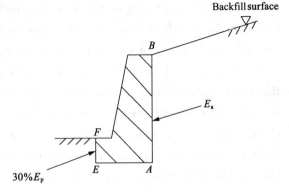

Figure 9-32 Earth pressure on retaining wall with double backs

It should be noted that for the permanent retaining wall (or retaining structure), when the displacement of wall tends to be stable, we need to consider whether the earth pressure previously calculated will change in a long-term condition and whether the backfill behind the wall is still in active or passive limit equilibrium state and how it will change. More scientific researches are needed to solve these problems.

Knowledge expansion

Separate and Combined Calculation of Water and Earth Pressure

The method of "separate calculation of water and earth pressure" is often used for sandy soil or silty soil, i. e., lateral pressure acting upon the retaining structure is the sum of earth pressure and hydrostatic water pressure. Moreover, for soil layer below the groundwater table, buoyant unit weight γ' and effective stress strength parameters of c' and ϕ' are adopted to calculate the earth pressure. This is because water is an isotropic material while soil is an anisotropic material.

If behind the retaining structure is filled with cohesive soil, we often use the method of "combined calculation of water and earth pressure". For soil layer below the groundwater table, saturated unit weight γ_{sat} and total stress strength parameters of c and ϕ are adopted. Normally, for clayey soil with large porosity or with large horizontal permeability coefficient, the separate calculation method of water and earth pressure is used.

When there is a steady flow of groundwater, the lateral pressure on retaining structures will be calculated according to the following principles using the method of separate calculation of water and earth pressure:

(1) The earth pressure acting on retaining structure is calculated using flow net method;

(2) Consider the influence of pore water pressure on earth pressure. For soil layer above the excavation surface, the earth pressure is calculated using hydrostatic pressure; for soil layer between excavation surface and bottom of embeded structure, the earth pressure is calculated using hydrostatic pressure of triangle distribution.

In China, the codes for design of foundation and local engineering constructions made specific regulations for calculation of earth pressure, including the separate calculation and combined calculation of water and earth pressure.

Exercises

[9-1] What is active earth pressure, at-rest earth pressure and passive earth pressure? Try to present examples in the actual engineering projects.

[9-2] Describe the conditions of occurrence for three typical earth pressures.

[9-3] Why we say the active earth pressure is the maximum value of the active limit equilibrium? While the passive earth pressure is the minimum value of the passive limit equilibrium?

[9-4] What assumptions are used in the Rankine's theory and Coulomb's theory about earth pressure? What error will it bring?

[9-5] How do Rankine's earth pressure theory and Coulomb's earth pressure theory establish the earth pressure calculation equation? In what conditions do they have the same results?

[9-6] What is the effect of groundwater on the stability of the retaining structure and what kind of engineering measures can be made to increase the stability?

[9-7] Try to compare the advantages and disadvantages of the Rankine's earth pressure theory and Coulomb's earth pressure theory.

[9-8] Try to illustrate some engineering applications of retaining structures and their characteristics.

[9-9] Figure 9-33 shows a retaining wall. It is 5m in height, and the surfare of wall back is vertical. Behind the wall is filled with sand and the groundwater table is 2m below the surface. If the unit weight of sand $\gamma = 16 \text{kN/m}^3$, the saturated unit weight of sand $\gamma_{sat} = 18 \text{kN/m}^3$, effective internal friction angle $\phi' = 30°$. Try to calculate the value, distribution and the resultant force of at-rest earth pressure and water pressure.

[9-10] Figure 9-34 shows a retaining wall. The wall back is vertical and smooth with a height of 10m and the backfill surface is horizontal. There is a continuous uniform surcharge load q of 20kPa. Backfill consists of two layers of non-cohesive soil, soil properties and groundwater table are shown in Figure 9-34, try to solve:

(1) The distribution of active earth pressure and water pressure;

(2) The total pressure force (the sum of earth pressure and water pressure);

(3) The acting point of total pressure force.

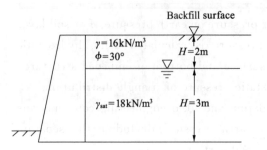

Figure 9-33 Schematic diagram of Exercise [9-9]

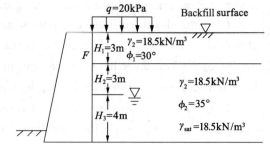

Figure 9-34 Schematic diagram of Exercise [9-10]

[9-11] As shown in Figure 9-35, calculate the active and passive earth pressure force acting on the retaining wall with Rankine's theory and plot the pressure distribution.

[9-12] As shown in Figure 9-36, assuming the wall back is smooth and vertical, calculate the active and passive earth pressure force acting on the retaining wall and plot the pressure distribution.

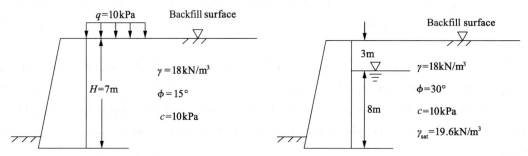

Figure 9-35 Schematic diagram of Exercise [9-11]   Figure 9-36 Schematic diagram of Exercise [9-12]

[9-13] Try to use Coulomb's theory and Culmann's method respectively to calculate the active earth pressure force acting on the retaining wall shown in Figure 9-37.

[9-14] Figure 9-38 shows a gravity retaining wall. There is a local distributed load on the backfill surface with a distance away from the wall. Try to calculate a limit value of d which will make distributed load have no impact on the wall.

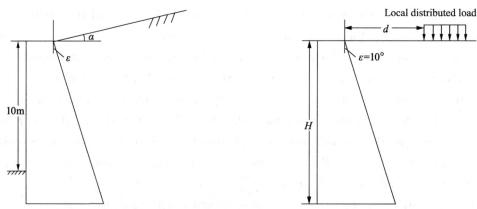

Figure 9-37 Schematic diagram of Exercise [9-13]   Figure 9-38 Schematic diagram of Exercise [9-14]

[9-15] Figure 9-39 shows a retaining wall, assuming the wall back surface is smooth.

Calculate the value, direction and action point of the earth pressure by Rankine's theory and Coulomb's theory, respectively.

[9-16] Figure 9-40 shows a retaining wall. The backfill conditions and properties are shown in the figure. Try to use Rankine's theory calculating the earth pressure on points A, B, C and find the point where the earth pressure is zero.

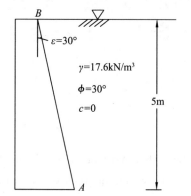
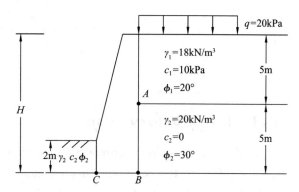

Figure 9-39 Schematic diagram of Exercise [9-15]   **Figure 9-40** Schematic diagram of Exercise [9-16]

Chapter 10　Slope Stability

10.1　Introductory case

　　A slope is an inclined ground surface which can be either natural or artificial. Slope stability refers to the condition that an inclined slope can withstand its own weight and external forces without experiencing displacement. The quantitative determination of the stability of slopes is necessary in a number of engineering activities, such as: ① the design of earth dams and embankments; ② the analysis of stability of natural slopes; ③ analysis of the stability of excavated slopes; ④ analysis of deep seated failure of foundations and retaining walls.

　　When the stability conditions are not met, the soil or the rock mass of the slope may experience downward movement which could be either slow or rapid. This phenomenon is known as slope failure or landslide. A landslide may be triggered by an earthquake, rainfalls that cause exceeding porewater pressure or degradation of the ground's mechanical properties. Slope failures systematically harm human infrastructure and cause numerous fatalities and significant economic loss every year. For example, The January 10, 2005 La Conchita landslide was the deadliest single event triggered by the 2004—2005 storm sequence. The landslide, which occurred about 130km northwest of Los Angeles, California, mobilized over 40000 cubic yards of wet debris into a large scale debris flow that flowed into a residential community at the foot of the slope, killing 10 persons and damaging or destroying 36 residences. Ten years earlier, in March 1995, a large rotational landslide had occurred in the same area following a period of heavy rainfall. Therefore, we need to study the knowledge about slope stability.

　　Slopes can be divided into two kinds including natural slopes and artificial slopes. The slopes formed due to natural geological processes are called natural slopes, such as slopes, river bank slope, etc. The stability of natural slope is determined by the engineering geological conditions, hydrogeological conditions and mechanical properties of rocks and soils. Slopes that are formed with excavation and compaction of soils are known as artificial slopes, such as foundations, tunnels, earth dams, embankments, etc. The stability of artificial slope is determined by the properties of the soil, the construction procedure, and the groundwater conditions. The composition of the slope is shown in Figure 10-1. According to the material properties, slope can be categorized into soil slope, rock slope and soil-rock mixed slope.

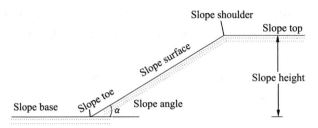

Figure 10-1 Schematic diagram of the slope composition

Since the slope surface is inclined, under the action of soil weight and other external forces, the rock and/or soil mass has a tendency to move downward. A landslide is defined as the movement of a mass of rock, debris, or soil down a slope. The main components of a landslide are shown in Figure 10-2. Slope instability is usually caused by the combined effect of shear stress increase and shear strength decrease. The factors that may cause an increase in the shear stress include: ① water infiltration or other external load that increase the unit stress of the slope soil; ② excavation that makes the slope steep; ③ other impact loads and vibrations caused by blasting or earthquakes; ④ seepage force. The factors that may cause the soil properties change with a reduced shear strength include: ① increase of pore water pressure; ② impact or vibration load or periodic load; ③ weathering of the rock; ④ seasoning check, freeze-thaw and softening factors, etc.

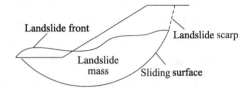

Figure 10-2 The components of a landslide

Before the slide of slope, it is generally observed a significant settlement at the top of slope with some possible cracks, and the ground near the toe has a larger lateral displacement and a slight bulge. With the development of the cracks at slope top and the lateral displacement at slope toe, some part of soils may suddenly slide along a certain sliding surface to form a landslide. For some soft soil slope, such as a pier slope built on coastal soft soil, due to the creep effect of the soil, the occurrence of a landslide may develop slowly. In contract, rock landslide is generally developed from micro-cracks to form a continuous sliding surface along the weak rock structural surfaces by sudden changes under external loads (such as rainfall to increase pore pressure, earthquake, etc). Compared with the artificial slope, the sliding of natural slope is sudden, and difficult to predict. In order to prevent the landslide effectively, we needs to design the artificial slope with the caution and have stable analysis to make it more reasonable. We should also take appropriate engineering measures to strengthen the soil and eliminate certain unfavorable factors. For natural slopes, we need to make a thoughtful geological survey and necessary investigation, then study the potential sliding surface and analyze the stability of slope. If necessary, we should take some preventive and reinforcement measures.

Geotechnical structures such as dams, embankments, etc., the length is far larger than the width and height, and the extent of the slope along the length is uncertain. Although there are resistances of both sides soils to the slope, the impact of the side soil resistance on soil stability is difficult to be determined. Therefore, for analyzing the stability of the slope, the effect of the resistance at both sides of the sliding soil body is usually not taken into account. Thus the stability analysis of the slope can be simplified as a plane strain problem. The previous studies showed that this simplification is feasible in most cases.

The main task of slope stability analysis is to determine the factor of safety for a slope and evaluate the stability. Different soil properties and analytical methods have different definitions on the factor of safety for the slope. For non-cohesive soils, the factor of safety is defined as the ratio of the maximum shear resistance force to the sliding force; for cohesive soils, considering the overall circular sliding stability analysis, the factor of safety for a soil element is defined as $F_s = \tau_f / \tau$. Since the factor of safety for the entire slope contains many soil elements, we can actually define to the ratio of the average shear strength to the average shear stress along the sliding surface or the ratio of the sliding resistant moment to the sliding moment. Thus, the commonly used factor of safety definition can be attributed to the average shear strength and the average shear stress over the entire sliding surface.

Determining the sliding surface and its location can be regarded as one of the tasks in slope stability analysis. The existing results showed that for homogeneous non-cohesive soil slopes, such as sand, pebbles, weathering gravel and other coarse aggregates, the sliding surface is often regarded approximately as a plane. For homogeneous cohesive soil slopes, the sliding surface is usually curved with a smooth surface, where the top radius of curvature is small and is often perpendicular to the slope top, while the bottom is relatively gentle. Empirically, if the shape of the calculated sliding surface is slightly different from that of actual one, the obtained factor of safety has little impact on the calculation results. For convenience, the sliding surface of the homogeneous clay soil slope is often assumed to be a cylindrical surface at failure, where the projection on the plane is a slip circle, namely as the sliding surface. For heterogeneous multi-layered soils or slopes with weak interlayers, such as earth dam or dam foundation, the presence of weak soil interlayers will enable the slopes tend to slide along the weak interlayer, where the sliding surfaces show a linear or curved sectional shape.

The location of sliding surface is unknown in general, unless there are significant weak parts in the soil, such as cracks, weak interlayer, ancient landslide, etc. Therefore, in calculation, we need to first assume a number of possible sliding surfaces, and then find the factor of safety on each potential sliding surface, in which the minimum value of the factor of safety corresponding to the sliding surface is used for slope stability analysis. For homogeneous soils, the location of the sliding surface is related to the nature of the soil, the falling gradient of slope, and the embeded depth of the hard soil or bedrock. The calculation results show that the minimum factor of safety for potential sliding surface can reflect the degree of stability for the actual slope as long as the strength parameters of soil are chosen properly. However, the calculation most dangerous sliding surface with minimum value of safety factor is sometimes different from the observed real sliding surface.

At present, the commonly used methods for slope stability analysis include: limit equilibrium analysis method, statistical comparison method, engineering geological comparison method and numerical analysis method. Among them, the limit equilibrium analysis method is the most used one. It assumes that the soil mass is a rigid body, and calculate the internal forces and moments at limit equilibrium state to obtain the slope stability. Statistical comparison method is used to compare the geological conditions, load conditions and other factors with similar slopes to obtain the statistical data, which can thereafter be combined with experiences to determine the slope stability. Numerical analysis method can be used to simulate the slope construction (fill or excavation) process. It can reflect the hydrogeological conditions of the slope and simultaneously obtain the displacement and stress at each point in the slope. Then we can analyze the stability of the slope. The most commonly used numerical analysis method is the finite element method. This chapter mainly introduces the limit equilibrium analysis method. Since the morphologies of the sliding surfaces for non-cohesive soil slope and the cohesive soil slope are different, thus the corresponding stability analysis methods are also different, which will be discussed in the following sections.

10.2 Stability analysis of non-cohesive soil slope

10.2.1 For non-cohesive soil slope without seepage flow

For homogeneous non-cohesive soil slopes, there is no cohesive force (cohesion) between the soil particles. No matter in a dry condition or saturated condition, as long as the soil element on the slope is stable, the entire slope is stable. Figure 10-3a shows a homogeneous non-cohesive soil infinite slope without seepage, which means that the slope stability is not affected by the boundary conditions. The slope angle is α. A soil mass element (parallelogram) is selected for analysis, the left and right sides are vertical lines, and the upper and lower lines are parallel to the slope surface. The weight of the soil element is W, which can be decomposed into two vectors including a normal force N perpendicular to the slope surface and a sliding force T downward parallel to the slope surface.

$$T = W\sin\alpha, \quad N = W\cos\alpha \qquad (10\text{-}1)$$

The force to prevent the soil from sliding is the shear force ($\overline{T}$) between the soil element and subsoil, which is equal to T when in a static equilibrium state, i.e., $T = \overline{T}$.

From force equilibrium, we also have $\overline{N}=N$. Therefore, the maximum value of $\overline{T}$, written as $\overline{T}_f$, is related to the soil internal friction angle ϕ,

$$\overline{T}_f = \overline{N}\tan\phi = W\cos\alpha\tan\phi \tag{10-2}$$

The factor of safety (F_s) for an infinite non-cohesive soil slope is defined as the ratio of maximum shear force to the downward sliding force

$$F_s = \frac{\overline{T}_f}{T} = \frac{W\cos\alpha\tan\phi}{W\sin\alpha} = \frac{\tan\phi}{\tan\alpha} \tag{10-3}$$

Thus, for an infinite slope in homogeneous non-cohesive soil, if the slope angle α is less than the internal friction angle ϕ, the soil mass is stable. When F_s is equal to 1, the soil is in the limit equilibrium state, and the slope angle α is equal to the internal friction angle ϕ for non-cohesive soil, which is called the angle of repose.

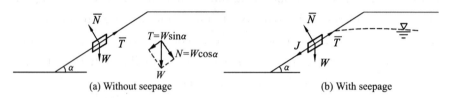

(a) Without seepage   (b) With seepage

Figure 10-3   Force analysis of non-cohesive infinite slope

10.2.2   For non-cohesive soil slope with seepage flow

If the water level in reservoir suddenly drops, certain seepage force will develop that adversely affect the stability. In this seepage condition, in addition to the soil weight W, there will be a seepage force J acting on the soil element, as shown in Figure 10-3b. Because the direction of seepage is parallel to the slope surface, the direction of the seepage force should be parallel to the slope surface. The total sliding force acting on the soil element is

$$T+J = W\sin\alpha + J \tag{10-4}$$

Since the maximum shear resistance of soil element is $T_f = W\cos\alpha\tan\phi$, the factor of safety is expressed as

$$F_s = \frac{T_f}{T+J} = \frac{W\cos\alpha\tan\phi}{W\sin\alpha + J} \tag{10-5}$$

For a small soil unit, we can use the unit buoyant weight γ' and the unit seepage force $j = i\gamma_w$. Since the direction of flow is downward along the slope surface, the hydraulic gradient is $i = \sin\alpha$. From Eq. (10-5), the factor of safety can be written as

$$F_s = \frac{\gamma'\cos\alpha\tan\phi}{(\gamma'+\gamma_w)\sin\alpha} = \frac{\gamma'\tan\phi}{\gamma_{sat}\tan\alpha} \tag{10-6}$$

where, γ_{sat} —saturated unit weight of soil.

Compared Eq. (10-6) with Eq. (10-3), we can find that the factor of safety for non-cohesive soil slope with seepage flow is multiplied by γ'/γ_{sat} than that without seepage flow, the value of which is close to 1/2. Therefore, the factor of safety for the non-cohesive soil slope without seepage will reduce almost half when the slope is with seepage flow.

【Example 10-1】 A infinite soil slope with homogeneous non-cohesive soil, the saturated unit weight of soil is $\gamma_{sat}=20\text{kN}/\text{m}^3$, the internal friction angle is $\phi=45°$. If the factor of safety of the slope is 1.2, so what the slope angle would be in the dry condition or fully saturated condition, or with a seepage flow? (Assuming that the internal friction angle of the soil does not vary with the water content)

【Solution】

By Eq. (10-3), when slopes are dry or fully saturated

$$\tan\alpha=\frac{\tan\phi}{F_s}=\frac{1}{1.2}=0.833$$

So $\alpha=39.8°$

When there is a seepage flow, by Eq. (10-6),

$$\tan\alpha=\frac{\gamma'\tan\phi}{\gamma_{sat}F_s}=\frac{10.2\times1}{20\times1.2}=0.425$$

So $\alpha=23°$

It can be seen from the calculation results that the slope angle with seepage is much smaller than that without seepage.

10.3 Stability analysis of cohesive soil slope

The common stability analysis method for cohesive soil slope can be divided into two kinds: ① friction circle method; ② slices method.

10.3.1 Friction circle method

The friction circle method of slope stability analysis is a convenient approach for both graphical and mathematical solutions. For cohesive soil, the soil mass will slide down as a whole along the failure surface due to the cohesive force between soil particles. Any soil unit on the slope cannot be used to represent the stability of the entire slope. For a plane strain problem, the soil above the sliding surface is regarded as a rigid body, and we can analyze the forces on the limit equilibrium state. The factor of safety of cohesive soil slope is defined by the ratio between the shear strength (τ_f) and the shear stress (τ) on the sliding surface,

$$F_s=\frac{\tau_f}{\tau} \tag{10-7}$$

For a homogeneous cohesive soil slope, the actual sliding surface is close to a cylindrical surface. It is generally assumed that the sliding surface is cylindrical and the section is circular for a plane strain problem. The factor of safety is the ratio of the moments of maximum resistance forces (M_f) to the moments of moving forces (M).

$$F_s=\frac{M_f}{M}=\frac{\tau_f\times\widehat{L}\times R}{\tau\times\widehat{L}\times R} \tag{10-8}$$

where, τ_f—soil shear strength on the sliding surface;

τ—shear stress on the sliding surface;

$\widehat{L}$—length of the failure arc;

R—radius of the failure arc.

Figure 10-4a shows a homogeneous cohesive soil slope, AC is the assumed sliding surface with the circle center O, and the radius R. The equilibrium of moments of forces must be satisfied when the sliding surface is stable. Note that the normal force on the sliding arc passes through the center of the circle, so

$$\frac{\tau_f \hat{L}}{F_s} R = Wd \qquad (10\text{-}9)$$

The factor of safety is

$$F_s = \frac{\tau_f \hat{L} R}{Wd} \qquad (10\text{-}10)$$

where, W—the weight of soil mass;

d—distance from center of arc to the action point of weight.

The shear strength of soil τ_f includes two parts, the cohesion c and friction $\sigma \tan \phi$. For saturated cohesive soils, if the factor of safety of slope is studied under undrained shear conditions, we have $\phi_u = 0$ and $\tau_f = c_u$. Thus Eq. (10-10) can be written as

$$F_s = \frac{c_u \hat{L} R}{Wd} \qquad (10\text{-}11)$$

Since the shear strength c_u is constant along the failure arc, Eq. (10-10) can be directly used to calculate the value of F_s. This slope stability method is commonly referred to as the analysis of $\phi_u = 0$. Besides, we may see a crack at the top of the slope surface before the slide of the slope, as shown in Figure 10-4b. The depth z_0 can be estimated by $z_0 = 2c/\gamma \sqrt{K_a}$. When $\phi_u = 0$, we have $K_a = 1$ and $z_0 = 2c_u/\gamma$. The occurrence of crack will reduce the length of the arc from AC to $A'C$. If there is water in the cracks, it also needs to consider the adverse effect of hydrostatic pressure acting on the crack.

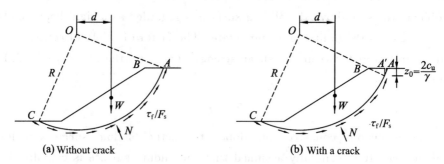

Figure 10-4 Sliding surface of a homogeneous cohesive soil slope

The above-obtained F_s is the factor of safety corresponding to a certain assumed sliding surface. In the practical engineering, it requires the minimum factor of safety corresponding to the most potential sliding surface. Thus, we usually need to assume a series of sliding surfaces, where a lot of calculations can be done by computer programs. The minimum factor of safety is calculated by programming of the possible sliding centers and the sliding radius, and obtain in the minimum value for the factor of safety corresponding to the most potential sliding surface. The calculation procedures are as follows:

(1) Determine the range of possible center points of sliding circles. Previous study showed that the sliding center of circle of the potential sliding surface will fall within the range of zone $ABCD$ for homogeneous slope. The procedures include: draw a vertical line and normal line through the middle point of slope surface; taking the middle point as the center, draw the circles with $1/4$ slope length (L) and $5L/4$ as radius respectively, where the four lines form the $ABCD$ zone.

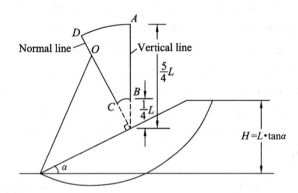

Figure 10-5　Determination of the center of circle for potential sliding surface

(2) For each center of circle, use Eq. (10-11) to calculate the factor of safety corresponding to a series of possible circles, where the minimum value is the factor of safety of slope and corresponding arc is the most unfavorable arc.

(3) Comparing the factor of safety obtained for each center of circle, the minimum value of safety factor is selected as the final factor of safety, and the corresponding sliding arc is the most unfavorable sliding surface.

In fact, there are numerous circles through each center. We can calculate a finite number of sliding radius according to the specific accuracy requirements. In the zone $ABCD$ in Figure 10-5, there will be an infinite number of sliding centers. For simplicity, we can specify the distance between two adjacent centers, calculate the factor of safety corresponding to a finite number of arcs, and take the minimum value as the final factor of safety.

At present, the commonly used friction circle method mainly include Fellenius method and Taylor method, and the detailed process can be found in the related literature. The friction circle method is mainly applied to homogeneous simple slope (slope top and base are horizontal, as shown in Figure 10-1). It should be pointed out that the overall slope analysis for cohesive soil can only be used for the total stress analysis under undrained shear condition, where $\phi=0$.

When $\sigma\tan\phi\neq 0$, the shear strength over the failure surface is not a constant, and the factor of safety cannot be calculated directly by Eq. (10-10). It is necessary to determine the normal stress on each point on the sliding arc (when the type of soil changes, we also need to determine the shear strength parameters c and ϕ to calculate the shear strength of soil on each point of the sliding surface. When the sliding surface is determined, the normal stress of each point can be calculated by the slices method described in the next section.

10.3.2 Slices method

From the Mohr-Coulomb failure criterion, the magnitude of shear strength τ_f of soil is related to the normal stress σ_n. If the value of σ_n at a given point on the sliding surface is unknown, then the shear strength at that point is also unknown. Therefore, we can divide the sliding soil into several soil slices, and the average normal stress and the shear strength at the bottom of the each soil slice can be obtained by an approximate method, which is known as the slices method. The basic assumption of this method is that each slice and soil below the sliding surface is rigid body without deformation, and the sliding surface is continuous. For any given assumed sliding surface, the sliding soil body has been subdivided into n slices, as shown in Figure 10-6a. The forces acting on the ith soil slice is shown in Figure 10-6b.

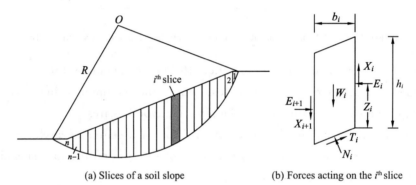

(a) Slices of a soil slope   (b) Forces acting on the ith slice

Figure 10-6   Schematic of slices method and the force analysis on a given slice

The forces and unknown quantities on i^{th} slice are given as follows:

1) Slice weight W_i. $W_i = \gamma_i b_i h_i$, where γ_i, b_i, h_i are the unit weight, width and height of the ith slice, respectively. They all have known values.

2) Normal force N_i and shear force T_i acting on the bottom of the soil, which are unknown. Assuming that N_i acts at the midpoint at the bottom of the ith slice, the direction of T_i is parallel to the bottom of the ith slice. Per the Mohr-Coulomb failure criterion, when the factor of safety is F_s, N_i and T_i have the following relationship.

$$T_i = (c_i l_i + N_i \tan \phi_i)/F_s \quad (i=1,2,3,\cdots,n) \tag{10-12}$$

where c_i, ϕ_i, l_i are the cohesion, internal friction angle and length of the ith slice. Eq. (10-12) indicates that the normal force N_i and the shear force T_i are linearly related to the factor of safety F_s and the shear strength of soil. Thus we have n independent unknown quantities for n slices.

3) Normal force E_i acting on the contact surface between two adjacent slices. The magnitude and action point of each normal force are unknown. Besides, for the normal

force acting on the right side of slice 1 and on the left side of slice n, as shown in Figure 10-6a, the value of normal force is 0 or equal to the known external force. So there is a total number of $(n-1)\times 2=2n-2$ unknown quantities.

4) Shear force X_i acting on the contact surface between two adjacent slices. The magnitude of each shear force is unknown, and the point of action is known. Besides, the shear force acting on the right side of slice 1 and the left side of slice n in Figure 10-6a is 0 or equal to the external forces. So there are $n-1$ unknown quantities.

5) Factor of safety F_s. When the sliding surface and the shear strength of soil and the external forces are determined, the shear stress and the shear strength at a given point on the sliding surface can be obtained and the related F_s can be calculated by Eq. (10-7). We assume that the factor of safety at any point on the failure surface is the same, then the factor of safety F_s is a single unknown quantity.

Table 10-1 lists the types of unknown quantity and the number of independent unknown quantities for n slices. We can find that the total independent unknown quantities are $4n-2$. If the slope stability analysis is given for a plane problem, for each soil slice, it can be divided into two orthogonal directions (such as the vertical and horizontal direction) static force equilibrium equation and one force of moment equilibrium equation. There are a total of $3n$ independent equations through n soil slices. Therefore, the number of total unknown quantity ($4n-2$) is more than the equation numbers ($3n$). Therefore, as long as the number of soil slices is greater than two, the slope stability analysis is a statically indeterminate problem. To solve the problem, we need to increase the number of equations or reduce the number of unknown quantities. Increasing the number of equations requires the use of soil stress-strain relationship considering the deformation characteristics of soil and use of elasto-plastic theory and other methods for calculation. Due to the complexity of the boundary conditions, only numerical solutions can be used. When adopting the limit equilibrium analysis method, we only consider the static force equilibrium of soil regardless of the deformation. Thus, we need to have additional assumptions and to reduce the number of unknown quantities and make the number of equations is no less than that of unknown quantities. Additional assumptions lead to different methods of analysis such as Swedish circle method, Bishop method, Janbu method and other slices methods. It should be noted that the factor of safety is different by using different methods even for the same problem. If the additional assumptions are more than $n-2$, the unknown quantities are less than $3n$ because the total unknown quantities are $4n-2$. Therefore, the solution about the forces acting on each slice cannot meet $3n$ static equilibrium conditions, which is called the approximate method; if there are exactly $n-2$ assumptions and the number of unknown quantity is equal to $3n$, the solution of the forces acting on each slice can meet all the $3n$ static equilibrium conditions, which is called the "exact" method. In fact, the so-called "exact" method only satisfies the static equilibrium condition, but not satisfy the condition of soil deformation, so it is still an approximate calculation method.

Table 10-1 Statistics of unknown quantities for *n* slices

Unknown quantity	Number of independent unknown quantity
Magnitude of N_i and T_i	n
Magnitude and action point of E_i	$2n-2$
Magnitude of X_i	$n-1$
Factor of safety F_s	1
Total	$4n-2$

10.3.3 Assumptions of commonly used slice methods

(1) Swedish circle method, also known as Fellenius method. It assumes that the slip surface is a circular arc regardless of the forces between the slices, which means $E_i = X_i = 0$ and can reduce $n-2$ unknown quantities.

(2) Bishop method. It assumes that the slip surface is a circular arc and the shear force between the slices, which means $X_i = 0$, and can reduce $n-1$ unknown quantities.

(3) Janbu method. It assumes that the slip surface can be any shape and the normal force between the slices is acting on the point at 1/3 of the height, which can reduce $n-1$ unknown quantities.

(4) Other slices methods. These methods assume that the slip surface can be any shape, the relationship between the normal force E_i and the shear force X_i is a function relationship, which can reduce $n-1$ unknown quantities. For example, imbalance thrust force method, Morgenstein-Price method, Spencer method, Sarma method, etc. probably will yield reasonable estimates of factor of safety for failure surfaces of any shape.

10.4 Swedish circle method

10.4.1 Basic assumptions and equations

The Swedish circle method of slices is the simplest basic method for slope stability analysis, which is known as the normal or ordinary method of slices. This method was first introduced by Fellenius in 1926 for analysis of slope stability of cohesive soil. In this method, the soil mass above the assumed slip circle is divided into a number of vertical slices of equal width, as shown in Figure 10-7. The location of the center of the failure arc is assumed. The forces between the slices are neglected and each slice is considered to

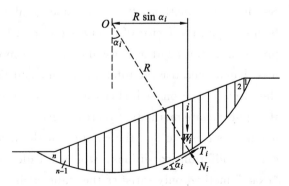

Figure 10-7 Force analysis of Swedish circle method

be an independent column of soil of unit thickness. Thus, we have the soil weight $W_i = \gamma_i b_i h_i$, the shear force T_i and the normal force N_i on the bottom of ith slice.

According to the static force equilibrium of ith slice, we can get

$$N_i = W_i \cos \alpha_i \tag{10-13}$$

If the factor of safety is F_s, according to Coulomb's strength theory, there is

$$T_i = \frac{1}{F_s} \times T_{fi} = \frac{c_i l_i + N_i \tan \phi_i}{F_s} \tag{10-14}$$

Based on the moment of force equilibrium at center O,

$$\sum (W_i R \times \sin \alpha_i - T_i R) = 0 \tag{10-15}$$

Substituting Eq. (10-13) into Eq. (10-14) and then into Eq. (10-15), we can get the following equation of Swedish circle method.

$$F_s = \frac{\sum (c_i l_i + W_i \cos \alpha_i \tan \phi_i)}{W_i \sin \alpha_i} \tag{10-16}$$

When the pore water pressure u on the sliding surface is known (Figure 10-8), Eq. (10-16) of the Swedish circle method can be rewritten in terms of effective stress.

$$F_s = \frac{\sum [c'_i l_i + (W_i - u_i b_i) \cos \alpha_i \tan \phi'_i]}{\sum W_i \sin \alpha_i} \tag{10-17}$$

Figure 10-8 Force analysis on i^{th} slice with pore water pressure

The Swedish circle method is an old and easy method, which is recommended for slope stability analysis in the related Chinese code. It should be noted that this method cannot satisfy all the static equilibrium conditions since it neglects the forces between slices, thus the calculated factor of safety is generally 10%~20% lower than those calculated with other strict methods. Specifically, when the central angle of the sliding arc is large and the pore water pressure is large, the calculated factor of safety may be half of the value calculated by more strict method.

10.4.2 Safety factor for layered soil slope with surcharge load

If the slope is made up of different soil layers, as shown in Figure 10-9, Eq. (10-16) can also be used to calculate the factor of safety with following requirements.

(1) We should calculate the soil slice weight by different layers and then sum them up. For example, if the soil has m layers, then

$$W_i = b_i (\gamma_{1i} h_{1i} + \gamma_{2i} h_{2i} + \cdots + \gamma_{mi} h_{mi}) \tag{10-18}$$

(2) The cohesive force c and the internal friction angle ϕ shall be determined based the location of the soil layer corresponding to the ith soil slice. Therefore, for the layered soil slope, the factor of safety F_s can be written as:

$$F_s = \frac{\sum [c_i l_i + b_i (\gamma_1 h_{1i} + \gamma_2 h_{2i} + \cdots + \gamma_{mi} h_{mi}) \cos \alpha_i \tan \phi_i]}{\sum b_i (\gamma_1 h_{1i} + \gamma_2 h_{2i} + \cdots + \gamma_{mi} h_{mi}) \sin \alpha_i} \tag{10-19}$$

If there is a surcharge load q acting on the slope surface or slope top, as shown in Figure 10-10, we need to add the surcharge load to the weight of ith soil slice right below the surcharge load, then the factor of safety of slope is

$$F_s = \frac{\sum [c_i l_i + (W_i + qb_i)\cos \alpha_i \tan \phi_i]}{\sum (W_i + qb_i)\sin \alpha_i} \qquad (10\text{-}20)$$

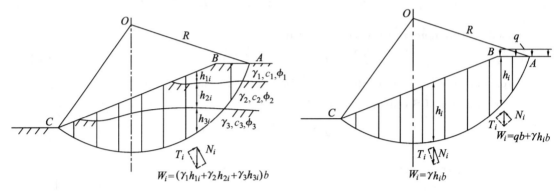

Figure 10-9 Calculation diagram of slope stability for layered soils

Figure 10-10 Calculation diagram of slope stability with surcharge load

10.4.3 Safety factor with seepage by unit weight substitution method

If the slope is partially immersed in water, as shown in Figure 10-11. According to Eq. (10-17), the weight of the soil slice in water should be calculated in terms of saturated unit weight. We need to consider the pore water pressure on the sliding surface (hydrostatic pressure) and the water pressure acting on the slope. Taking the slope below the water table EF as a free body (wedge CFE), we have the force of the static pore water pressure on the sliding surface (the resultant force is P_1) and the water pressure on the slope surface CF (the resultant force is P_2).

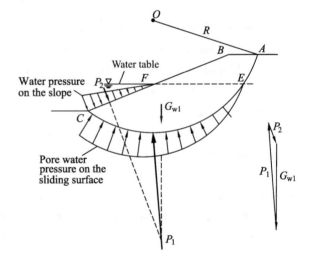

Figure 10-11 Calculation diagram of slope stability for part of slope immersed in water

Besides, there are the self-weight of pore water and soil buoyant force (the resultant force is equal to the weight of the same volume of water for the free body (wedge CFE), namely G_{w1}) acting at the position of the gravity center for the free body. Since it is static water, the three forces form an equilibrium force system. i.e., the resultant force of P_1 and P_2 is equal to the G_{w1}, and the direction is opposite. Thus, the effect of water pressure on the sliding soil body can be reflected by the buoyant force acting on the soil below the water

Chapter 10 Slope Stability

table. Therefore, the weight of soil slice below the water table is calculated with the buoyant unit weight.

Besides, if the slope partially immersed in water is composed of layered soil, the factor of safety can be calculated using Eq. (10-19) by replacing the unit weight γ_i with the buoyant weight γ_i' for the ith slice below the water table.

When the reservoir stores or drains water, or the embankment slope is with relatively high groundwater level, or the foundation pit is under drainage, etc., it will produce seepage flow through the slope and generate the seepage force. We should consider the effect of seepage on the slope stability analysis.

Based on the theory of seepage, we can use the flow net (Figure 10-12a) to obtain the seepage force. First, effective unit weight or buoyant unit weight if the soil mass is below the water table can be used to combine the seepage force to reflect the effect of seepage on slope stability. According to the flow net, the average hydraulic gradient i_i of each grid can be determined, and the seepage force $J_i = \gamma_w i_i a_i$ of each grid can be obtained. γ_w is the unit weight of water and a_i is the area of the grid. The force J_i acts on the centroid of the grid which is parallel to the direction of the streamline. If d_i is the distance between the sliding center and the action point of J_i, then the moment of seepage force of ith grid generated is $J_i d_i = \gamma_w i_i a_i d_i$. The total moment of sliding force generated by the seepage force in the entire sliding soil body is the sum of each moment of the seepage force acting on the grid, that is, $\sum J_i d_i = \sum \gamma_w i_i a_i d_i$. If we assume that the shear stress is uniformly distributed caused by the seepage force on the sliding surface and does not account the sliding resistency effect, the shear stress is $\sum J_i d_i / R \hat{L}$, where R is the radius of the sliding surface and $\hat{L}$ is the length of the sliding surface. If this expression of shear stress is used in Eq. (10-7), we can get the factor of safety for the slope under the action of seepage. Noted that the weight of soil below the water table should be calculated using the buoyant unit weight.

When we use flow net to calculate the seepage force, the accuracy can be guaranteed if the grid is plotted correctly. But the calculation is very tedious since drawing the flow net is also difficult in some cases. Therefore, the method of directly solving the seepage force by flow net to obtain the factor of safety has not been widely adopted in practical engineering. Instead, the commonly used method is called "replacement method". Replacement method uses the weight (G_{w2}) of the same volume of water for the area ($lmnl'$) enclosed between the seepage line and above the water level to replace the seepage force, and calculate the corresponding moment of seepage force. The procedures are described as follows:

As shown in Figure 10-12b, taking the pore water in the area enclosed between the sliding surface and the seepage line as a free body, the forces acting on it include:

① The resultant force of the pore water pressure on the sliding surface is P_w, which is pointing towards the center of circle;

② The water pressure on the slope surface nC, and the resultant force is P_2;

③ The resultant force of the pore water pressure and the buoyant force by soil particles within the range of nCl' is G_{w1}, which is pointing vertical downward;

④ The resultant force of the pore water pressure and the buoyant force by the soil particles within the range of $lmnl'$ is G_{w2}, which is pointing vertical downward and the distance between the line of G_{w2} and the center of the circle is d_w;

⑤ Soil resistance force to seepage T_j, where the distance to the center of the circle is d_j.

In the steady flow conditions, these forces form a equilibrium system. Taking the moment of each force through the center of circle, we have the moment of P_w is zero since the action line of P_w goes through the center. Besides, from the force analysis of the slope when soil is partially submerged in water, the moments of P_2 and G_{w1} can offset each other with respect to the circle center O, thus we can obtain: $T_j d_j = G_{w2} d_w$. Note that T_j and and the resultant of seepage force have the same magnitude and opposite direction. So the moment of seepage force on the sliding center of circle can be replaced by the moment of G_{w2}.

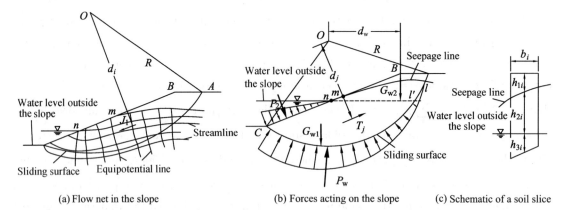

(a) Flow net in the slope (b) Forces acting on the slope (c) Schematic of a soil slice

Figure 10-12 Calculation diagram of slope stability with seepage flow

Since it is assumed that the sliding resistant effect of the seepage force is not taken into account, the moment of G_{w2} on the sliding center O can be calculated by summing the moment of force on each slice, i.e., $G_{w2} d_w$ is equal to $\sum \gamma_w h_{2i} b \sin\alpha_i R$. If this expression is taken to the moment equilibrium equation of the entire sliding soil body, the expression on the factor of safety with steady seepage can be obtained as

$$F_s = \frac{\sum [c_i' l_i + b_i(\gamma_i h_{1i} + \gamma_i' h_{2i} + \gamma_i' h_{3i}) \cos\alpha_i \tan\phi_i]}{\sum (\gamma_i h_{1i} + \gamma_i' h_{2i} + \gamma_i' h_{3i}) b_i \sin\alpha_i + \sum \gamma_w h_{2i} b_i \sin\alpha_i} \qquad (10\text{-}21)$$

Obviously, the second item in the denominator in Eq. (10-21) is the shear force caused by the seepage. Combined the two items in the denominator, we can get the final expression on the factor of safety F_s.

$$F_s = \frac{\sum [c_i' l_i + (\gamma_i h_{1i} + \gamma_i' h_{2i} + \gamma_i' h_{3i}) b_i \cos\alpha_i \tan\phi']}{\sum (\gamma_i h_{1i} + \gamma_{sat i} h_{2i} + \gamma_i' h_{3i}) b_i \sin\alpha_i} \qquad (10\text{-}22)$$

where, γ_i—natural unit weight of ith soil slice;

γ_{sati}—saturated unit weight of ith soil slice;

γ'_i—buoyant unit weight of ith soil slice.

h_{1i}, h_{2i} and h_{3i} are respectively the height of the soil above the seepage line, the height between the water level and the seepage line, and the height below the water table for i^{th} soil slice, as shown in Figure 10-12c.

【Example 10-2】 Figure 10-13 shows a homogeneous cohesive soil slope with the height of 20m, and the slope ratio is 1 : 2. The soil cohesion c is 45kPa, the internal friction angle ϕ is 7°, and the unit weight of soil is 20kN/m³. Try to calculate the factor of safety using the Swedish circle method.

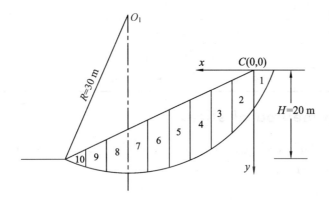

Figure 10-13 Soil slope profile of 【Example 10-2】

【Solution】

(1) Divide the slope into several slices and arc. As shown in Figure 10-13, in the possible sliding range, select one circle center O_1 and take the radius $R=30$m;

(2) Sliding soil mass is divided into 10 slices and we can number them thereafter;

(3) Measure the height h_i, breadth b_i, length l_i, and calculate the value of $\sin\alpha_i$, $\cos\alpha_i$ and W_i for each slice. The results are listed in Table 10-2. By Eq. (10-16), calculate the factor of safety corresponding to the center O_1 and radius $R=30$m;

(4) Select a different radius for the center O_1 to obtain the minimum factor of safety corresponding to O_1;

(5) In the potential sliding range, select different circle centers O_2, O_3, ⋯, and repeat the above calculation steps to obtain the minimum factor of safety, which is the final factor of safety.

Table 10-2 Swedish circle method result (center O_1 and radius of 30m)

Number of soil slice	h_i/m	b_i/m	l_i/m	$\sin \alpha_i$	$\cos \alpha_i$	W_i/kN	c_i/kPa	Results
1	5.65	4.5	9.28	0.8746	0.4848	508.5	45	$\sum c l_i = 2456.8$ $\sum W_i \cos \alpha_i \tan \phi = 930.1$ $\sum W_i \sin \alpha_i = 2845.0$ $F_s = \dfrac{2456.8+930.1}{2845.0} = 1.19$
2	10.64	4.5	6.73	0.7431	0.6691	957.6		
3	12.74	4.5	5.63	0.6018	0.7986	1146.6		
4	13.53	4.5	5.10	0.4695	0.8829	1217.7		
5	13.31	4.5	4.76	0.3256	0.9455	1204.2		
6	12.46	4.5	4.58	0.1908	0.9816	1121.4		
7	10.90	4.2	4.21	0.0698	0.9976	915.6		
10	10.61	4.9	4.91	−0.0698	0.9976	850.6		
9	5.60	4.5	4.60	−0.2079	0.9781	504		
10	2.40	4.5	4.79	−0.3420	0.9396	216		

10.5 Bishop method of slices

The Bishop method of slices also assumes that the sliding surface is a circular surface, as shown in Figure 10-14a, which takes into account the forces acting on the slice sides and assumes that the factor of safety on the bottom of each slice is the same and equal to the factor of safety for the entire failure surface. Bishop adopted the effective stress method to derive the solution, and this method can also be used for the total stress analysis.

10.5.1 Effective stress analysis

Taking the ith soil slice for analysis, the forces acting on the ith slice include: the weight of slice W_i, the shear force T_i, the effective normal force N'_i, the pore water pressure $u_i l_i$ acting on the bottom of the slice, the normal forces E_i, E_{i+1} and inter-slice shear forces X_i and X_{i+1} acting on both sides of the slice, and $\Delta X_i = X_{i+1} - X_i$, as shown in Figure 10-14b.

According to the theory of the shear strength, if the factor of safety is F_s, the shear force T_i and the normal force N'_i have the following relationship:

$$T_i = \frac{1}{F_s}(c'_i l_i + N'_i \tan \phi'_i) \tag{10-23}$$

Since the force equilibrium is on the vertical direction of the ith soil slice, there is

$$W_i + \Delta X_i - T_i \sin \alpha_i - N'_i \cos \alpha_i - u_i l_i = 0 \tag{10-24}$$

Substituting Eq. (10-23) into Eq. (10-24), we have

$$N'_i = \frac{1}{m_{ai}}\left(W_i + \Delta X_i - u_i b_i - \frac{1}{F_s} c'_i l_i \sin \alpha_i\right) \tag{10-25}$$

where

$$m_{ai} = \cos \alpha_i + \frac{\tan \phi'_i}{F_s} \sin \alpha_i \tag{10-26}$$

In Eq. (10-25), $b_i = l_i \cos \alpha_i$, b_i is the width of the ith soil slice.

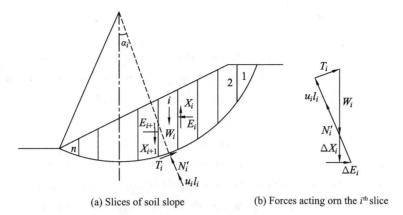

(a) Slices of soil slope   (b) Forces acting orn the ith slice

Figure 10-14   Schematic diagram of Bishop method of slices

The overall moment equilibrium about the center O is satisfied, and the forces acting on the slice sides will offset each other. The action line of normal force N_i on the sliding surface of ith slice is through the center of the circle, so

$$\sum W_i x_i - \sum T_i R = \sum W_i R \sin \alpha_i - \sum T_i R = 0 \qquad (10\text{-}27)$$

Substituting Eq. (10-25) into Eq. (10-23), and then into Eq. (10-27), we can obtain

$$F_s = \frac{\sum \frac{1}{m_{\alpha i}}[c'_i b_i + (W_i - u_i b_i + \Delta X_i)\tan \phi'_i]}{\sum W_i \sin \alpha_i} \qquad (10\text{-}28)$$

Eq. (10-28) is the equation for calculating the factor of safety by Bishop method of slices. Since ΔX_i is unknown, in order to solve the problem, Bishop assumed $\Delta X_i = 0$ and proposed the simplified Bishop method to calculate the factor of safety as

$$F_s = \frac{\sum \frac{1}{m_{\alpha i}}[c'_i b_i + (W_i - u_i b_i)\tan \phi'_i]}{\sum W_i \sin \alpha_i} = \frac{\sum A_i}{\sum W_i \sin \alpha_i} \qquad (10\text{-}29)$$

It has been proved that the effect of this simplification on the factor of safety is only about 1%, and the calculated results meet the engineering requirements. Thus, the simplified Bishop method of slices is widely used in the engineering applications.

10.5.2 Total stress analysis

Bishop method of slices on factor of safety can also be written in the form of total stresses

$$F_s = \frac{\sum \frac{1}{m_{\alpha i}}[c_i b_i + (W_i + \Delta X_i)\tan \phi_i]}{\sum W_i \sin \alpha_i} \qquad (10\text{-}30)$$

The simplified Bishop method of slices for total stress analysis calculating the factor of safety is

$$F_s = \frac{\sum \frac{1}{m_{\alpha i}}(c_i b_i + W_i \tan \phi_i)}{\sum W_i \sin \alpha_i} \qquad (10\text{-}31)$$

where, $m_{ai} = \cos\alpha_i + \dfrac{\tan\phi_i}{F_s}\sin\alpha_i$.

When we use the simplified Bishop's equation, there is a factor of safety F_s on the right side of Eq. (10-31), so it is necessary to perform the iteration calculation. It is assumed that F_s is equal to 1 at first, then m_{ai} can be calculated, and a new value of F_s can thereafter be obtained by Eq. (10-29) or Eq. (10-31). If the calculated F_s is not equal to 1, then we can use the F_s to calculate the new m_{ai} and new F_s, repeat iterating until the final F_s is very close to the previous one. The accuracy of the solution can normally be satisfied through three to four times iteration, and the result is always convergent.

It should be pointed out that for the soil slices with negative α_i values, we should find whether m_{ai} will approach zero. In this case, the simplified Bishop may not be used because N_i will tend to infinity, which is obviously unreasonable. According to the recommendations of some existing studies, when the value of m_{ai} for any given slice is less than or equal to 0.2, there will be a larger error in the calculation of F_s. We need to use other methods instead. In addition, when the value of α_i is very large, we can get $N_i < 0$, then we can use $N_i = 0$ instead.

The simplified Bishop method of slices assumes that all the values of ΔX_i are equal to zero, which will reduce $n-1$ unknown quantities. According to the equilibrium of the vertical force of each soil slice and the moment equilibrium for the entire sliding soil body, we can obtain the factor of safety F_s. It can be seen that the simplified Bishop method cannot satisfy all the equilibrium conditions, thus it is not a strict method, the calculation error is about 2% to 7%.

The Swedish method and the Bishop method are the most commonly used methods of slices in slope stability analysis in general engineering applications. Some softwares that contains those methods built-in are used for the numerical calculation of slope stability. Those sofwares are introduced in the knowledge expansion part.

【Example 10-3】 A slope had the same size with that in [Example 10-2], as shown in Figure 10-13. The soil unit weight γ is 20kN/m³, cohesion c' is 15kPa, and the internal friction angle ϕ' is 45°. The pore water pressure u_i on the bottom surface of the ith slice can be obtained by $\gamma h_i \overline{B}$, where h_i is the height of the ith slice, and the pore pressure coefficient $\overline{B}$ is equal to 0.5. Try to use simplified Bishop method of slices to calculate the factor of safety for the slope.

【Solution】
Since $u_i = \gamma h_i B$, subsitituting this into Eq. (10-29), we have

$$F_s = \dfrac{\sum \dfrac{1}{m_{ai}}[c'_i b_i + (1-\overline{B})W_i \times \tan\phi'_i]}{\sum W_i \sin\alpha_i} = \dfrac{\sum A_i}{\sum W_i \sin\alpha_i}$$

Make a list to calculate the factor of safety, where the calculation processes are shown in Table 10-3. The factor of safety obtained for the sliding arc is 1.661 through iterative calculation. It should be noted that this is just the result of one potential sliding arc, and we can assume a number of sliding surfaces and use the same method to find the smallest value of F_s. Noted that the potential sliding surface with the minimum F_s obtained by the Bishop method is not necessarily the same one obtained by the Swedish method.

Table 10-3 Results of Bishop simplified method of slices for [Example 10-3] (center O_1 and radius of 30m)

Number of soil slice	h_i/m	b_i/m	l_i/m	$\sin \alpha_i$	$\cos \alpha_i$	W_i/kN	First trail $F_s=1.0$	
							$m_{\alpha i}$	A_i
1	5.65	4.5	9.28	0.8746	0.4848	508.5	1.3594	236.69
2	10.64	4.5	6.73	0.7431	0.6691	957.6	1.4122	386.84
3	12.74	4.5	5.63	0.6018	0.7986	1146.6	1.4004	457.58
4	13.53	4.5	5.10	0.4695	0.8829	1217.7	1.3524	500.11
5	13.38	4.5	4.76	0.3256	0.9455	1204.2	1.2711	526.79
6	12.46	4.5	4.58	0.1908	0.9816	1121.4	1.1724	535.82
7	10.90	4.2	4.21	0.0698	0.9976	915.6	1.0674	487.91
10	8.68	4.9	4.91	−0.0698	0.9976	850.64	0.9278	537.64
9	5.60	4.5	4.60	−0.2079	0.9781	504	0.7702	414.83
10	2.40	4.5	4.79	−0.3420	0.9396	216	0.5976	293.67

Number of soil slice	Second trail $F_s=1.536$		Third trail $F_s=1.642$		Forth trail $F_s=1.659$		Notation	
	$m_{\alpha i}$	A_i	$m_{\alpha i}$	A_i	$m_{\alpha i}$	A_i		
1	1.0542	305.22	1.0174	316.25	0.8777	137.34		
2	1.1529	473.87	1.1216	487.07	0.9965	226.29		
3	1.1904	538.32	1.1651	550.01	1.0954	643.61	First trail	$F_s=1$
4	1.1885	569.06	1.1688	578.67	1.1498	701.85	Calculated Second trail	$F_s=1.536$ $F_s=1.536$
5	1.1575	578.51	1.1438	585.43	1.1501	719.05	Calculated Third trail	$F_s=1.642$ $F_s=1.642$
6	1.1058	568.09	1.0978	572.24	1.1129	710.78	Calculated Forth trail	$F_s=1.659$ $F_s=1.659$
7	1.0430	499.31	1.0401	500.72	1.0447	677.71	Calculated	$F_s=1.661$
10	0.9522	523.88	0.9551	522.27	0.9483	614.79	Final results	$F_s=1.661$
9	0.8428	379.11	0.8515	375.22	0.8235	505.14		
10	0.7170	244.78	0.7313	239.97	0.6679	302.45		

10.6 Slope stability analysis for non-circular failure surface

For non-cohesive soil slopes, the sliding or failure surface is generally a plane, which is different from that of a generally circular failure surface for homogeneous cohesive soil slopes. In the practical engineering projects, for example, the well-graded compacted earth dam or the earth-rockfill dam, the slices method is often used to calculate the dam slope stability. However, when there is a clear weak interlayer in the slope, such as clay core wall dam, weak interlayer in the backfill slope, construction of soil embankment along an inclined rock surface, rock and soil mass with relatively developed fissures or old landslides, etc., where the sliding surface will occur in the weak layer and the potential failure surface is not a circular surface in these cases. The Swedish method and Bishop method of slices are no longer applicable. Instead, some slope stability analysis used for non-circular failure surface are proposed, such as Janbu method and imbalance thrust force method.

10.6.1 Janbu method

Janbu method is a general method of slices developed on the basis of limit equilibrium. It requires satisfying equilibrium of forces and moments acting on individual slices (only moment equilibrium at last uppermost slice is not satisfied). The slices are created by dividing the soil above the slip surface by dividing planes, as shown in Figure 10-15a. For a given slope with known slip surface, Janbu assumed that the action point of the horizontal force is known. The existing analysis results showed that the action point of the force acting on the slice has little influence on the factor of safety, and we can assumed that it acts on the 1/3 height from the bottom of the soil slice. The connection line through the action points on each slice is called the thrust line. Take the ith soil slice for example, the forces acting on it are shown in Figure 10-15b, where h_{ti} is the distance between the action point and the bottom of the slice, α_{ti} is the angle between the thrust line and the horizontal line. h_{ti} and α_{ti} are known quantities.

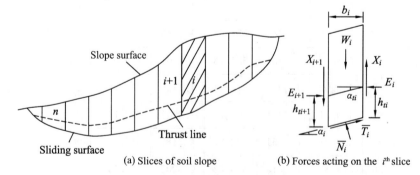

Figure 10-15 Schematic diagram of Janbu method of slices

Based on the equilibrium of vertical forces acting on the ith slice, we can get

$$\overline{N_i}\cos \alpha_i = W_i + \Delta X_i - \overline{T_i}\sin \alpha_i \tag{10-32}$$

or

$$\overline{N_i} = (W_i + \Delta X_i)\sec \alpha_i - \overline{T_i}\tan \alpha_i \tag{10-33}$$

Then according to the equilibrium of the horizontal forces,

$$\Delta E_i = \overline{N}_i \sin \alpha_i - \overline{T}_i \cos \alpha_i \tag{10-34}$$

Substituting Eq. (10-23) into Eq. (10-34),

$$\Delta E_i = (W_i + \Delta X_i) \tan \alpha_i - \overline{T}_i \sec \alpha_i \tag{10-35}$$

Based on the equilibrium of the moment acting on the middle point of the ith slice, we have

$$E_{i+1}\left(h_{ti+1} - \frac{b_i}{2}\tan \alpha_{ti}\right) = X_{i+1} \cdot \frac{b_i}{2} + X_i \cdot \frac{b_i}{2} + E_i\left(h_{ti} + \frac{b_i}{2}\tan \alpha_{ti}\right) \tag{10-36}$$

Since $h_{ti+1} = h_{ti} + \Delta h_{ti}$, and $X_{i+1} = X_i + \Delta X_i$; $E_{i+1} = E_i + \Delta E_i$, so we get

$$E_{i+1}(h_{ti} + \Delta h_{ti}) - E_i h_{ti} = \frac{b_i}{2}\tan \alpha_{ti}(2E_i + \Delta E_i) + \frac{b_i}{2}(2X_i + \Delta X_i) \tag{10-37}$$

Ignore the higher order items and re-arrange Eq. (10-37), then

$$X_i b_i = -E_i b_i \tan \alpha_{ti} + h_{ti} \Delta E_i \tag{10-38}$$

or

$$X_i = -E_i \tan \alpha_{ti} + h_{ti}\frac{\Delta E_i}{b_i} \tag{10-39}$$

By the boundary condition: $\sum \Delta E_i = 0$ and Eq. (10-39), we can get

$$\sum (W_i + \Delta X_i)\tan \alpha_i - \sum \overline{T}_i \sec \alpha_i = 0 \tag{10-40}$$

With the definition of factor of safety and the Mohr-Coulomb failure criterion,

$$\overline{T}_i = \frac{\tau_{fi} l_i}{F_s} = \frac{c_i b_i \sec \alpha_i + \overline{N}_i \tan \phi_i}{F_s} \tag{10-41}$$

Through solving Eq. (10-33) and Eq. (10-36), we can get

$$\overline{T}_i = \frac{1}{F_s}[c_i b_i + (W_i + \Delta X_i)\tan \phi_i]\frac{1}{m_{\alpha_i}} \tag{10-42}$$

where

$$m_{\alpha_i} = \cos \alpha_i + \frac{\sin \alpha_i \tan \phi_i}{F_s} \tag{10-43}$$

Substituting Eq. (10-42) into Eq. (10-40), the factor of safety calculated by Janbu method is

$$F_s = \frac{\sum [c_i b_i + (W_i + \Delta X_i)\tan \phi_i]\frac{1}{\cos \alpha_i m_{\alpha_i}}}{\sum (W_i + \Delta X_i)\tan \alpha_i} \tag{10-44}$$

In the calculation process of Janbu method, we need to use the iterative method when calculating the factor of safety and the inter-slice force X_i and E_i at the same time. The steps are as follows:

① Assuming $\Delta X_i = 0$, it is equivalent to the simplified Bishop method. Then use the Eq. (10-44) to calculate the factor of safety. In this case, it is necessary to iterate over F_s. First we can assume that $F_s = 1$, calculate the m_{α_i} and substitute into Eq. (10-44) to get the new F_s. If the new F_s is larger than the assumed value and the difference is large, then we can use the new F_s to repeat the interation until the new calculated F_s is close to the first approximation of F_s. Then we can use this F_s to calculate the $\overline{T}_i$ of ith soil slice.

② Substituting $\overline{T}_i$ into Eq. (10-35), calculate the ΔE_i and E_i of ith soil slice, then calculate the value of ΔX_i by Eq. (10-39).

③ Repeat step ① with the new ΔX_i to obtain the second approximation of F_s and recalculate the $\overline{T}_i$ of ith soil slice.

④ Repeat steps ② and ③ until F_s converges within a given error.

Janbu method can basically satisfy all the force equilibrium conditions, hence it is one of the "strict" methods. However, the assumption of the thrust line must meet reasonable requirements, i. e., no tension and shear failure occur between two slices. Most of the commercial softwares about the slope stability analysis include the Janbu method. It should be noted that in some cases, the results may have convergence problem, thus we need to consider other available methods.

10.6.2 Imbalance thrust force method

Some slopes in the mountain areas are often with rock surface, and most of the slope instability problems occur along these interfaces, forming a polyline sliding surface. For rock slopes, the failure surface often occurs along the fault or fracture, which is also generally with a polyline sliding surface. Stability analysis of those slopes can be done using the imbalanced thrust force method. This method is a limit state method. It builds upon the equation of equilibrium of forces acting on individual slices and does not consider the moment equation of equilibrium.

As shown in Figure 10-16, the soil above the slip surface can be divided into several slices. We assume that the direction of force between two slices is parallel to the bottom line of soil slice.

For a given soil slice based on the equilibrium of forces perpendicular and parallel to the bottom of the previous soil slice, we can get

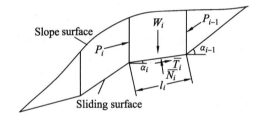

Figure 10-16 Schematic diagram of imbalance thrust force method

$$\overline{N}_i - W_i \cos \alpha_i - P_{i-1} \sin(\alpha_{i-1} - \alpha_i) = 0 \quad (10\text{-}45)$$

$$\overline{T}_i + P_i - W_i \sin \alpha_i - P_{i-1} \cos(\alpha_{i-1} - \alpha_i) = 0 \quad (10\text{-}46)$$

By the definition of factor of safety and the Mohr-Coulomb failure criterion, we have

$$\overline{T}_i = \frac{c_i l_i + \overline{N}_i \tan \phi_i}{F_s} \quad (10\text{-}47)$$

Solving Eq. (10-46) and Eq. (10-47), we can eliminate $\overline{T}_i$, $\overline{N}_i$ and get the following equation

$$P_i = W_i \sin \alpha_i - \left(\frac{c_i l_i + W_i \cos \alpha_i \tan \phi_i}{F_s} \right) + P_{i-1} \psi_i \quad (10\text{-}48)$$

where ψ_i is the transfer coefficient, which can be expressed as

$$\psi_i = \cos(\alpha_{i-1} - \alpha_i) - \frac{\tan \phi_i}{F_s} \sin(\alpha_{i-1} - \alpha_i) \quad (10\text{-}49)$$

Eq. (10-48) and Eq. (10-49) are implicit expressions to get factors of safety for the imbalance thrust force method.

When using the imbalance thrust force method, it is usually assumed that $F_s = 1$. Then from the uppermost slice, we can calculate the thrust force P_i onto the next slice, until the last slice thrust force P_n is zero.

The Eq. (10-48) was simplified to the following explicit form per Chinese code GB 50007—2011 "Code for design of building foundation".

$$P_i = F_s W_i \sin \alpha_i - (c_i l_i + W_i \cos \alpha_i \tan \phi_i) + P_{i-1} \psi_i \quad (10\text{-}50)$$

where the transfer coefficient ψ_i is changed to the following equation:

$$\psi_i = \cos(\alpha_{i-1} - \alpha_i) - \tan \phi_i \sin(\alpha_{i-1} - \alpha_i) \quad (10\text{-}51)$$

In fact, Eq. (10-50) and Eq. (10-51) are obtained by the Eq. (10-48) and Eq. (10-49) when $F_s = 1$. Existing studies have shown that the calculated factor of safety using imbalance thrust force method no matter with explicit or implicit solution is greater than that using simplified Bishop method. Moreover, the error is lager when using explicit solution. When the factor of safety is equal to 1, the implicit solution and explicit solution is equivalent; when the factor of safety is not equal to 1, there will be a difference between the factor of safety calculated by the two solutions.

Although the implicit solution is superior to the explicit solution, it still has some disadvantages. This is because of the assumption that the inter-slice thrust force is parallel to the bottom line of the previous soil slice, the calculated factor of safety is affected by the angle of the sliding surface. Existing studies have shown that for smooth and continuous sliding surfaces, implicit solutions can be used unconditionally; for rough surfaces consisting of broken lines, the use of implicit solutions should be limited and the inclined angle at all turning points on the sliding surface should be less than 10°; when the angle of slope at the turning point exceeds 10°, we should eliminate the angle effect.

When using the imbalance thrust force method, the shear strength index c, ϕ can be determined by the test or local experiences. In addition, because the soil slices cannot withstand the tension, if any soil slice thrust force P_i is negative, the P_i is no longer transfer to the next slice, where the value of P_{i-1} on the next slice is taken as zero. Imbalance thrust force method is also commonly used to calculate the thrust force on each slice and the last slice based on a given value of F_s. Then determine whether or how to set up a retaining structure to withstand the thrust force. According to the nature of landslide and the impact on the construction, the allowable factor of safety is in the range of 1.05~1.25.

[Example 10-4] As shown in Figure 10-17, the height of slope is 10m. A soft soil layer locates 3m below the slope toe. The length of slope is 15m, and the unit weight of the soil is 20kN/m³. The cohesion c is 15kPa, and the internal friction angle ϕ is 45°. The undrained strength of soft soil layer is 12kPa, and the internal friction angle ϕ_u is zero. Try to calculate the factor of safety for the slope using the imbalance thrust force method.

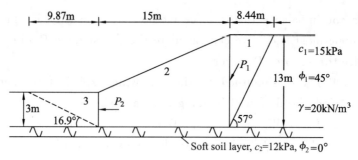

Figure 10-17 Schematic diagram of [Example 10-4]

[Solution]

Using Eq. (10-48) and Eq. (10-49) of the imbalance thrust force method, the calculation process and result are listed in Table 10-4.

Table 10-4 Calculation process and result of imbalance thrust force method (unit of P_i is kN/m)

Number of soil slice	W_i (kN/m)	$F_s=1$ P_i	$F_s=3$ P_i	$F_s=2.5$ P_i	$F_s=2.8$ P_i	$F_s=2.85$ P_i	Result
1	1097.2	90.1	643.5	588.2	623.7	628.9	
2	2400.0	−130.9	290.5	248.3	275.4	279.4	$F_s=2.105$
3	296.1	−611.1	17.8	−52.5	−7.5	−0.9	

Knowledge expansion

Some Softwares that Used for Numerical Analysis of Slope Stability

(1) GEO5—It is a package of programs designed to solve various geotechnical problems. Each program is used to analyse a different geotechnical task but all modules communicate with each other to form an integrated suite. The programs are based on traditional analytical methods and the Finite Element Method (FEM). GEO5 is capable of solving the following geotechnical tasks: slope stability analysis, foundations design, retaining wall design, sheet pile wall, soil settlement, underground structures etc. This program can be used to perform slope stability analysis (embankments, earth cuts, anchored retaining structures, MSE walls, etc.). The slip surface is considered as circular (Bishop, Fellenius/Petterson, Janbu, Morgenstern-Price or Spencer methods) or polygonal (Sarma, Janbu, Morgenstern-Price or Spencer methods).

(2) GEOSLOPE-SLOPE/W—It is the leading slope stability software for soil and rock slopes. SLOPE/W can effectively analyze both simple and complex problems for a variety of slip surface shapes, pore-water pressure conditions, soil properties, and loading conditions. SLOPE/W supports a comprehensive list of material models including Mohr-Coulomb, undrained, high strength, impenetrable, bilinear, anisotropic

strength, SHANSEP, spatial Mohr-Coulomb and more. SLOPE/W can be used to analyze almost any slope stability problem you will encounter in the geotechnical, civil, and mining engineering projects.

(3) SLIDE—It is comprehensive slope stability analysis software available, complete with finite element groundwater seepage analysis, rapid drawdown, sensitivity and probabilistic analysis and support design. All types of soil and rock slopes, embankments, earth dams and retaining walls can be analyzed. Flows, pressures and gradients are calculated based on user defined hydraulic boundary conditions. Seepage analysis is fully integrated with the slope stability analysis or can be used as a standalone module. Slide offers no less than 17 different material strength models for rock and soil including Mohr-Coulomb, anisotropic and generalized Hoek-Brown. Support types include tieback, end anchored, soil nail, micro pile and geotextile. Back analysis allows you to determine the required support force for a given safety factor. Advanced search algorithms simplify the task of finding the critical slip surface with the lowest safety factor.

(4) TSLOPE-2D/3D—They use the same model for slope stability analysis. They can be used for all types of slope stability problems including open pit mines, landfills, and any slope with complex geological structure and variable surface topography.

(5) SVSLOPE—It represents the new standard in 2D/3D slope stability analysis. Users can perform classic limit equilibrium slope analysis of soil or rock slopes by the method of slices or newer stress—based methods. Comprehensive 2D slip surface searching and pore-water pressure conditions and innovative 3D spatial analysis allow modeling at new levels.

(6) PLAXIS 2D/3D—They perform two or three-dimensional analysis of deformation and stability in geotechnical engineering and rock mechanics. Design and perform advanced finite element analysis of soil and rock deformation and stability, as well as soil structure interaction and groundwater and heat flow.

Exercises

[10-1] What are the main factors controlling slope stability?

[10-2] Why we say that all limit equilibrium analysis methods for calculating factor of safety are approximate methods? If the shear strength index is assumed to be true, is the calculated safety factor higher or lower than the true value?

[10-3] What are the main differences between the simplified Bishop method and the Swedish method? Why is the calculation of Bishop method more safe than that of the Swedish method for the same slope project?

[10-4] What are the differences between the imbalance thrust force method and the

Janbu method? Can they be used for the circular sliding slope analysis?

[10-5] Why are the allowable factors of safety for slope is different for different projects?

[10-6] What are the similarities and differences in controlling the factors of the slope stability in case of the earth-rock dam at construction period, water storage period and reservoir water level dropping period?

[10-7] If you find that a slope may be unstable, what treatments can be done to increase its stability?

[10-8] A gravel slope with the saturated unit weight γ_{sat} is 19kN/m³, the internal friction angle ϕ is 32°, and the slope ratio is 1 : 3. What is the factor of safety when the slope is dry or fully saturated? Whether the slope can remain stable when there is a seepage flow? If the slope ratio changes to 1 : 4, how is the stability?

[10-9] A homogeneous cohesive soil slope with the height of 20m and slope ratio of 1 : 3. The cohesion c is 10kPa, the internal friction angle ϕ is 20°, and the unit weight of soil γ is 18kN/m³. Assuming that the sliding surface is through the slope toe, radius R is 55m. Try to calculate the factor of safety for the slope using Swedish method.

[10-10] Slope profile imformation is the same as described in Exercise [10-9], if the effective strength index c' is 5kPa, ϕ' is 38° and unit weight γ is 18kN/m³. The pore pressure coefficient $\overline{B}$ is 0.55, the sliding surface is the same as in the Exercise [10-9]. Try to calculate the factor of safety using Bishop method in the construction period.

[10-11] The slope and soil properties are the same as in Exercise [10-9]. The effective strength index and the total stress index are the same, but there is a stable seepage, where the seepage line and the location of the water level are shown in Figure 10-18. The saturated unit weight of soil γ_{sat} is 19kN/m³. Try to calculate the factor of safety.

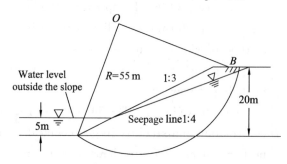

Figure 10-18　Schematic diagram of Exercise [10-11]

[10-12] A homogeneous excavation slope with the height is 10m, and slope ratio of 1 : 2. The unit weight γ is 18kN/m³, the internal friction angle ϕ is 25°, and the cohesion c is 5kPa. There is a soft soil layer located 3m below the slope base, where the cohesion c is 10kPa, and the internal friction angle ϕ is 5°. Try to calculate the factor of safety using the imbalance thrust force method.

References

[1] Atkinson J H. The Mechanics of Soils and Foundations[M]. 2th ed. Boca Raton: CRC Press, 2017.

[2] Bell F G. Fundamentals of Engineering Geology [M]. London: Butterworth-Heinemann, 1983.

[3] Blyth F G H, de Freitas M H. A Geology for Engineers[M]. 7th ed. London: Butterworth-Heinemann, 2005.

[4] Budhu M. Soil Mechanics Fundamentals[M]. 2th ed. New York: John Willey & Sons, 2015.

[5] Chen Z H. Rock Slope Stability Analysis[M]. Beijing: China Water Power Press, 2005 (in Chinese).

[6] Coduto D P. Geotechnical Engineering Principles and Practices[M]. New Delhi: Prentice Hall of India Private Limited, 2002.

[7] Craig R F. Craig's Soil Mechanics[M]. 7th ed. New York: Spon Press, Taylor & Francis Group, 2004.

[8] Das B M. Advanced Soil Mechanics[M]. 5th ed. Boca Raton: CRC Press, 2019.

[9] Das B M. Principles of Geotechnical Engineering[M]. New York: John Willey & Sons, 2002.

[10] Guo Y. Soil Mechanics [M]. Beijing: China Building Industry Press, 2014 (in Chinese).

[11] Ishibashi I, Hazarika H. Soil Mechanics Fundamentals and Applications[M]. 2th ed. Boca Raton: CRC Press, 2015.

[12] Jiang D Y, Zhu H H. Slope Stability Analysis and Landslide Treatment Plant[M]. Chongqing: Chongqing University Press, 2005 (in Chinese).

[13] Kaliakin V. Soil Mechanics: Calculations, Principles, and Methods[M]. London: Butterworth-Heinemann, 2017.

[14] Knappett J, Craig R F. Craig's Soil Mechanics[M]. 9th ed. Boca Raton: CRC Press, 2019.

[15] Lang Y H. Soil Mechanics [M]. Beijing: Peking University Press, 2012.

[16] Li G X, Zhang B Y, Yu Y Z. Soil Mechanics[M]. 2th ed. Beijing: Tsinghua

University Press, 2013 (in Chinese).

[17] Liu S Y. Soil Mechanics[M]. 5th ed. Beijing: China Building Industry Press, 2020 (in Chinese).

[18] Liu Z Y. Engineering Geology[M]. 2th ed. Beijing: China Electric Power Press, 2016 (in Chinese).

[19] McCarthy D F. Essentials of Soil Mechanics and Foundations Basic Geotechniques [M]. 6th ed. New Jersey: PrenticeHall, 2002.

[20] Mitchell J K, Soga K. Fundamentals of Soil Behavior[M]. 3th ed. New York: John Wiley & Sons, 2005.

[21] Miu H J, Su L J, Li H Z, et al. Soil Mechanics [M]. Xi'an: Xi'an Jiaotong University Press, 2015.

[22] Miu L C. Soil Mechanics[M]. Beijing: Metallurgical Press, 2004.

[23] Neilson J L. Engineering Geology of Melbourne [M]. London: Routledge Press, 2018.

[24] Pells P J. Engineering geology of the Sydney Region[M]. London: Routledge Press, 2018.

[25] Powrie W. Soil Mechanics: Concepts and Applications[M]. 3th ed. Boca Raton: CRC Press, 2013.

[26] Ranjan G, Rao A S R. Basic and Applied Soil Mechanics[M]. New Delhi: New Age International Publishers, 2000.

[27] Roy W. Basic Soil Mechanics[M]. Upper Saddle River: Prentice Hall, 2001.

[28] Shi B. Engineering Geology[M]. Beijing: Science Press, 2019 (in Chinese).

[29] Shi J Y. Soil Mechanics[M]. Beijing: China Communications Press, 2004.

[30] Shi Z M. Engineering Geology[M]. 3th ed. Beijing: China Building Industry Press, 2018 (in Chinese).

[31] Sun J Q. Engineering Geology[M]. Wuhan: Wuhan University Press, 2000 (in Chinese).

[32] Terzaghi K, Peck R B, Mesri G. Soil Mechanics in Engineering Practice[M]. New York: John Wiley & Sons, 1996.

[33] Waltham T. Foundations of Engineering Geology[M]. 3th ed. New York: T & F Books UK, 2009.

[34] Wang G R. Engineering Geology[M]. Beijing: China Machine Press, 2017 (in Chinese).

[35] William P. Soil Mechanics: Concepts and Applications[M]. Boca Raton: CRC Press, 2013.

[36] Xia J Z. Soil Mechanics and Engineering Geology [M]. Hangzhou: Zhejiang University Press, 2012 (in Chinese).